BACTERIAL GENOMES

Bacterial Genomes

Trees and Networks

Aswin Sai Narain Seshasayee

https://www.openbookpublishers.com

All external links were active at the time of publication unless otherwise stated and have been archived via the Internet Archive Wayback Machine at https://archive.org/web

Any digital material and resources associated with this volume will be available at https://doi.org/10.11647/OBP.0446#resources

ISBN Paperback: 978-1-80511-495-6
ISBN Hardback: 978-1-80511-496-3
ISBN Digital (PDF): 978-1-80511-497-0
ISBN Digital ebook (EPUB): 978-1-80511-498-7
ISBN HTML: 978-1-80511-499-4
DOI: 10.11647/OBP.0446

Cover images by Prerana Sudarshan and David Goodsell, assembled by Ashitha Arun, Inder Raj Singh and Madhumitha K. Cover design: Jeevanjot Kaur Nagpal.

Contents

About the author

Aswin Sai Narain Seshasayee is a researcher and Associate Professor at the National Centre for Biological Sciences (NCBS), a centre of Tata Institute of Fundamental Research (TIFR) in Bangalore, India. His lab is interested in fundamental aspects of the function and evolution of bacterial genomes and gene regulatory networks. His career in the sciences started off with a Bachelors of Technology at Anna University, Chennai, India, during which a lot of time left alone to explore and break things in the bioinformatics laboratory of Professor Gautam Pennathur and in the experimental microbiology and protein engineering laboratory of Professor Sankaran encouraged him to take up research for a career. He then pursued research as an intern, a PhD student and then briefly a postdoc with Nicholas Luscombe at EMBL-European Bioinformatics Institute, Hinxton, Cambridge, UK (and St John's College and Girton College, University of Cambridge). He has been with NCBS since December 2010, his research here funded over the years by the Department of Atomic Energy (Govt of India) core support to TIFR and NCBS, Department of Biotechnology, Department of Science and Technology and Science and Engineering Research Board (all Govt of India), CEFIPRA and DBT-Wellcome Trust India Alliance. Beyond science, he enjoys making music, painting watercolour landscapes and reading classic crime and fantasy fiction and popular history.

List of figures and tables

Preface

Bacteria are the most dominant form of cellular life on Earth. Not just in terms of numbers, but also in terms of their capabilities that allowed other life forms to emerge a few billion years ago, and continue to help sustain life to this day. To fear them as pedlars of disease would be to do great disservice to their incomparable contribution to making life on Earth habitable; after all it is only a minuscule minority of bacteria that are pathogenic.

Bacteria are hardy. They are everywhere. In the human body, they contribute to processes that we take for granted and we are often ignorant of their role (a quick Google search for human microbiomes[1] is instructive). They are present in extreme environments—from hydrothermal vents on the one hand to arctic permafrost on the other. Of course, they are also present in large numbers in the soil and in fresh and salt water. One would not be too far off the mark to claim that any catastrophe that drives most life forms to extinction would leave these organisms relatively unscathed. The only exception would be events that create conditions that are inimical to the very foundations of life.

Central to the bacterial dominance of our planet is their insatiable ability to adapt. Imagine the first bacterial cell living in a very ancient world of water, devoid of oxygen, and lacking even the most common source of energy necessary to modern-day life forms—i.e., sugars—spreading outwards and eventually colonising every conceivable habitat on Earth, and along the way oxygenating it and sowing the seeds for many other life forms to emerge and evolve. We are justifiably in awe (and at times in despair when concerned about the environment) of the great journey of *Homo sapiens* out of Africa some 50,000–100,000 years ago, culminating in our habitation of all continents including Antarctica. This was enabled by our outsize brains, which provided us with the wherewithal to invent new tools, create societies, cities, civilisations, religions, all the way through to rockets and artificial intelligence. Bacteria are not so endowed. They are single cells and what they have achieved is a product of their chemical capabilities. Underlying this is their genetic material or their genome, its ability to change and stabilise in a manner best suited to their circumstances. This is what we call genetic evolution, which has, over several billion years, caused the emergence of a whole diversity of simple and complex bacterial individuals and species starting from a presumably primitive ancestor.

1 A small piece of advice: it might be best to give AI summaries produced by Google a pass for the time being.

 https://doi.org/10.11647/OBP.0446.00

Talking about evolution over such epochs creates a striking narrative. Yet, this is merely an accumulation of more humble changes that happen every day to permit bacteria to adapt from one day's circumstance to the next. These circumstances might be the external environment, or they can themselves be genetic. One genetic change can require that subsequent genetic changes must occur, and this also constitutes adaptation. The genome merely specifies a recipe that needs to be acted upon for anything to happen. How this recipe is interpreted by the cellular machinery and how this reading is regulated is yet another contributor to adaptation, one that can effect physiological switches in a matter of minutes, in contrast to genetic changes which take generations to establish.

This book is about bacterial genetic diversity, the processes that establish such diversity, and how this is driven by the need for bacteria to adapt to circumstances. Here we use a very broad definition of adaptation (from the Cambridge Dictionary): "the process of changing to suit different conditions," and not merely evolutionary adaptation that necessarily requires genetic change for adaptation; according to our use of this word, even physiological changes can be adaptive. Thus, tightly regulated cellular processes that read the information contained in the genome to produce life-sustaining molecules such as proteins are also central to adaptation, but mostly over a timescale different from that over which genetic changes operate. However, the genome and its reading are intertwined. Regulation of the processes that read the genome are also evolvable and in a manner that helps bacteria adapt. More than that, one drives the other. The way the genome is organised helps facilitate its seamless reading by the cellular machinery, and this organisation might have evolved in response to the need to ensure that the genome is read efficiently.

Finally, from a human standpoint, what bacteria are capable of doing is one thing, and what we know of it is another. Human knowledge of bacteria is nearly 350 years old and, like technological progress, has accelerated over time, reaching a crescendo over the last few decades.

The book is about bacterial adaptation. It begins by presenting a selection of people and ideas that have helped us understand bacteria. Much of this work, especially in the 20th century, was driven by the medical need to combat infectious diseases. Additionally, the ease with which some bacteria could be grown and manipulated in the laboratory allowed them to serve as models for studies of the fundamental processes that drive life. A whole body of research was facilitated by the plasticity and adaptability of bacteria, and in turn helped us understand bacterial adaptation. After nearly 300 years of scratching our heads, we finally developed an understanding of the place of bacteria in the universal tree of life only in the final quarter of the 20th century, before establishing how we can know what the genome encodes.

The book then takes a more technical turn and discusses how bacterial genomes span two orders of magnitude in size, but remain compact and rich in information. We will also see how natural selection, or 'survival of the fittest' as espoused by Charles Darwin, underlies the evolution of the bacterial genetic material, and how the bacterial genome differs from much of the mammalian genome in this respect. We explore how the genome changes, how it expands and how it contracts, and how all these facilitate adaptation. We will then attempt to understand the processes the cell uses to interpret its own genetic material and produce molecules that determine its traits, how these processes are regulated, how such regulation has played a role in shaping the genome, and how various contemporary researchers have addressed these questions.

Chapter 1 discusses the history of our study of bacteria, starting from their first microscopic observation in the 17th century. It then discusses how we learnt, mostly in the 19th century, about the vast array of functions they perform, from causing disease to running biogeochemical cycles. Medical microbiology came into its own with the discovery of antibiotics in the first half of the 20th century, simultaneously highlighting the role of microbial competition in the environment. The first indication of bacterial resistance to antimicrobial agents as early as in the 1920s also brought to light their metabolic flexibility and adaptability.

Chapter 2 first discusses the fascinating story of the discovery of viruses that prey on bacteria and how research (once again) demonstrating the flexibility of bacterial traits led to the stunning revelation that DNA and not protein was the genetic material of cellular life. This led to a cascade of discoveries that created the field of molecular biology, which not only helped us to understand our phylogenetic relationships with bacteria, but also culminated in the complete sequencing of the first cellular genomes towards the end of the 20th century.

In Chapter 3, we introduce the bacterial genome. We discuss how they span two orders of magnitude in size and at the smaller end of the spectrum can approximate the theoretical minimal genetic requirement for cellular life to exist. Even the largest bacterial genome is compact and information-rich, in contrast to 'junk'-filled genomes of the so-called higher eukaryotes like ourselves. We see how this is a reflection of the degree to which Darwinian natural selection operates on the genomes of various organisms.

In Chapter 4, we ask what the forces are that determine the range of bacterial genome sizes. Gene loss underlies parasitic lifestyles, whereas genome growth is driven by the remarkable phenomenon of horizontal gene transfer by which DNA is transferred, not vertically from parent to progeny, but from one organism to another unrelated one. We ask how the prevalence of these processes compare with the humbler mutation that changes one, or just a few, monomeric units of the DNA polymer at a time, and how these processes reflect the need for bacteria to adapt to their circumstances.

Finally in Chapter 5, we introduce the process of transcription by which the genetic material is read to eventually produce proteins that do the cell's work. We talk about how this process is regulated and how this enables rapid adaptation to changing situations. Genetic evolution by mutation or DNA loss or horizontal transfer meets physiological adaptation when genetic changes act on regulators of transcription. We ask how this plays out, especially in the face of collateral damage that any mutation affecting regulators of many processes can suffer. Then we conclude by asking how transcription and replication are factors in determining how the very organisation of the bacterial genome is determined.

Bacterial adaptation is an extraordinarily vast field. A search of the Pubmed database of scholarly biomedical literature with the key "(gene OR genetics OR genome) AND bacteria AND adaptation" returns nearly 60,000 journal articles! This will include influential papers, run-of-the mill works that are the bread and butter of scientists like me, as well as hidden gems that have unfortunately not received as much recognition as they deserve. It is impossible for anyone to read, absorb and write about the entirety of this vast literature. It would be extremely challenging to even represent all the many dimensions of bacterial adaptation that these papers encompass. I have taken a decision of my own volition, reflecting my interest in this field, to limit the ambit of this book to the principles underlying the organisation of the genome, the way it is read by the cell, how these two talk to each other and how all these things together result in bacterial adaptation.

I thank Ganesh Muthu for helping to compile and organise the figures for this book. Nitish Malhotra and Meghna Nandy also suggested figures for some sections. I thank Sunil Laxman for reading most of this manuscript and providing detailed comments and suggestions while also taking the liberty of correcting any minor typos he might have found, making life that much easier for me! I thank Anjali Vinodh, a young high school student at the time, for her comments on the first two chapters of this book. I thank the editorial team at OBP, Alessandra Tosi and Annie Hine in particular, for helping get the book in the shape it is in now. Thanks to Jeevanjot Singh for creating a lovely design for the book cover working with visual material created by Prerana Sudarshan, Ashitha Arun, Inder Raj Singh and Madhumitha Krishnaswamy in my lab. Many thanks to all the scientists and illustrators who made their work available under an open access licence, allowing me to embellish the contents of this book with their work. Working on a book does make demands, not only of the author, but also of his or her family, who may not be invested in his or her research interests. Many thanks to my father, Gayathri, my better half, and Harini and Hamsini for putting up with me.

Most of the first draft of the book (except the sections that almost wrote themselves) were written by hand with fountain pens, especially Indian handmade pens. A shout out to Gama Pens and to Mr. Pandurangan and Mr. Kandan at Ranga Pens for crafting these gorgeous pens!

Finally, I dedicate this book to the memory of Mrs Nagalakshmi, my mother and a teacher to many hundreds more!

1. All creatures great and small

1.1. Bacteria: numerous and diverse

The forms of life that we see around us constitute but a tiny proportion of all that lives. In other words, most of life on Earth—and possibly any that we may eventually find on say, Mars—are too small to be visible to the naked eye. They are microscopic. These microorganisms or microbes, despite their small form, have an outsized impact on our lives. These organisms often find a place in discourse as the causative agents of a variety of infectious diseases, the source of much morbidity and mortality. However, disease-causing microbes form a small minority of the vast diversity of microbes on our planet. Many microbes residing in our bodies, for example in the gut, perform reactions that enable routine digestion and many other bodily functions that are being increasingly attributed to the activities of these tiny denizens. Outside our own bodies and in the greater world around us, these microbes catalyse many unique reactions that sustain various essential biogeochemical cycles that breathe life into our planet; these include the fixation of atmospheric nitrogen gas into biochemically usable forms in the roots of leguminous plants, a process without which there would be no food to sustain us. To cut this long and much-articulated story short, there would never have been any large life forms in existence had microbes and their inventory of chemical capabilities not evolved. Any future catastrophe that somehow selectively annihilates microbes will, as a consequence, destroy most, if not all, other forms of life.

While it is not inappropriate to use a single word 'microbe' or 'microorganism' to refer to these invisible drivers of life, the fact is that microbes encompass organisms from all kingdoms of life: viruses, whose status as living organisms might be debatable; bacteria, single-celled beings representing enormous biochemical capabilities; the enigmatic archaea which, together with bacteria are often referred to as 'prokaryotes'; and tiny relatives of larger life forms including bread- and wine-making fungi and the plant-like algae, which, in contrast to the prokaryotes, are 'eukaryotes' with highly compartmentalised cell structures not seen in prokaryotes. The book will focus on bacteria, which, as a very large and diverse kingdom of life, are the most numerous cellular[1] life forms on our planet.

How large and diverse is the bacterial kingdom of life? In other words, how many bacteria are there on Earth? Or, what proportion of biomass on this planet is

1 The adjective cellular leaves out viruses.

 https://doi.org/10.11647/OBP.0446.01

bacterial? As one might guess, the answers to these questions are not straightforward to obtain. Let us begin with a complex, but in comparison to the bacterial question a straightforward problem: enumerating humans of various genders, ethnicities etc. on our planet. Most major economies have robust mechanisms for recording births and deaths, and decadal census-taking efforts are also common. These are mammoth exercises, and the creation of accurate population registers is critical to the delivery of essential welfare schemes. That said, such enumerations are not without their errors and uncertainties.[2] For example, the US census of 1940 undercounted ~3% of its population. This may not sound too bad, even if it could equal the population of an entire state, considering the magnitude of the exercise. However, the accuracy of census data may not be equal across geographies and demographics. The same US census undercounted African Americans by as much as 15%. Conscious efforts since then have brought down these errors to less than a quarter of a percentage, and not more than one percent for any demographic.[3]

Surveys of wild animal populations deal with far greater uncertainty. These surveys identify and count a small sample of the true population by trapping them in life or on camera, or even more indirectly by identifying their spoor. For example, Project Tiger, the famous effort for tiger conservation in India, estimated the tiger population in 2018 to 2,967, but with a range of 2603–3346.[4] This is for a single, large, 'visible' species. New species of small animals such as frogs are even now being discovered, making estimates of total animal populations rather uncertain. Going down the size scale, uncertainty increases exponentially when we talk about insect populations. It is believed that there might be 2–30 million species of insects; note the already wide range of possibilities here and that the range is only an informed guess. About a million species have been described and the current best estimate is around 5–6 million species.[5] These amount to 10^{18} individuals, or over 100 million insects for every human.[6] That the uncertainty in these numbers is large will not be surprising if we realise that most insect species are unknown and counting tiny, though visible to the naked eye, critters directly is well-nigh impossible.

Enumerating microorganisms, invisible to the naked eye and found in every nook and cranny including the most inhospitable and inaccessible environments, is a different ball game altogether. It is so forbidding to conceive and implement that

---

2 Dan Bouk, 'Error, Uncertainty, and the Shifting Ground of Census Data', HDSR, 26 May 202. https://hdsr.mitpress.mit.edu/pub/5zxrvthz/release/6

3 US Census Bureau Press Release CB22-CN.02, 10 March 2022. https://www.census.gov/newsroom/press-releases/2022/2020-census-estimates-of-undercount-and-overcount.html

4 Y.V. Jhala, Q. Qureshi and A.K. Nayak AK, 'Status of tigers, copredators and prey in India 2018', National Tiger Conservation Authority, Government of India and Wildlife Institute of India, Dehradun, 2020. http://moef.gov.in/wp-content/uploads/2020/07/Tiger-Status-Report-2018_For-Web_compressed_compressed.pdf

5 N.E. Stork, 'How Many Species of Insects and Other Terrestrial Arthropods Are There on Earth?', *Annual Review of Entomology* 63 (2018), 31–45. https://doi.org/10.1146/annurev-ento-020117-043348

6 The Smithsonian Museum Information Sheet 18, 1996. https://www.si.edu/spotlight/buginfo/bugnos

the first landmark global survey of prokaryotes (the subset comprising bacteria and archaea), to my knowledge, was published by Whitman, Coleman and Wiebe only in 1998, well over 300 years after these organisms were first observed under a microscope.[7] In their seminal work, Whitman and co-workers compiled data on local densities of prokaryotes in various sample habitats and extrapolated these to estimate that the total number of prokaryotes is of the order of 10^{30}–10^{31}. Assuming that bacteria account for 90% of all prokaryotes, the total bacterial population would be 10^{29}–10^{30}. There are 100 billion–1 trillion bacteria to every insect. Large uncertainties arise in these estimates from the assumption that the samples used to anchor these projections are representative of the whole population. Even within each sample, the counting exercise uses various microscopic techniques under the assumption that these techniques work equally well for all prokaryotes, including those we may know nothing about. Neither assumption is entirely valid, and therefore these estimates are only the best possible under the circumstances.

Most bacteria (~10^{29}) are found in the sub-surface—defined as spaces more than eight metres below ground in land, and marine sediments ten centimetres or more below the ocean floor. The uncertainty in population estimates is greatest for these habitats and, keeping this in mind, Whitman and colleagues consider their projections an underestimation. The open ocean and topsoil are other habitats supporting large prokaryotic populations, about ten times less than in the sub-surface. Even the inhospitable polar regions house a million times as many prokaryotes as there are insects on the entire planet. The number of bacterial individuals on the average human body—one to ten times the number of human cells—is often used to tout the importance of bacteria in our lives. Given that the total human population is only under nine billion, it becomes obvious that an anthropocentric view of the bacterial world does little justice to their dominance of our planet and then role in the sustenance of all life.

Another approach to estimating populations of various organisms involves quantifying their biomass, which can be complementary to the difficult problem of accurately counting microbial cells directly. The total biomass of any large, multi-cellular organism will be many orders of magnitude greater than that of a single-celled bacterium or an archaeon. Therefore, the dominance of bacteria (and microbes in general) is considerably reduced when measured in terms of their contribution to biomass. A recent study by Bar-On, Philips and Milo, published in 2018, estimated the total carbon contained in different kinds of organisms and used this as a measure of biomass.[8] These researchers showed that bacteria, their small size notwithstanding, contribute 15% of the total of biomass carbon on Earth. They are second only to plants, which make up a whopping 80% of all biomass carbon. That all other forms of life together contribute a measly 5% is a sobering thought.

7 W.B. Whitman, D.C. Coleman and W.J. Wiebe, 'Prokaryotes: the unseen majority', *Proc. Natl. Acad. Sci. USA* 95 (1998), 6578–6583. https://doi.org/10.1073/pnas.95.12.6578

8 Y.V. Bar-On, R. Philips and R. Milo, 'The biomass distribution on Earth', *Proc. Natl. Acad. Sci. USA* 115 (2018), 6506–6511. https://doi.org/10.1073/pnas.1711842115

Whitman and colleagues' 1998 census, described above, estimated prokaryotic carbon to be 60–100% of that contained in plants. Bar-On and colleagues' more recent work, based on updated data, considerably lowers this estimate. However, a large proportion of plant biomass would be accounted for by non-living material such as tree bark, making the contribution of bacteria to total 'living carbon' higher than the 15% projected by Bar-On and colleagues. While plant biomass, living or otherwise, dominated bacterial biomass, bacteria account for ~70% of aquatic carbon. Biological material also contains substantial nitrogen (N) and phosphorus (P) and prokaryotic N and P may be ten times that in plants. Once again, it is important to emphasise that hazarding these estimates represent heroic efforts, and that projections for census population sizes or biomass of organisms that are microscopic and mostly inhabit difficult-to-access sites are subject to great uncertainty. For instance, Bar-On and colleagues estimate that their calculation of bacterial biomass carries an uncertainty that is seven to eight times as much as that for plant biomass. Taken together, bacteria, and more broadly prokaryotes and microorganisms, form the most dominant forms of life in terms of their sheer numbers and contribute large proportions of the planet's biological carbon, nitrogen and phosphorus.

That bacteria are found in such a wide range of habitats throughout the planet should already suggest but not necessarily establish the great diversity of capabilities this kingdom of life encompasses. Although the large number of prokaryotes could—in a non-existent zero-error life-producing factory—represent so many identical copies of a single sort of bacterium (as it was often believed till the second half of the 19th century), the fact is that not all members, even of the same species, are equal. Let us assume that one every 10^6 reproduction events that bacteria perform produces progeny that differs substantially from its parent and that each reproduction event produces one child bacterium. Whitman and colleagues estimate that prokaryotes in the upper 200m of the open ocean reproduce every 6–25 days. Compare this to our own generation time of 20–30 years. If each of the 10^{28} prokaryotes in this habitat were to reproduce every 15 days on average, there would be over 2×10^{29} reproductions every year and over 2×10^{31} per century. We then expect ~10^{25} new variants to be generated every 100 years in the upper layers of our oceans. Larger organisms, with their smaller populations and often much longer generation times, would produce far fewer variants in the same timeframe even though reproduction involving two parents would generate a child different from its parents pretty much every time. Multicellular organisms also include many cellular reproduction events within their bodies, but such somatic events would not be passed down generations. These complexities in the manner in which a single-celled bacterium and a multicellular insect or human or plant reproduces notwithstanding, it is abundantly clear that prokaryotes would best larger, multicellular life forms in the numbers game, representing not only the most numerous, but also the most diverse forms of free-living life on Earth.

1.2. Animalcules

When and how did we become aware of the very existence of microorganisms? How did our attitudes and approaches to their study evolve subsequently? We must go back in time to Renaissance and post-Renaissance Europe, and what is sometimes referred to as the scientific revolution, to begin to answer this. Though the early discovery of microbes and its immediate fallout are not of any urgent relevance to the more molecular scope of this book, I will spend some time on this subject as it refers to a key period in the development of science, and also because of its sheer improbability and as a tribute to an odd enthusiast's indefatigable spirit and energy, fortitude and unmatched skill. Hence a brief detour into the history of science.

The discovery of microorganisms was by direct observation under the microscope in the Netherlands in the late 17th century. Though ancient Romans may have developed techniques for shaping glasses, it was the Arab scholar ibn-Heitam (also spelt ibn-Haytham) who is credited with having proposed, in the 10th–11th centuries, that smoothened lenses could aid vision. This idea was put into practice later in the 13th century in Italy.[9] A 14th-century painting by Tomasso de Moderna shows a biblical scholar working on a text wearing spectacles.[10] Scholarship in optics accumulated over centuries resulting in the principles of telescopes being established in the 16th century and formally invented in the first decade of the 17th century. It was around this time that Galileo and a group of intrepid astronomers "turned telescopes towards the heavens", revolutionsing astronomy.[11]

Galileo was responsible, albeit indirectly, for the first microscopic observations, whose details survive to this day. In 1624, Galileo wrote to Federico Cesi, who had founded an early scientific society in Rome of which Galileo was a member, describing his microscope and inviting Cesi to observe "thousands upon thousands of tiny details".[12] Cesi and his colleague Francesco Stelluti published their microscopic observations on the anatomy of a bee in a book called *Apiarium* in 1625.[13] Only six physical copies of this book are known to exist today, making each an invaluable original record of the beginnings of microscopy.

Progress in microscopy led Henry Power in England to write—in 1661—an ode to the microscope[14] in which he hoped that the microscope would, before long, allow

9 'The history of glasses', *Zeiss*, 12 November 2021. https://www.zeiss.co.in/vision-care/better-vision/understanding-vision/the-history-of-glasses.html

10 'File:Tommaso da modena, ritratti di domenicani (Ugo di Provenza) 1352 150cm, treviso, ex convento di san niccolò, sala del capitolo.jpg', Wikimedia Commons. https://commons.wikimedia.org/wiki/File:Tommaso_da_modena,_ritratti_di_domenicani_(Ugo_di_Provenza)_1352_150cm,_treviso,_ex_convento_di_san_niccol%C3%B2,_sala_del_capitolo.jpg

11 'Galileo and the Telescope', Library of Congress. https://www.loc.gov/collections/finding-our-place-in-the-cosmos-with-carl-sagan/articles-and-essays/modeling-the-cosmos/galileo-and-the-telescope

12 'Galileo's Instruments', ETH Zurich. https://library.ethz.ch/en/locations-and-media/platforms/virtual-exhibitions/galileo-galilei/galileos-works/galileos-instruments.html

13 Kerry Grens, '*Apiarium*, 1625', *The Scientist*, 1 March 2015. https://www.the-scientist.com/foundations/apiarium-1625-35869

14 T. Cowles, 'Dr. Henry Power's poem on the microscope', *Isis* 21 (1934), 71–80.

us "to see the magnetical effluviums of the lodestone, the solary atoms of light, the springy particles of air, the constant and tumultuous motion of the atoms of all fluid bodies." Power also published his microscopic observations of the eyes of a spider as part of the voluminous series *Philosophy in Three Books* in 1664.[15] The magnum opus of microscopy published in this period was Robert Hooke's *Micrographia*.[16] This book contains gorgeous drawings of the eye of a fly, the ultrastructure of the stitches in a piece of cloth, as well as observations of a mould, a microbe that forms structures that are in fact visible to the naked eye. *Micrographia* also introduced the term 'cell' to describe the elements that make up the honeycomb-like structure of cork, a plant material.

Micrographia was the first publication of the Royal Society of London, formed in 1660 as a "College for the promoting of Physico-Mathematical Experimental Learning."[17] Robert Hooke, the author of *Micrographia*, was the Curator of Experiments in the Royal Society. Another important figure in the story of the discovery of microbes and bacteria was Henry Oldenburg, the first Secretary of the Royal Society and, in 1664–65, the founding Editor of the *Philosophical Transactions of the Royal Society*, the world's first and longest-running scientific journal. Henry Oldenburg was a vocal advocate of communicating research outputs broadly, far beyond the scope of private correspondences. As the Introduction to the inaugural edition of the *Philosophical Transactions* says, "solid and useful knowledge may be further entertained, ingenious endeavours and undertakings cherished, and those, addicted to and conversant in such matters, may be invited and encouraged to search, try and find out new things, impart their knowledge to one another and contribute what they can to the grand design of improving Natural knowledge, and perfecting all philosophical arts, and sciences".[18]

Oldenburg was a voracious communicator, maintaining written correspondences with a wide range of natural philosophers from around Europe. So worried was he, given the tense political climate in Europe at that time, of his correspondences being misconstrued as treasonous by the authorities that he often used the name Grubendol—an anagram of Oldenburg—to sign his letters. This did not serve him well, for he found himself incarcerated in the Tower of London in 1667 for suspected treason. Luckily however, he stayed there for only three months. He lived for ten years after his release from prison, during which time he was introduced to Antonie van Leeuwenhoek, who would soon become the preeminent microscopist of his time and unsurpassed for another century and a half, to the Royal Society. Among Oldenburg's correspondents was a physician called Regenerus de Graaf from Delft, Netherlands. In 1673, de Graaf wrote to Oldenburg introducing "a certain most ingenious person here,

15 https://archive.org/details/b30331171_0002

16 Robert Hooke, *Micrographia, or, Some physiological descriptions of minute bodies made by magnifying glasses with observations and inquiries thereupon* (London: Martyn & Allestry, 1665). https://archive.org/details/mobot31753000817897

17 'History of the Royal Society', The Royal Society. https://royalsociety.org/about-us/who-we-are/history/

18 'An Introduction to this tract', *Phil Trans Royal Soc.* 1 (1665), 1–2. https://doi.org/10.1098/rstl.1665.0002

named Leeuwenhoek who has devised microscopes which far surpass those which we have hitherto seen".[19] He referred to an enclosed letter from Leeuwenhoek, containing a "sample of his work", which is now presumed lost.

Leeuwenhoek was a draper by profession and had held various administrative posts in the municipality of Delft. While commending Leeuwenhoek in a letter to Robert Hooke, the noted physicist Constantijn Huygens noted that Leeuwenhoek was a "person unlearned in sciences and languages, but of his own nature exceedingly curious and industrious."[20] This is a sentiment Leeuwenhoek himself admitted to in his letter to Oldenburg that followed the latter's response to de Graaf's introduction of Leeuwenhoek. Leeuwenhoek claimed that he had "no style, or pen, wherewith to express my thoughts properly... I have not been brought up to languages or arts but only to business." In the same letter he also expressed his reluctance to write about his work to a broad audience, for he did not "gladly suffer contradiction or censure from others."[21] Oldenburg's letter, as a reply to which Leeuwenhoek had made this assertion, however appears to have appeased the Dutchman who then gladly noted that his "memoir ... did not displease the Royal Society". He enclosed drawings made by others more capable than he in art, certifying, "my observations and thoughts are the outcome of my own unaided impulse and curiosity alone."[22] His early letters primarily described his observations of fluids in plants and animals, everyday objects and living things like cloth, flies and tree bark. These were largely in response to observations of similar objects published for others by the Royal Society, but at times achieving hitherto unseen levels of detail. It was finally in 1674 that Leeuwenhoek stirred the proverbial hornet's nest.

In a letter dated September 1674,[23] Leeuwenhoek noted that the water in a lake called Berkelse Mere was clear in winter but then "becomes whitish" in summer with "little green clouds floating in it." He further noted the presence of "very many little *animalcules*", which he described as "little creatures ... above a thousand times smaller than the smallest ones I have ever yet seen upon the rind of cheese". He expressed his wonder at seeing these animalcules move in the water: "... so swift and so various upwards, downwards and round about that (it was) wonderful to see." This account was met with disbelief. His status as an unlettered man, "unlearned in sciences and languages" did not help his cause. Several of his letters from this period went unpublished. Leeuwenhoek wrote to Hooke a few years later complaining that he suffered "many contradictions"; and that it was often said that he was telling "fairy tales about the little animals."

---

19 Most of the material on Leeuwenhoek in this text, including various quotes and translations of Leeuwenhoek's Dutch to English, is from C. Dobell, *Antony van Leeuwenhoek and his "Little Animals"* (New York: Harcourt, Brace & Co., 1932). https://archive.org/details/antonyvanleeuwen00dobe

20 Dobell, *Antony van Leeuwenhoek*, p. 43.

21 Ibid., p. 42.

22 Ibid.

23 N. Lane, 'The unseen world: reflections on Leeuwenhoek (1677) "Concerning little animals"', *Phil Trans Royal Soc. B.* 370 (2015), 20140344. https://doi.org/10.1098/rstb.2014.0344

In a now-famous letter of 1676, published in 1677, Leeuwenhoek described his studies observing animalcules in rainwater that had stood in an earthen pot for a few days, in addition to those in well water and water collected from the sea coast.[24] He provided detailed descriptions of these creatures: "bodies consisted of 5, 6, 7 or 8 very clear globules", "animalcules whose figure was an oval", "the circumference of one [...] is not so great as the thickness of a hair on a mite" are short excerpts from his copious and graphic descriptions. His descriptions of these animalcules were sufficiently detailed for the microbiologist, Clifford Dobell, researching Leeuwenhoek in the early 20th century, to make positive identifications of several as specific members of the protozoa, a type of eukaryotic microbe.[25]

Leeuwenhoek, "with great wonder", then made his momentous discovery of what we now know are bacteria frolicking in "1/3 ounce of whole pepper" left to stand and soften in water for three weeks. He described bacteria as "wee animals", 100 of which, if "stretched out one against another could not reach to the length of a grain in coarse sand; [...] ten thousand of these living creatures could scarce equal the bulk of a coarse sand grain."[26] His adjectives for these organisms included "round ones", "a little animal that was three of four times as long as broad", "little eels, which were even smaller than the very tiny eels spoken of before", "exceeding little animalcules, to which, because of their littleness, no shape can be given." He regularly noted that these organisms moved rapidly, pointing to their vitality. His studies went beyond descriptions of the appearance and behaviours of animalcules but also demonstrated the antimicrobial property of vinegar in a systematic manner.

Whereas his 1674 letter on animalcules in Berkelse Mere only attracted disbelief and derision, his careful descriptions in 1676 created quite a sensation. Several meetings of the Royal Society were held expressly to confirm Leeuwenhoek's claims.[27] Robert Hooke and one Nehemiah Grew were tasked with attempting to reproduce Leeuwenhoek's observations. They failed, but were, in response, urged to build better microscopes and examine pepper water with these. While the Royal Society was holding these deliberations, Leeuwenhoek recruited "two clergymen, one notary and eight other trustworthy persons who had seen the animalcules moving in pepper water"[28] to provide testimony bolstering his claims. Eventually, on 15 November 1677, Hooke reported seeing "divers very small creatures swimming up and down" in several "liquors" and "even in Raine itself and that they had various shapes and differing motions."[29] Leeuwenhoek stood vindicated. And thus came to pass the discovery of microbes and bacteria by the draper Antonie van Leeuwenhoek of Delft.

A little over two years later, Leeuwenhoek was elected Fellow of the Royal Society, a title considered a great honour to this day. Around this time, Leeuwenhoek had also

24 As translated in Dobell, *Antony van Leeuwenhoek*, pp. 118, 123, and 157.
25 Dobell, *Antony van Leeuwenhoek*.
26 Ibid., p. 133.
27 J.B. Stein, 'On the trail of Van Leeuwenhoek', *The Scientific Monthly* 32 (1931), 116–134.
28 Ibid., p. 123.
29 Ibid.

published results, which, when reproduced some 240 years later by Martinus Beijerinck, established that anaerobic bacteria, which do not require oxygen for respiration, had been discovered by Leeuwenhoek a century before the very discovery of oxygen. Louis Pasteur, who has been credited with their discovery, thus rediscovered these bacteria centuries after Leeuwenhoek's primary work. By the 1670s, Leeuwenhoek was considered "the great man of the century" as claimed by Constantijn Huygens in a letter to his illustrious brother Christiaan. Leeuwenhoek was even invited to meet the Tsar, Peter The Great, who spent "no less than two hours" with him and "on taking leave shook Leeuwenhoek by the hand."[30]

At the end of the 17th century, Robert Hooke, speaking about microscopes, asserted that they "are now reduced to a single Votary which is Mr. Leeuwenhoek."[31] This was probably because Leeuwenhoek's skill in lens making—irrespective of whether it was inspired by Hooke or not—was so unmatched that his observations with single-lens simple microscopes would prove superior to those made using the more complex compound microscopes favoured by Hooke and others. It might have even required a certain fortitude to work with single-lens microscopes, something not many might have possessed. For instance, in 1678 Hooke considered single-lens microscopes "offensive to my eye and to have much strained the sight."[32] Leeuwenhoek also owned a great ability to prepare samples for microscopic examination, as attested to by some of his original specimens that had been left untouched but remarkably well preserved at the Royal Society for 300 years until their discovery by Brian Ford.[33] That said, Leeuwenhoek, despite living to his nineties, did not pass his knowledge down to the next generation, expressing that he "never had any desire to teach" while complaining that "work[ing] with all one's soul was not attractive to most young people."[34] It might be that it was his reluctance to teach that ensured that microscopic—and indeed all—study of microorganisms would languish till the middle of the 19th century, but for the occasional treatise. For example, the 1720 work of one Dr. Benjamin Marten posited that the ultimate cause of consumptive disease "may possibly be some certain species of animalcula or wonderfully minute living creatures that [...] are inimical to our nature."[35]

30 As recorded by one van Loon—who knew Leeuwenhoek personally—and quoted in Dobell, *Antony van Leeuwenhoek*, p. 55.

31 Robert Hooke's 1692 lecture on the history of the microscope and the telescope: https://lensonleeuwenhoek.net/content/hooke-single-votary-mr-leeuwenhoek

32 H. Gest, 'The Discovery of Microorganisms by Robert Hooke and Antoni van Leeuwenhoek, Fellows of the Royal Society', *Notes and Records of the Royal Society of London* 58 (2004), 187–201, p. 197. https://doi.org/10.1098/rsnr.2004.0055

33 B.J. Ford, 'The van Leeuwenhoek specimens', *Notes and Records of the Royal Society of London* 36 (1981), 37–59. https://doi.org/10.1098/rsnr.1981.0003

34 Stein, 1931, p. 126.

35 W. Bulloch, *The History of Bacteriology* (Oxford: Oxford University Press, 1938), p. 34. https://archive.org/details/b2982378x/

1.3. The golden age of microbiology: chemistry, biology and ecology

The idea that microorganisms—or at least, particles invisible to the naked eye—cause disease is old, having been expressed in some form or another in the ancient world and in the mediaeval Middle East and Europe. However, it found formal experimental proof only in the late 19th century. We owe this proof to Louis Pasteur and Robert Koch, the two most famous microbiologists of the 19th century and indeed of all time. Louis Pasteur was the son of a poorly educated French tanner. Despite showing limited scholastic ability as a child, Pasteur found himself studying chemistry at the prestigious École Normale Supérieure in Paris. Not long after finishing college, he accepted a professorship in chemistry at the University of Strasbourg (1849) and later at Lille (1854). His contributions to microbiology stemmed from his expertise in chemistry and a deep interest in solving practical problems using science. In his own words, "There are no such things as pure and applied science. There is only science and the application of science."[36]

Pasteur's first major scientific discovery, arrived at even before starting his professorship in Strasbourg as a young man, was in pure chemistry. After much painstaking work, he isolated the two forms, one being the mirror image of the other, of the same molecule: tartaric acid. Such variants of the same compound are called enantiomers or stereoisomers. Ten years later, he showed how biological processes can selectively produce one and not the other enantiomer of tartaric acid, something not easily achieved in the laboratory. This finding would lead Pasteur to show how wine fermentation is the result of the action of microorganisms. This would in turn lead him to famously demolish the idea of the spontaneous generation of life and contribute to proving the germ theory of disease.

Why Pasteur moved on from pure chemistry to microbiology around 1860 is a matter of conjecture.[37] A matter of fact, however, is that the wine industry around Lille was suffering from poor production quality because the fermentation processes used there often produced lactic acid instead of the desired ethanol. Pasteur was made aware of this problem by one Mr Bigo, whose commerce involved producing ethanol by fermenting beetroot. Pasteur was, at that time, a professor at the University of Lille at which the local minister for public instruction had exhorted the faculty to "appropriate to yourselves the special applications suitable to the real wants of the surrounding country".[38] Pasteur, with his own fascination for and commitment to the application of science to practical problems, would have needed no additional encouragement to take up the task of solving the local wine industry's problem. Pasteur noticed that

36 A. Ullmann, 'Spontaneous Generation', *Brittanica*. https://www.britannica.com/biography/Louis-Pasteur/Vaccine-development

37 The interested reader may refer to the following article for a discussion: J. Gal, 'The discovery of biological enantioselectivity: Louis Pasteur and the fermentation of tartaric acid, 1857—a review and analysis 150 years later', *Chirality* 20 (2007), 5–19. https://doi.org/10.1002/chir.20494

38 L. Ligon, 'Louis Pasteur: a controversial figure in a debate on scientific ethics', *Seminars in Paediatric Infectious Diseases* 13 (2002), 134–141. https://doi.org/10.1053/spid.2002.125138

visually different kinds of microorganisms populated broths producing ethanol and lactic acid. He further observed that fermentation was accompanied by the production of one and not the other enantiomer of certain chemical compounds, an indication of biological activity. Over several years, he isolated microbes responsible for the proper fermentation of various starting materials to ethanol. He also showed that heating these mixtures would sterilise them and stop fermentation, thus establishing the basis for pasteurisation.

Though microorganisms such as yeasts—a type of fungi—had been known to be present in wine for some time, the theory that microbial life forms arose spontaneously from inanimate matter was rife at the time, and yeast were believed to be a product, and not the cause, of fermentation. Fermentation itself was supposed to be caused by 'vibrations' of some kind. Pasteur showed that the yeast responsible for fermenting grapes into wine grew on the skin of the fruit; the sterile fleshy part of the grape alone would not undergo fermentation, which demonstrated that these microbes did not arise spontaneously from unrelated components in the broth. He later used the ability of a juice to undergo fermentation as evidence for the presence of living microorganisms in a series of experiments designed to deliver a "mortal blow" to the prevalent "doctrine of spontaneous generation."[39] He concluded emphatically in 1864: "There is no circumstance known in which it can be affirmed that microscopic beings came into the world without germs, without parents similar to themselves".[40] Of course, while this may be true under the conditions prevalent on Earth then and now, life would have originally arisen at least once from inanimate bodies, if not on primordial Earth, somewhere in the Universe.

Pasteur went on to prove, in a variety of contexts, the germ theory of disease, i.e., the theory that certain diseases are caused by microorganisms. Of particular importance was his work on anthrax, which he summarised thus in 1878:

> To demonstrate experimentally that a microscopic organism is the cause of a disease [...] subject the microbe to the method of cultivation out of the body [...] Having cultivated it a great many times in a sterile fluid, each culture being started with a minute drop from the preceding, we then demonstrated that the product of the last culture was capable of further development and of producing anthrax with all its symptoms. Such is—as we believe—the indisputable proof that anthrax is a bacterial disease.[41]

In the early 1880s, Pasteur demonstrated the efficacy of immunisation in preventing cholera in chicken and anthrax in sheep. Later, he performed human vaccination preventing the development of rabies in a child who grew up to be eternally grateful to the memory of Pasteur. These works were a significant reiteration and development of the experiments performed some 80 years earlier by Edward Jenner to develop

39 Ibid.

40 R. Valerie-Radot, *The Life of Louis Pasteur*, trans. by R. L. Devonshire (New York: McClure, Phillips and Co., 1902). https://archive.org/details/lifeofpasteur02vall/page/n5/mode/2up

41 Pasteur's Papers on the Germ Theory, Historic Public Health Articles. https://biotech.law.lsu.edu/cphl/history/articles/pasteur.htm

smallpox vaccination, which in turn had helped substantiate empirical knowledge on immunisation against smallpox from mediaeval India, the Middle East and North Africa.[42]

It was in the early 1880s, following his work on anthrax, that Pasteur came into conflict with the younger German physician-microbiologist Robert Koch. Koch became a physician at a young age in 1866. In 1873, five years before Pasteur's work on anthrax was read before the French Academy of Sciences, Koch, of his own initiative, had begun observing the blood of sheep that had died of anthrax under the microscope. He noticed that these specimens contained rod-shaped bacteria first described 10 years earlier by a French scientist named Davine. A year later, Koch described the life cycle of the anthrax bacterium as one that alternates between a rod-shaped cellular form and a dormant spore form. His discovery of the spore form, which is found in the soil, explained how sheep could become sick on consuming spore-infested soil. Koch showed that bacteria from the blood of an infected animal could be used to cause disease in an otherwise healthy sheep. This finding led him to propose what are called 'Koch's Postulates', a set of diagnostic criteria that must be satisfied before proving that a microorganism was the cause of a disease. He achieved great fame in the 1880s by isolating and identifying the causative agent of tuberculosis in man and cattle.

Koch and Pasteur, both patriotic individuals living during times of conflict between their nations, met under cordial circumstances in London in 1881. Pasteur applauded the younger Koch's demonstration of methods for growing bacteria on solid media and staining them for viewing under the microscope. Soon after their meeting however, Koch and his students criticised Pasteur's work on anthrax as "little which is new, and that which is new is erroneous."[43] Though Pasteur responded to Koch's criticism, the latter replied aggressively, leading to much acrimony between the two. Koch found the quality of evidence that Pasteur had used in his work as insufficient to substantiate the claims made. Pasteur's work might not have met the exacting criteria prescribed by Koch's Postulates, but he did get it right. Scientific arguments apart, national and linguistic factors almost certainly played their part in the Pasteur-Koch debate, a cautionary note against the idealistic view of science one sometimes hears in conversations today.

Pasteur's scientific integrity has also recently come under scrutiny. He might not have given due credit to co-workers and others. For instance, Jean-Joseph Henri Toussaint had developed a vaccine for anthrax before Pasteur did and had also been given credit for the discovery by Koch. Though Pasteur appears to have admitted that Toussaint was first to the discovery in his private correspondence, he never gave credit to Toussaint in public. He might have even suppressed experimental data that went against what he wanted to prove; however, it did help his cause that he was

---

42 J.Z. Holwell, *An Account of the Manner of Inoculating for the Small Pox in the East Indies* (London: Becket & De Hondt, 1767). https://archive.org/details/accountofmannero00holw

43 A. Ullmann, 'Pasteur-Koch: Distinctive ways of thinking about infectious diseases', *Microbe* 2 (2007), 383–387.

generally proven correct in his conclusions. Pasteur, in sum, has been revealed as "a great imperfect man who was both a strong diligent and driven researcher and teacher and a sometimes secretive and deceptive brute."[44]

Koch was a physician, and his interest in microbiology arose from its possible medical relevance. Pasteur, on the other hand, was interested more broadly in science, which nonetheless led him to make seminal contributions to medicine and medical microbiology. Though their work, thanks in part to its obvious impact on human health, found great prominence in public discourse and is regularly presented in school textbooks today, our understanding of the role of microbes and bacteria on our planet goes well beyond medical microbiology into the realm of what is referred to as general microbiology. This is apparent from our discussion earlier, showing how direct interaction between the animal or human world and microbes is but a miniscule component of the microbial world. Among those who contributed pioneering work to general microbiology in the 19th century were Sergei Winogradsky and Martinus Beijerinck.

Sergei Winogradsky[45] was born to a wealthy family in Kyiv, Ukraine. He found school and graduation in his hometown tedious and uninspiring and moved to the famed University of St Petersburg to study music. He then shifted to the natural sciences division, which hosted distinguished scientists such as Dmitri Mendeleev, the creator of the periodic table, and Elie Metchnikoff, who would later win a Nobel Prize for his work in immunology. After three years of study, Winogradsky pursued research work in the laboratory of the botanist Andrei Famintsyn. Here, he worked with a species of yeast that had also attracted the attention of Pasteur in his fermentation days. Winogradsky developed methods for isolating this yeast in pure culture by serial dilutions, and performed experiments that would later in 1953 be described by Nobel prize-winner Selman Waksman as "among the first careful investigations ever made on the influence of controlled environment on the growth of microorganisms in pure culture, under well-defined conditions."[46] Soon the unfavourable political conditions in St Petersburg, and its inhospitable weather, which was disagreeable to his wife, forced Winogradsky westwards to join Anton deBary's laboratory in Strasbourg in 1885.

In Strasbourg, Winogradsky worked on the bacterium *Beggiatoa*, which was known to accumulate granules of sulphur in its cell body. The eminent microbiologist Ferdinand Cohn had shown that several organisms that grow in sulphur springs gather these granules. Instead of working with the *Beggiatoa* already collected in the

---

44 Donald P. Francis, 'Book Review—The Private Science of Louis Pasteur by Gerald L. Geison', *The Body*, 1 February 1996. https://www.thebody.com/article/book-review-private-science-louis-pasteurby-gerald-l-geison

45 The material on Sergei Winogradsky was primarily sourced from the following review of his life and work: M. Dworkin, 'Sergei Winogradsky: a founder of modern microbiology and the first microbial ecologist', *FEMS Microbiology Reviews* 36 (2012), 364–379. https://doi.org/10.1111/j.1574-6976.2011.00299.x

46 Ibid., p. 366.

lab, Winogradsky travelled to sulphur springs in Germany and Switzerland to collect filamentous bacterial masses growing there. After quickly confirming that cells freshly isolated from these springs were indeed loaded with sulphur granules, Winogradsky asked, "Is the sulphur in the cells of *Beggiatoa* produced by the reduction of sulphate or by the oxidation of hydrogen sulphide?"[47] After demonstrating that cells grown in the lab in the presence of hydrogen sulphide grew well and accumulated sulphur granules, whereas those grown in calcium sulphate quickly died, Winogradsky concluded, "development of the sulphur granules occurs only during hydrogen sulphide oxidation; it is impossible to conclude that the process occurs at one time by the reduction of sulphate and at another by the oxidation of hydrogen sulphide"[48].

Winogradsky would not end the story here. He moved on to ask why: "Why does *Beggiatoa* need so much sulphur? [...] is it assimilated or excreted?" He proved that the sulphur is assimilated by being "further oxidised to its highest oxidation state, sulphuric acid."[49] He had thus shown that in these organisms, sulphur is the "sole respiratory source", the equivalent of "carbohydrates in other organisms."[50] Winogradsky had discovered chemolithotrophy, i.e., the ability of an organism to derive energy by oxidising inorganic compounds, a phenomenon very foreign to our own capabilities as *Homo sapiens*, but one that is essential to sulphur and nitrogen cycles and hence the very sustenance of life on our planet.

Winogradsky then sought to return to his native Kyiv, but failed to secure a job there or in St Petersburg. He therefore moved instead to Zurich to study nitrification, the process by which ammonia is oxidised to nitrate. This would, in principle, be similar to the oxidation of hydrogen sulphide to sulphate. The French scientist Jean Jacques Schoesling had shown in 1868 that the reverse process, i.e., the release of nitrogen gas from nitrates, occurs in some biological fluids. Such a process, if one-way and widespread, would be disastrous, for example by depleting the soil of essential nitrates. Pasteur had earlier proposed that nitrification did occur and was in fact a biological process. Later in 1877, Schloesing and colleagues demonstrated nitrification in sewage. Winogradsky entered this field with the objective of establishing the occurrence of biological nitrification with the experimental rigour espoused by Robert Koch. He isolated the bacteria responsible for nitrification initially as a mixed culture, though he believed that he had one type of organism in pure culture. Later, following or in parallel with the work of the English chemist Robert Warrington, Winogradsky proved that nitrification of ammonia to nitrate was a two-step process with nitrite being the intermediate, and that each step was carried out by a different type of bacteria. Then he moved away from pure cultures to study bacteria in their complex natural environment, the soil in this case. He demonstrated how the interaction between the pair of nitrifying bacteria with other, presumably biotic, components of

47 Ibid., p. 368.
48 Ibid.
49 Ibid.
50 Ibid., p. 369.

the soil was essential for the complete conversion of ammonia to nitrate. This would be the second example of chemolithotrophy—the first being the result of his earlier work in Strasbourg—that Winogradsky established. In the case of nitrifying bacteria however, Winogradsky additionally showed that inorganic carbonate, derived from carbon dioxide, was the source of cellular carbon; construction of biological material from inorganic carbon had previously been known only to be achieved with the help of solar energy during photosynthesis.

In 1891, after politely declining an offer from Louis Pasteur to join his institute, Winogradsky returned to St Petersburg where he isolated for the first time a free-living, anaerobic bacterium capable of nitrogen fixation, i.e., the conversion of nitrogen gas into ammonia, one of the forms of nitrogen that can be assimilated by plants. This, along with the 1888 work of Hermann Hellriegel and Hermann Wilfarth demonstrating symbiotic nitrogen fixation in the roots of leguminous plants, 'completed' the nitrogen cycle. In 1902, Winogradsky did finally join the Pasteur Institute where he was a vocal advocate of microbial ecology, exhorting microbiologists to study microbes in conditions similar to their natural habitat, something he had done, in part, in his earlier studies. He is sometimes considered as the first microbial ecologist, although one can argue that Leeuwenhoek himself had taken an ecological approach to discovering microbes and that Winogradsky's contemporary, Martinus Beijerinck, another great microbiologist who pursued his research in Delft, had also taken a similar approach.

Martinus Beijerinck was born in Amsterdam in the Netherlands.[51] The failure of his family's tobacco business forced them to move to the provinces in the year of his birth. Poverty meant that the young Martinus was schooled at home till the age of 12. Later, at secondary school he was greatly influenced by his botany teacher, one Mr. van Eeden. His precocious talent and interest in botany was revealed when, at the age of 15, Beijerinck won a contest for his collection of 150 species of plants. With help from his brother and uncle, he entered the Delft Polytechnical School where he studied chemistry. At Delft, he encountered J.H. Van 't Hoff, later the winner of the first Nobel Prize in chemistry, who would become his life-long advisor. Following the completion of his studies, Beijerinck took up a couple of teaching jobs in which he did not particularly distinguish himself except as a demanding teacher who would often reveal his impatience with his students by shouting at them. In 1885, he joined an alcohol factory in Delft where he set up an outstanding microbiology laboratory. The Dutch government would later set up a special position in the chemistry department at the Delft Polytechnical School, which he accepted in 1985 to mark his return to academic research.

Whereas Winogradsky was the first to isolate an anaerobic, free-living, nitrogen-fixing bacterium, Beijerick discovered an aerobic counterpart. Beijerinck also isolated *Rhizobium*, famous now for its symbiotic nitrogen-fixing ability in the roots of

51 The material on Martinus Beijerinck was primarily sourced from the following review of his life and work: K-T. Chung and D.H. Ferris, 'Martinus Willem Beijerinck (1851–1931): Pioneer of general microbiology', *ASM News* 62 (1996), 539–543. https://doi.org/10.1142/9789813200371_0016

leguminous plants. Beijerinck 'completed' the sulphur cycle by demonstrating the reduction of sulphate to hydrogen sulphide, the reverse of the process described by Winogradsky, thus discovering the first known sulphate-reducing bacterium. The evolutionary invention of this ancient reaction is a biochemical landmark, one that would make the synthesis of essential amino acids and some oxidation-reduction reactions in metabolism possible. In fact, bacteria would have used this series of reactions to eventually produce the cellular energy currency adenosine triphosphate (ATP) in an ancient, pre-oxygenated world, and of course, they still do so in their select environs, for example in anoxic deep waters. Diffusion of hydrogen sulphide upwards, towards water layers not devoid of oxygen, would allow its oxidation back to sulphate by bacteria in the manner described by Winogradsky.[52] Thus, Beijerinck takes his place in the biologists' hall of fame for being one of the discoverers of biological nitrogen fixation, and for describing some of the microbial drivers of two of the major biogeochemical cycles on our planet.

Beijerinck, like his contemporaries discussed earlier, was also among the first to establish methods for culturing specific microorganisms from their natural environments, something he did not emphasise enough in his writings. However, he did describe his ecological approach in a speech as:

> the study of [...] the relation between environmental conditions and the special forms of life corresponding to them [...] It leads us to investigate the conditions for the development of organisms that have (somehow) come to our attention; [...] to the discovery of living organisms that appear under predetermined conditions, because they alone can develop, or because they are more fit and win out over their competitors.[53]

These seminal contributions to microbiology are in addition to what I believe Beijerinck is best known for today: showing that the causative agent of tobacco mosaic disease was smaller than a bacterium, suggesting that it was capable of replicating in a living plant cell, and calling it a virus. He made no impact on medicine, having developed a dislike for medical microbiology, presumably after his request to meet and discuss science with Robert Koch had been ignored. Unlike the other great microbiologist from Delft, Beijerinck left a mighty legacy in the form of researchers trained in what is now called the Delft School of Microbiology, many of whom would make major discoveries improving general microbiology in the 20th century.

Thus, thanks to the work of a young Pasteur, Winogradsky and Beijerinck (among others not discussed in this short summary), general microbiology found itself in a golden era in the late 19th century. However, the first "*schism*" (as a present day microbiologist puts it)[54] appeared in microbiology - creating the sub-discipline, medical

52 C.T. Walsh, *The Chemical Biology of Sulfur* (London: Royal Society of Chemistry, 2020). https://doi.org/10.1039/9781839161841-00005

53 C.B. van Niel, 'The Delft School and the rise of general microbiology', *Bacteriology Reviews* 13 (1949), 161–174, pp. 163–164. https://doi.org/10.1128/br.13.3.161-174.1949

54 R. Kolter, 'The History of Microbiology—A Personal Interpretation', *Annual Review of Microbiology* 75 (2021), 1–17. https://doi.org/10.1146/annurev-micro-033020-020648

microbiology - thanks to the prominent work of Pasteur and Koch. This schism would have profound effects on the progress of microbiology in the 20th century, which also saw the development of multiple silos or specialisations within microbiology that were often impervious to knowledge exchange.

1.4. Antibiotics: microbial competition and metabolic plasticity

Medical microbiology found great prominence in the first half of the 20th century, primarily in relation to antibiotic discovery and deployment. Antimicrobial agents, in the form of moulds and herbs, had been known in the ancient and mediaeval world across civilisations. Later, the botanist and herbalist Thomas Parkinson documented the use of moulds against infections in his 1640 book *Theatrum Botanicum*.[55] In the 19th century, supernatants of cultures of the versatile bacterium *Pseudomonas* were used as an antimicrobial agent with mixed success by Rudolf Emmerich and Oscar Loew. The "modern era of antimicrobial chemotherapy" began in the laboratory of Paul Ehrlich in the first decade of the 20th century.[56]

Paul Ehrlich was born in a province of Prussia and at a young age showed an interest in staining biological samples for microscopic examination. After studying medicine at various centres in Europe, Ehrlich joined Robert Koch's Berlin Institute of Infectious Diseases and later the University of Gottingen in the 1890s. His search for antimicrobials was based on the premise that the existence of chemicals that could selectively bind to and stain bacteria implied that some of these molecules could be used in antibacterial therapy. Driven by his observation that the arsenical compound atoxyl was effective against the protozoan parasite *Trypanosoma*, his laboratory embarked on an endeavour to synthesise various derivatives of atoxyl and test them for antimicrobial activity. In 1907, Alfred Bertheim in Ehrlich's lab synthesised arsphenamine, also known as Salvarsan. Sahashiro Hata, a research assistant in the lab, showed that Salvarsan was active against *Spirochaetes*, a class of bacteria.

Ehrlich was well aware of Koch's embarrassment in the 1890s when tuberculin, a substance Koch had hastily claimed was an effective anti-tuberculosis drug, turned out to be a failure, a setback that also revealed damaging lapses in the process Koch had used while testing the substance's activity. Ehrlich ensured that arsphenamine went through a rigorous clinical testing exercise before being established as a drug for treating the bacterial disease syphilis. The drug, however, because of its chemical instability when exposed to air, needed careful handling to ensure safe administration. This, and its unpleasant side-effects, meant that Salvarsan would be set aside when penicillin came along.

The serendipitous discovery of penicillin by Alexander Fleming is the stuff of

---

55 John Parkinson, *Theatrum botanicum* (London: Tho. Cotes, 1640). https://archive.org/details/gri_33125008297760

56 K. Gould, 'Antibiotics: from pre-history to the present day', *Journal of Antimicrobials and Chemotherapy* 71 (2016), 572–575. https://doi.org/10.1093/jac/dkv484

legend. Fleming was a British bacteriologist working at St Mary's hospital in London. He was interested in discovering antibacterial agents that were non-toxic to humans. He had in 1921 discovered the antibacterial action of lysozyme, a protein. But this molecule showed only modest activity against pathogens of human interest. Then came the now-famous culture plate of the bacterium *Staphylococcus* which also grew a contaminating mould that caused the bacteria to die. As Fleming himself claimed in this Nobel Lecture of 1945, the discovery of this activity was not "a result of serious study of the literature and deep thought that valuable antibacterial substances were made by moulds [...] My only merit is that I did not neglect the observation and that I pursued the subject as a bacteriologist."[57] Fleming isolated the mould in pure culture and showed that the culture supernatant even diluted "1,000 times [...] would inhibit the growth of *Staphylococci*." Other moulds that Fleming had tested did not show such antibacterial action. Fleming named the antibacterial agent penicillin after the fungus *Penicillium* that produced it. Importantly from medical point of view, the diluted Penicillium culture supernatant "when tested in human blood [...] had no more toxic effect on the leucocytes[58] than the original culture medium in which the culture had been grown." It showed no toxicity when injected into animals.

Fleming published his results in 1929 and referred to them a couple of times in the next few years, but otherwise his work was largely ignored. It is probably a reflection of the growth of microbiology and our own understanding of microbial behaviour in the late 19th and early 20th centuries that Fleming would recollect later in his Nobel Lecture: "to my generation of bacteriologists the inhibition of one microbe by another was commonplace and it is seldom that an observant clinical bacteriologist can pass a week without seeing in the course of his ordinary work very definite instances of bacterial antagonism." A mould that killed bacteria was just another example of microbial antagonism.

Two years before Fleming published his discovery of penicillin, Oswald Avery—a key figure in the identification of DNA as the genetic material—had recruited René Dubos, a soil chemist and bacteriologist, to the Rockefeller Institute in New York. At that time—thanks in part to the work of the German physiologist Emil von Behring—serum therapy was the treatment of choice for several infections, including bacterial pneumonia caused by a bacterium commonly referred to as *Pneumococcus*. But a certain type of pneumococci produced a capsule made of polysaccharides (linear or branched chains of simple sugar molecules) as a protective sheath around itself, making it impervious to serum therapy. Dubos was assigned the task of discovering an agent that could destroy the pneumococcal capsule. Dubos approached this using the principles of microbial ecology, which conceptually are the same whether the microbe is in the environment or in the human body.

57 Alexander Fleming, *Nobel Lecture* (1945). https://www.nobelprize.org/uploads/2018/06/fleming-lecture.pdf

58 A type of blood cell.

In a 1931 publication describing their work,[59] Dubos and Avery summarised their approach as follows:

> [Since the] capsular polysaccharides of *Pneumococcus* [...] have many of the properties of hemicelluloses an attempt was made in a natural environment for organisms possessing the capacity to split complex substances of this type. It was thought that those locations where large amounts of [...] hemicelluloses accumulate and undergo decomposition were the most likely to harbour the desired organisms.

Material from such environments, including soil from the cranberry bogs of New Jersey that eventually yielded positive results, were fed a defined medium containing purified pneumococcal polysaccharides as the sole source of carbon.

> It was possible that the materials used for inoculation contained organisms potentially capable of decomposing the specific substance [...] [This] material used for inoculation was of course a mix of a great variety of microbial species. A medium containing the specific substance as the sole source of carbon rendered conditions favourable only for these organisms capable of utilising the specific substance itself or the products of its decomposition.

In other words, bacteria that are capable of utilising the polysaccharide or its degradation products, even if in a small minority, would be able to grow and outcompete all other bacteria types in the soil community and become the dominant member. This led to the discovery of an enzyme that could degrade the pneumococcal capsule.

This experiment was truly in the spirit of the ecological approach to microbiology and enrichment cultures, which allows the selection of organisms of desired properties from a complex microbial community. These approaches had been vocally advocated by Winogradsky and, if a little less vocally, by Beijerinck in the decades past. Dubos radically applied this approach from general microbiology to antimicrobial agent discovery, thus making a pioneering contribution to medical microbiology. This approach was later called by Dubos "one of the most important biological laws I have ever been in contact with."[60] By the time the capsule-degrading enzyme discovered by Dubos started showing promise of becoming a viable treatment, Gerhard Domagk at Bayer's Institute of Pathology and Bacteriology had discovered, by chemical screening similar to Ehrlich's approach 30 years earlier, a sulfonamide drug, called Prontosil, effective against some pathogenic bacteria. The subsequent advent of several sulfonamide drugs made serum therapy passé, rendering Dubos' discovery of academic interest.

Sulfonamides came with the risk of toxicity when used at high doses, which encouraged Dubos to search for gentle antibacterial agents, again deploying his ecological approach. This time he fed soil samples mixtures of whole bacteria

---

59 R. Dubos and O.T. Avery, 'Decomposition of the capsular polysaccharide of pneumococcus type III by a bacterial enzyme', *Journal of Experimental Medicine* 54 (1931), 51–71. https://doi.org/10.1084/jem.54.1.51

60 Via M. Honigsbaum, 'Antibiotic antagonist: the curious career of Rene Dubos', *Lancet* 387 (2015), 118–119. https://doi.org/10.1016/s0140-6736(15)00840-5

instead of purified polysaccharides for many months. Unlike polysaccharides, whose degradation would immediately produce nutrients that bacteria could feed on, the selective pressure that a mix of whole bacteria would impose on a community of soil bacteria is more complex. It is possible that Darwin's principle of natural selection, or survival of the fittest, operated at multiple levels. The presence of these new bacteria can induce a small subset of the soil bacterial community to start producing antibacterial agents of broad specificity, eliminating most of the soil community as well as the added bacteria. This would require that the antibiotic producers themselves be resistant to the toxic substance they release. This might then favour the small subset of antimicrobial producers by removing competition for limiting nutrients that the added bacteria might have introduced. Additionally, dead and lysed bacteria release their contents, which can become food for living bacteria. Irrespective of the mechanism by which selection occurred, Dubos finally isolated a bacterium that produced a proteinaceous substance that could inhibit the growth of many pathogenic bacteria. One of the components of this antibacterial substance was gramicidin, which became the first clinically adopted antibiotic of natural origin, albeit one of limited use. Its limitation arose from toxicity when administered intravenously. But it found its role as a topical agent, helping treat wounds and ulcers during the Second World War. Despite its limitations, the discovery of gramicidin had two major consequences.

The first fallout of the discovery of gramicidin was a rekindling of interest in Fleming's penicillin. Though Fleming had not investigated penicillin's therapeutic potential extensively, but for a few trials with ordinary yet favourable outcomes, he was "convinced that before it could be used extensively it would have to be concentrated and some of the crude culture fluid removed."[61] As bacteriologists and not chemists, Fleming's team tried but failed to achieve this. But the success of gramicidin led Ernst Chain, Howard Florey and "the forgotten man of penicillin"[62] Normal Heatley to start using Fleming's cultures of the penicillin-producing mould to successfully concentrate penicillin and achieve its large-scale production. The rest, as they say, is history.

The discovery of gramicidin encouraged Selman Waksman to adopt Dubos' ecological approach to finding antimicrobials, resulting in the discovery of streptomycin in the early 1940s. This watershed moment in antibiotic history was the culmination of "24 years of detailed early research on soil microorganisms, something new that others had only minimally tackled."[63] In fact, a young Dubos himself had decided to pursue study of soil microbiology only after a "chance meeting"[64] with Waksman in 1924. Waksman studied a group of bacteria called the Actinomycetes, which in Waksman's time were considered to be neither bacteria nor fungi, at the Rutgers Agricultural School. In 1923, Waksman and his student Robert Starkey showed that

61 Fleming, *Nobel Lecture*.

62 P. Brack, 'Norman Heatley: the forgotten man of penicillin', *Biochemist* 37 (2015), 36–37. https://doi.org/10.1042/bio03705036

63 B. Woodruff, 'Selman A. Waksman, Winner of the 1952 Nobel Prize for Physiology or Medicine', *Applied and Environmental Microbiology* 80 (2014), 2–8. https://doi.org/10.1128/aem.01143-13

64 Honigsbaum, 2015.

certain actinomycetes in soil could kill other bacteria in another example of microbial antagonism: "Certain actinomycetes produce substances toxic to bacteria [...] around an actinomycete colony, upon a plate, a zone is found free from bacterial growth."[65] This, similar to the later discovery of penicillin, did not lead to any immediate progress in antibiotic discovery till the arrival of gramicidin on the scene in 1939.

Soon after Dubos' discovery had come into the limelight, Waksman reportedly told his new student Boyd Woodruff, "Woodruff, drop everything. My former student René Dubos has discovered a way to find antibacterial agents produced by soil microorganisms. And found an antibiotic. I am impressed. We must discover a better one."[66] The influence of Dubos' work on Waksman's becomes even more apparent from Waksman's description of it as "the stimulus which flooded with bright light the whole previously unilluminated field of the study and application of antibiotics."[67]

Woodruff fed soil with types of bacteria different from those used as feed by Dubos and discovered a potent antibacterial agent, which was named actinomycin. As the name suggests, this agent was produced by an actinomycete. Unfortunately however, actinomycin proved to be toxic. This was followed by the discovery of several antibacterials of actinomycete origin in Waksman's lab; however, none was sufficiently safe for therapeutic use. Finally a student, Albert Schatz, discovered streptomycin, which was then shown by researchers at the Mayo Clinic to be active against *Mycobacterium tuberculosis*, which causes TB in humans. This dramatic application of streptomycin ensured its enduring success.

Bacteria are broadly classified into gram positive and gram negative bacteria, with many known bacteria falling into one of the two classes. Gram positive bacteria have a thick cell wall composed of a molecule called peptidoglycan. Gram negative bacteria carry a thin peptidoglycan layer, which is protected by an outer layer comprising lipids and sugars. While penicillin was effective against a class of gram positive bacteria, "streptomycin is active against a variety of gram negative and gram positive bacteria, as well as upon gram positive bacteria which have become resistant to penicillin."[68] The discovery of streptomycin provided fresh impetus to antibiotic discovery. "The discovery of streptomycin as a product of a rather obscure group of microorganisms, the actinomycetes, led to the study of these organisms as producers of other chemotherapeutic substances. Following streptomycin, there came in rapid succession, chloramphenicol, chlortetracycline, neomycin, oxytetracycline and more recently erythromycin and others."[69] The golden age of antibiotics to kill infectious bacteria was well and truly underway. Infectious diseases were dead... long live infectious diseases!

---

65 S.A. Waksman and R.L. Starkey, 'Partial sterilization of soil, microbiological activities and soil fertility', *Soil Science* 16 (1923), 343–358. https://doi.org/10.1097/00010694-192310000-00003 via Woodruff 2014 referenced above.

66 Woodruff, 2014..

67 S.A. Waksman, *Frontiers in Medicine 99–119* (New York: Columbia University Press, 1951). Via C. Moburg, 'Early antibiotic from a Cranberry bog', *Nature* 518 (2015), 303. https://doi.org/10.1038/518303a

68 S. Waksman, *Nobel Lecture* (1952). https://www.nobelprize.org/uploads/2018/06/waksman-lecture.pdf

69 Ibid.

Alexander Fleming, after discovering penicillin, noticed that sensitivity to its antibacterial action varied across bacterial species.[70] He wrote: "penicillin contains bacterio-inhibitory substance which is very active towards some microbes while not affecting others." This, he suggested, would be useful to selectively isolate bacteria that are insensitive to penicillin from a community of otherwise sensitive bacteria:

> It sometimes happens that in the human body a pathogenic microbe might be difficult to isolate because it occurs in association with others which grow more profusely and which mask it. If in such a case the first microbe is insensitive to penicillin and the obscuring microbes sensitive, then by the use of this substance these latter can be inhibited while the former are allowed to develop normally.

At this time, the ability to resist penicillin's action was thought to be a property of a bacterial species as a whole and not one that might apply to only some individuals or 'strains' the same species.

Ten years later, after the discovery of sulfonamide drugs, Fleming showed that strains of the same species of pneumococci could vary in their sensitivity to these drugs.[71] He also demonstrated the insidious phenomenon by which an originally sensitive population of a bacterial species—pneumococci in this instance—developed resistance on continuous exposure to sub-lethal concentrations of a sulfonamide. This study prompted Fleming to note that the "primary dosage of the drug should be bold, so that the infection can be dealt with before the cocci have had a chance to become fast [...] Insufficient dosage [...] will merely result in the infecting microbes acquiring an increased resistance to the drug, so a later increase in dosage may still be insufficient."[72] This is an idea he would reiterate later in his Nobel Lecture of 1945.

The late 1930s and early 1940s, during which a slew of antibiotic discoveries were made, also saw the publication of a series of works, like Fleming's cited above, demonstrating antibiotic resistance and also attempting to dissect its underlying mechanisms. For example, in 1939, Colin Macleod and Giuseppe Daddi, starting with a population of sensitive pneumococci, produced in the laboratory strains that were resistant to sulfonamides by transferring the bacteria repeatedly in increasing, yet sub-lethal concentrations of the drug. Every transfer would have activated the phenomenon of natural selection, favouring strains that could outcompete their contemporaries at the concentration of the antibiotic in which they were grown, progressively yielding strains that were more and more resistant to the drug.[73] Outside the laboratory, cases

---

70 A. Fleming, 'On the antibacterial action of cultures of a Penicillium with special reference to their use in the isolation of B. influenzae', *The British Journal of Experimental Pathology* 10 (1929), 223–236. https://doi.org/10.1093/clinids/2.1.129

71 See C.L. Moberg, 'René Dubos, a harbinger of microbial resistance to antibiotics', *Perspectives in Biology and Medicine* 42 (1999), 559–580. https://doi.org/10.1089/mdr.1996.2.287, for an excellent review of early research on antibiotic resistance with special emphasis on the contribution of René Dubos.

72 I.H. Maclean, K. Rogers and A. Fleming, 'M and B 693 and pneumococci', *Lancet* 1 (1939), 562–568. https://doi.org/10.1016/s0140-6736(00)73741-x

73 C.M. Macleod and G. Daddi, 'A "Sulfapyridine-Fast" Strain of Pneumococcus Type I', *Experimental Biology and Medicine* 41 (1939), 69–71. https://doi.org/10.3181/00379727-41-10575p

of clinical antibiotic resistance were also becoming known. For example, at least one case of treatment failure as a result of resistance was reported in a 1943 clinical trial of penicillin. This persuaded Howard Florey to comment, "there are reasons for believing that organisms will sometimes develop resistance to penicillin during administration."[74] During the Second World War, an epidemic of sulfonamide-resistant Streptococci spread from a group of sailors to military personnel and civilians around the world. Richard Krause, writing about this decades later, would note, "Let me remind you that this occurred just before penicillin was widely available. Fortunately, penicillin was waiting in the wings. But it was a close call. Had penicillin not become available just at that time, there would have been epidemics of streptococci for which we had no treatment."[75]

Progress was also being made in understanding the molecular basis of antibiotic resistance. In 1940, Ernst Chain and Edward Abraham, following up on Fleming's observation that some bacteria such as *B. coli*[76] were insensitive to penicillin, "made an extract of *B. coli*" by "crushing a suspension of the organisms" and found a "substance destroying the growth-inhibitory property of penicillin" in it.[77] They proved that this substance was an enzyme and named it penicillinase. They noted that "a number of bacteria contain an enzyme active on penicillin." Little might they have known then that this enzyme and its relatives would make its presence felt half a century later, making penicillin useless to our unceasing efforts to conquer bacterial infections.

While these studies point to increasing awareness of antimicrobial resistance at least among microbiologists in the late 1930s and early 1940s, Fleming had demonstrated bacterial resistance to the modest antimicrobial agent lysozyme as early as in 1922. In their paper, Fleming and Allison say (quote rearranged):

> It was shown [...] than one type of coccus, which was called *Micrococcus lysodeikticus*, was especially suitable for experimental purposes with this lytic agent[78] as it was particularly susceptible to the lytic action [...] It was found however that if the culture plates (containing a small amount of lysozyme [...] embedded deep in the centre) were kept for two or three weeks there developed in the zone of inhibition a few isolated colonies of a microbe which appeared identical to *M. lysodeikticus* [...] It was thought that these colonies were produced from certain of the cocci which were especially resistant to the lysozyme.[79]

In the 1940s, René Dubos, appreciating the metabolic plasticity of bacteria (i.e., the ability of bacteria to change their physiological states, for example, to gain the ability

74 Via Moberg, 1999.

75 R. Krause, *The Restless Tide: The Persistent Challenge of the Microbial World* (Washington D.C.: National Foundation of Infectious Diseases, 1981). https://doi.org/10.1016/0167-5699(83)90082-8

76 Now called *E. coli* for *Escherichia coli*, but it has been called *Bacterium coli* and *Bacillus coli* at various times in the past.

77 E. Chain and E. Abraham, 'An Enzyme from Bacteria able to Destroy Penicillin', *Nature* 146 (1940), 837. https://doi.org/10.1038/146837a0

78 Lysozyme.

79 A. Fleming and V.D. Allison, 'Further Observations on a Bacteriolytic Element found in Tissues and Secretions', *Proceedings of the Royal Society B* 93 (1922), 142–151. https://doi.org/10.1098/rspb.1922.0051

to metabolise a new type of sugar in response to its availability), raised several red flags about the possibility of antibiotic resistance growing into a serious matter. In 1942, Dubos, commenting on the variation in sensitivity across bacteria to antibacterial agents and the chemistry behind antibiotic action, wrote:

> It is unknown whether the differential sensitivity is due to differences in cellular metabolism or to differences in permeability determined by the nature of the cell wall. In the analysis of the mode of action of antibacterial agents, it may be profitable to keep in mind that susceptible species often give rise by 'training' to variants endowed with great resistance to these agents. In some cases, drug resistance may be due to changes in metabolic behaviour, for instance the ability to use a metabolic channel not blocked by the inhibitor [...] It is also possible [...] that resistance may develop from a change in cell permeability.[80]

Again in 1944, at a symposium titled 'Wartime Advances in Medicine', Dubos warned[81]:

> the development of drug-fastness has been recognised *in-vivo* [...] There exists the possibility [...] that as a result of the widespread use of sulfonamides in therapy and especially for prophylaxis, there may develop in the population strains of pathogenic agents which have become resistant to these drugs [...] the problem should not be ignored, and it is to be hoped that laboratories throughout the land will find it possible to maintain a permanent survey in order to follow the shift in susceptibility of the different pathogenic agents to the drugs in common use.

His hopes were misplaced!

Dubos developed his ideas on antimicrobial resistance in his 1945 book, *The Bacterial Cell.*[82] In particular, he asked whether "resistant bacteria always occur in small numbers during normal growth in the presence of the drug, or whether they are produced only as a response to the presence in the medium of the substance with reference to which resistance develops." Referring to a 1942 work by Horsfall that had shown that in an original population of susceptible bacteria, a "small but constant proportion" are capable of growing in the presence of sulfathiazole, a sulfonamide, Dubos concludes: "During growth in the presence of sulfathiazole, there occurs a selection of the few individuals endowed with higher resistance to the drug, which are normally present in any susceptible culture." In a famous piece of work, Salvador Luria and Max Delbrueck had shown that the resistance of *E. coli* to viruses[83] that kill bacteria arises predominantly by natural selection of mutants existing prior to the exposure of the bacteria to the virus.[84] This is classical natural selection or survival of the fittest,

80 R. Dubos, 'Microbiology', *Annual Review of Biochemistry* 11 (1942), 659–678. https://doi.org/10.1146/annurev.bi.11.070142.003303

81 R. Dubos, 'Trends in the study and control of infectious diseases', *Proceedings of the American Philosophical Society* 88 (1944) 208–213, p. 209.

82 R. Dubos, *The Bacterial Cell* (Cambridge, M.A.: Harvard University Press, 1945), p. 25. https://archive.org/details/in.ernet.dli.2015.271563/mode/2up

83 Called Bacteriophage. See Chapter 2.

84 S.E. Luria and M. Delbrueck, 'Mutations of bacteria: from virus sensitivity to virus resistance', *Genetics* 28 (1943), 491–511. https://doi.org/10.1093/genetics/28.6.491

as proposed by Charles Darwin nearly a century earlier, applied to microorganisms. Dubos also recognised that, though resistance is often specific to the toxic agent under which selection had occurred, broad specificity resistance to multiple antibiotics was also not unknown.

Dubos, rather unfairly, is far from being the most recognisable face of the golden age of antibiotics in discussions today. One may therefore not be surprised that his extensive writings on antibiotic resistance, its mechanisms and potential clinical impact flew under the radar. However, both Fleming and Waksman did raise the spectre of resistance to penicillin and streptomycin respectively, not just in scholarly writings, but also in their more popular and accessible Nobel Lectures of 1945 and 1952.

In his 1945 Nobel Lecture, Fleming said:

> Penicillin is to all intents and purposes non-poisonous so there is no need to worry about giving an overdose and poisoning the patient. There may be a danger, though, in underdosage. It is not difficult to make microbes resistant to penicillin in the laboratory by exposing them to concentrations not sufficient to kill them, and the same thing has occasionally happened in the body. The time may come when penicillin can be bought by anyone in the shops [...] Here is a hypothetical illustration. Mr. X. has a sore throat. He buys some penicillin and gives himself, not enough to kill the streptococci but enough to educate them to resist penicillin. He then infects his wife. Mrs. X gets pneumonia and is treated with penicillin. As the streptococci are now resistant to penicillin the treatment fails. Mrs. X dies. Who is primarily responsible for Mrs. X's death? Why Mr. X whose negligent use of penicillin changed the nature of the microbe. Moral: if you use penicillin, use enough.

By the time Waksman won his Nobel Prize in 1952, streptomycin-resistant *Mycobacterium tuberculosis* was known. In his Nobel Lecture, Waksman echoed Fleming's views thus:

> The problem of variation in sensitivity of different strains of the same organism to streptomycin and the increasing resistance of the bacteria on prolonged contact with it are of considerable theoretical and practical importance [...] Freshly isolated cultures of tubercle bacilli from patients with pulmonary tuberculosis are uniformly sensitive to streptomycin [...] The organism develops resistance to streptomycin in-vitro at a rapid rate [...] Streptomycin fastness may be a major factor in the failure of streptomycin therapy in certain infections. When the disease condition presents physical barriers to the penetration of streptomycin the organisms are exposed to sub-lethal concentrations thus resulting in increased resistance.

Darwin's theory of natural selection predicts that antibiotic resistance had always been inevitable. And as the above discussion amply shows, the discoverers of antibiotics themselves, among others, also unearthed cases of antimicrobial resistance in the laboratory and also in clinical settings and wrote extensively about this. Today we are fighting widespread antibiotic resistance, even leading to doomsday predictions of an impending 'post-antibiotic' age. Why were these early warnings not acted upon? I do not know. Perhaps the incredible early success of antibiotics in treating some of the most serious infectious diseases, and the consequent euphoria surrounding a famous victory of humanity against nefarious agents of infections, blinded us to the possibility of a minor nuisance of that time exploding into a major health crisis. We have abused antibiotics. We buy them off the shelf from pharmacies that are happy to sell them even in the absence of a doctor's prescription, to self-treat viral colds and allergies. We take the liberty of stopping a prescribed course of antibiotics midway. Both of these habits have the potential to create sub-lethal concentrations of the antibiotic in the body, thus increasing the risk of resistance.[85] Polluting the environment with antibiotic wastes also leads to the selection of resistance among environmental bacteria, which can then transfer this ability to pathogenic ones; this phenomenon will take up much of this book. Prophylactic use of antibiotics (as warned by Dubos) as well as their use in animal husbandry to prevent infections are also problematic. The term 'antimicrobial stewardship', often used today, refers to the responsible use of these life-saving drugs. It is revealing that the first recorded use of this term, at least in the scientific literature and commentary, was only in the mid-1990s.[86] This was more than 50 years since the early warnings about resistance were made, but only a few years after cases of antibiotic-resistant bacteria started flooding the popular press.[87] So much for vision!

A hundred years of general and medical microbiology leading up to the 1950s taught us several important lessons on bacteria and microbes in general. We learnt that microbes are responsible for a wide variety of biogeochemical, industrial and disease processes. For example, the sulphur cycle, which plays an indispensable role in the formation of fossil fuels underground, is driven by bacteria, as is the nitrogen cycle that feeds us. Industrial processes such as fermentation, including those producing wine, bread, curd and *idlis*, are products of microbial action. A variety of diseases, such as pneumonia and tuberculosis, are of bacterial origin. We learnt that microbes, just like us, live in communities and that interactions between members of such communities play critical roles in their lives, and in ours indirectly. For instance, members of the soil community, other than the bacterial species specifically performing the reactions that convert ammonia to nitrate as part of the nitrogen cycle, contribute to the efficient completion of nitrification. We are still discovering new bacterial lineages that perform reactions of biogeochemical cycles such as nitrification, denitrification and nitrogen fixation from poorly explored parts of our planet such as the deep ocean, as revealed by papers published while this book was being finished.[88] Community living also activates microbial antagonism, a phenomenon we exploited to discover antibiotics safe for therapeutic use. These emphasise the need to study microbes in their natural environments to truly understand their behaviour. Most importantly from the

85 The relationship between antibiotic dosage and its concentration in the body can, however, be complex. See Bradley J. Langford and Andrew M. Morris, 'Is it time to stop counselling patients to "finish the course of antibiotics"?', *Can Pharm J* 150:6 (2017), 349–350. https://www.doi.org/10.1177/1715163517735549

86 O.J. Dyar, B. Huttner, J. Schouten and C. Pulcini, 'What is antimicrobial stewardship?', *Clinical Microbiology and Infection* 23 (2017), 793–798. https://doi.org/10.1016/j.cmi.2017.08.026

87 A.A. Salyers and D.D. Whitt, *Revenge of the Microbes: How Bacterial Resistance is Undermining the Antibiotic Miracle* (Washington, DC: ASM Press, 2005). https://doi.org/10.1128/9781555817602

88 Y. Huang, X. Zhang, Y. Xin, J. Tian and M. Li,. 'Distinct microbial nitrogen cycling processes in the deepest part of the ocean', *mSystems* 28 (2024), e0024324. https://doi.org/10.1128/msystems.00243-24

perspective of this book, early studies with antibiotics established that bacteria possess an enormous ability to rapidly adapt to new challenges.

In the mid-20th century, our understanding of how bacteria perform these reactions was within the realm of biochemistry, an extension of the already advanced field of chemistry. For example, antibacterial agents forming a small part of complex bacteria- or fungi-growing media could be isolated and characterised. Another example is the identification of the enzyme penicillinase as the agent responsible for penicillin resistance in some bacterial species. But how are these molecules produced, and how does the ability to produce interesting molecules, and to rapidly adapt to changing circumstances, evolve? Simultaneously with the explosive growth of medical microbiology in the first half of the 20th century came the second "schism", in Roberto Kolter's words, in microbiology: the emergence of what would be termed molecular microbiology.

2. The molecules of bacteria and of life

2.1. A molecular biology primer

DNA is the genetic material of all known cellular life forms. 'DNA' abbreviates deoxyribonucleic acid. It contains the information that, in large part, determines the characteristics of all organisms, more so of microorganisms in which brain-based mechanisms of learning and teaching do not operate. It is the material on which evolutionary change manifests itself.

DNA is a polymer of complex monomeric units called nucleotides. Each nucleotide is made of three components. The information content of DNA is contained in what is called the base. The base is connected to a deoxyribose sugar group, which in turn is attached to a short chain of phosphates to form a nucleotide. Adjacent nucleotides are linked to each other through chemical bonds between the phosphate of one nucleotide and the deoxyribose sugar of the next, forming a sugar-phosphate backbone. The bases protrude from this backbone. If all nucleotides were identical, the DNA would carry very little information. This obviously is not the case. There are four types of nucleotides each differing from the others only in the type of base it carries. These four types of bases in DNA are called adenine, cytosine, guanine and thymine,[1] abbreviated to A, C, G and T respectively. 'A' and 'G' are larger molecules called purines whereas the smaller 'C' and 'T' are called pyrimidines.

Though there are only four kinds of bases and therefore four kinds of nucleotides, these can be arranged in sequence in myriad ways. Even a short stretch of DNA with 10 monomeric units can take one of 4^{10} ($>$ 1 million) possible sequences. The genetic material of a cellular life form generally contains more than a million nucleotides, creating an almost infinite variety of possible sequences. The number of ways in which a 1-million nucleotide sequence (more commonly called a base sequence) can be created from an alphabet of a mere four monomeric units has a value with over 600,000 digits. Most such polymers would be nonsensical and incapable of defining life as we know it, but the very small fraction that works is good enough to produce the entire diversity of life on our planet.

DNA adopts what is called a double-helical structure. Two strands of the DNA polymer are wound around each other, held together by forces called hydrogen bonds

1 Not to be confused with the vitamin thiamine.

 https://doi.org/10.11647/OBP.0446.02

between atoms of the bases, which protrude into the helix. Such hydrogen bonds in DNA follow what is called the base pairing rule. 'A' forms two hydrogen bonds with 'T', and 'G' forms three hydrogen bonds with 'C'. Other pairs that can form under certain conditions would usually be considered as lesions (or damage) in the DNA. The base pairing rule ensures that the sequence of bases on one strand of the DNA strictly determines that of the other. Because of the predictability of base pairing, the sequences of the two strands are said to be complementary to each other.

Before a cell can divide and produce two daughter cells, the DNA must replicate to ensure that each daughter cell can receive its complete genetic material. The molecular machinery that performs DNA replication unwinds the DNA, reads the sequence of each strand, synthesises a new DNA strand complementary to each parental DNA strand that was read, and leaves each parent strand paired with its own newly synthesised daughter strand. The process of replication is highly accurate but, as with even the best machines, it does make the occasional mistake. Sometimes, such errors in replication can be left uncorrected. Uncorrected errors, when not lethal to the cell bearing them, help generate genetic variation on which the forces of evolution can operate.

The sum total of genetic material contained in a cell is called its genome. Most of an organism's genome is partitioned into chromosomes. Most bacteria carry a single chromosome in the form of a long circular chain. Other bacteria, including some of the actinomycetes that were brought into the limelight by the discovery of streptomycin, carry linear chromosomes. Eukaryotes have their genomes partitioned into multiple linear chromosomes. This, however, is not a property unique to eukaryotes; some bacteria carry multiple chromosomes. Replication of a chromosome is normally regulated such that it is complete before the cell bearing it divides into two, allowing each daughter cell its own full set of chromosomes. Many bacteria also contain one or more extrachromosomal DNA, which are typically two to three orders of magnitude shorter in sequence length than a typical chromosome: these are called plasmids. Replication of a plasmid often occurs independent of chromosomal DNA replication and cell division.

A chromosome is not merely a long, unorganised sequence of nucleotides with the property of defining life. It is, as the clichéd description goes, like a book with each sentence or chapter serving a specific purpose. Just as a book has many meaningful sentences, a chromosome has several short blocks called genes, each of which might define a function: for example, a gene may define a step in the synthesis of a complex molecule like streptomycin from simpler building blocks. Many functions would require multiple steps to be carried out in series, and each step may be encoded by one gene or a group of genes working in concert. An average bacterial gene contains ~1,000 nucleotides, but spans a wide range. In many eukaryotes, a gene could be split into multiple segments which are stitched together when required, but this is extremely rare in bacteria and will find little mention in this book.

A piece of DNA, whether it be a gene or not, can be gained or lost, and when present can undergo changes in its sequence, for example, through errors during replication. These processes drive evolution. Depending on the strength of selection, i.e., the contribution to population growth and survival, positive or negative, of the outcome of these processes, these changes can be maintained or lost from a population of organisms. For example, some of these processes would have been involved in the evolution of antibiotic resistance in bacteria. The strength of selection imposed by a piece of DNA is not the same across organisms: this depends on the lifestyle of the organism, as well as its population size. Whereas the former is intuitive to understand, the latter will be discussed in Chapter 3. In short though, when the size of the population is small, even a piece of DNA with slightly detrimental effects on the organism can be retained in the population. This can explain some differences between the genomes of bacteria and eukaryotes.

A gene is functional not in the sense that it catalyses a reaction, but because it encodes information that allows an active principle to be produced. In this sense, it is more like a recipe book than the final dish. The first step that a cell takes in reading the information contained in a gene is transcription. A system of proteins transcribes the sequence of bases that make a gene into another polymer of nucleotides called RNA, or ribonucleic acid. RNA is chemically similar to DNA except in the following ways.

1. Whereas the sugar component of the DNA is deoxyribose, that in RNA is ribose.
2. RNA does not contain thymine as one of its bases; instead, it uses a related pyrimidine base called Uracil, abbreviated to 'U'.
3. Whereas DNA is double-stranded, RNA is single-stranded but with the possibility of intrastrand base pairing, which can produce complex 3D structures and make the RNA capable of performing a variety of chemical reactions. The sequence of the RNA produced by transcription is the same (but for U replacing T) as that of one strand of the gene encoding it; this strand is called the coding strand. The opposite strand, whose sequence is complementary to that of the RNA, is called the template strand. Spaces between genes on the DNA, usually called intergenic DNA, are typically short in bacterial genomes and contain sequence patterns that determine when and to what extent an adjacent gene is transcribed.

In many cases, the RNA acts as an intermediate that is further read to produce proteins in a process called translation. Such RNAs are called messenger RNAs, or mRNAs. Proteins are polymers comprising chains of amino acid monomers. There are 20 standard amino acids encompassing a range of physicochemical properties. Every sequence of three bases on the mRNA, called a codon, serves as an instruction to the translation machinery to add a specific amino acid to a growing protein chain or to end translation of that mRNA. There are 64 possible codons, of which three act as stop signals. Each of the remaining 61 codons code for one of the 20 possible amino acids. This mapping of codons to amino acids or a stop signal is called the genetic code.

One sometimes sees the genome or a gene being referred to as a genetic code in the popular press. Whilst this is correct English, it is bad jargon and can confuse the biologist. In this book we will stick to the biologist's definition of the genetic code. As the wide gulf between the numbers of codons and amino acids indicates, there is considerable redundancy built into the genetic code, with multiple codons mapping to the same amino acid. This redundancy implies that all variations introduced into a gene, for example as errors during replication, will not result in a change in the amino acid sequence and therefore the properties of the protein produced.

In several other cases, the RNA is an end in itself and serves as the final functional product. The most prominent and abundant among these are ribosomal RNA (rRNA) and transfer RNA (tRNA), which help in the translation of mRNAs. Various other non-coding RNAs have been described more in eukaryotes than in bacteria; these help in regulating translation and mRNA stability among other roles.

This brief primer recapitulates and summarises material readily found in molecular biology textbooks and is pretty much all the essential basic biology one needs to know to race through most of the rest of this book.

2.2. Viruses that prey on bacteria

The proof that DNA is the genetic material was the culmination of disparate work that brought about what Roberto Kolter[2] designates as the second "schism" in unified microbiology, called molecular microbiology. The beginnings of molecular microbiology, which is the beginning of all molecular biology, can be defined only in retrospect. This is quite unlike medical microbiology, which began with the definite goal of identifying causative agents of disease. Today, the term molecular microbiology is commonly used to identify studies that take a molecular genetic approach to microbiology. By molecular genetics we mean the genetic material, and its decoding by mean processes and substances defined in the above primer on the cellular machinery. One can make the argument that molecular microbiology began as an offshoot of medical microbiology, but in obscurity and, soon enough, controversy.

Molecular microbiology began without knowing it, thanks to a 1915 paper by Frederick Twort, a "son of a country doctor and a hard-working mother."[3] He had in the previous decade worked on sugar fermentation in bacteria, proposing that the ability to metabolise a sugar was not constant for a bacterium, and that bacterial species could be induced to metabolise new sugars by exposure to the sugar over extended periods of time.[4] In this way he had anticipated René Dubos, who would, a few decades

---

2 R. Kolter, 'The History of Microbiology—A Personal Interpretation', *Annual Review of Microbiology* 75 (2021), 1–17. https://doi.org/10.1146/annurev-micro-033020-020648

3 D. Duckworth, 'Who discovered bacteriophage?', *Bacteriological Reviews* 40 (1976), 793–802. https://doi.org/10.1128/mmbr.40.4.793-802.1976

4 F.W. Twort, 'The fermentation of glucosides by bacteria of the Typhoid-coli group and the acquisition of new fermenting powers by Bacillus dysenteriae and other microorganisms', *Proceedings of the Royal Society of London B* 79 (1907), 329–336. https://doi.org/10.1098/rspb.1907.0025

later, emphasise the metabolic plasticity of bacteria while discovering antibiotics and raising red flags about antibiotic resistance. In his 1915 work,[5] carried out at the Brown Institution in London, Twort described the results of his "attempts [...] to demonstrate the presence of non-pathogenic filter-passing viruses." He did not find any and then tried to find ways to grow viruses in artificial media. Today we know that he was at the very outset doomed to fail, for viruses cannot multiply in the absence of an amenable host cell that they can infect and manipulate. But his efforts would nevertheless lead to an important incidental discovery.

While trying to culture material called vaccinia, a substance used in smallpox vaccination, on artificial media, Twort observed that colonies of bacteria of the type micrococcus had grown on it. This on its own would have been unremarkable but for the following observation:

> Inoculated agar tubes, after 24 hours at 37°C, often showed watery-looking areas, and in cultures that grew micrococci, it was found that some of these colonies could not be subcultured but if kept they became glassy and transparent. On examination of these glassy areas nothing but minute granules [...] could be seen. [6]

He also observed "that when a pure culture of the white or yellow micrococcus isolated from vaccinia is touched with a small portion of one of the glassy colonies, the growth at the point touched soon starts to become transparent or glassy, and this gradually spreads over the whole growth, sometimes killing all the micrococci and replacing these by small granules." And that this "disease" could be "transmitted to pure cultures of the micrococcus" indefinitely. But "the transparent material will not grow by itself on any medium." The emergence of these glassy areas started as "transparent points" that would quickly encompass the entire bacterial growth. This antibacterial agent acted only on the micrococcus; its activity was much reduced against the related bacterium staphylococcus, and it was inactive against the more distant tubercle bacillus. Thus, Twort had discovered an agent that could not only kill specific bacteria, but one that could also multiply exclusively on its prey.

Twort is now credited with the discovery of a virus that can kill bacteria, but he himself was unsure about the nature of the killing agent he had discovered. In his paper, he offered a range of possible explanations. It is worth noting that the very nature of ultramicroscopic viruses was a great mystery in Twort's time. A virus was "*contagium vivum fluidum*"[7], as Martinus Beijerinck had defined them while studying the causative agent of tobacco mosaic virus. In other words, a virus was merely any "live reproducing organism that differed from other organisms", and that could pass "through a minute filter that would not allow the passage of bacteria."[8] The first images of a virus would

5 F.W. Twort, 'An investigation on the nature of ultramicroscopic viruses', *Lancet* 186 (1915), 1241–1243. https://doi.org/10.1016/s0140-6736(01)20383-3

6 Ibid.

7 R.R. Wagner and R.M. Krug, 'Virus', *Encyclopaedia Britannica*. https://www.britannica.com/science/virus

8 Ibid.

not be available for another 25 years after the publication of Twort's paper. Given this state of knowledge, it was perfectly par for the course that Twort wondered if the antibacterial agent he had discovered was "a minute bacterium that will only grow on living material, or it may be a tiny amoeba." He believed that "the transparent material contains an enzyme", which somehow "increases in quantity when placed on an agar tube containing micrococcus."[9] Considering the fact that proteins were the leading candidates for genetic material at that time, an enzyme (a protein) with the power of replication does not sound all that ridiculous.[10] Twort suggested in his paper that this enzyme might be a part of the micrococcus itself. And finally, he remarked that the possibility of the antibacterial agent being an ultramicroscopic virus "has not been definitely disproved." After expressing these hypotheses and his doubts as to which of these might be the correct one, Twort concluded, "I regret that financial considerations have prevented my carrying these researches to a definite conclusion, but I have indicated the lines along which those more favourably situated can proceed."

Twort's paper was neglected and it stayed under the radar for over five years. But viruses that prey on bacteria did not, thanks to the work of the mercurial French-Canadian researcher Felix d'Herelle. In 1917 d'Herelle, working at the Pasteur Institute hospital, "isolated from stools of different individuals convalescent from bacterial dysentery [...] an invisible microbe with antagonistic properties against the Shiga bacillus."[11] By opening his short report with these words, he left no room for uncertainty by claiming that the antibacterial agent he had found was a "microbe." He showed that the invisible microbe had multiplied in the "lysed Shiga culture." He also noticed, consistent with Twort's observations, that the killing action initiated at specific points on the bacterial culture. However, unlike Twort, d'Herelle took this to be "visible proof that the antagonistic activity is produced by a living germ [...] no chemical substance would be able to concentrate on precise points."

d'Herelle also demonstrated that the action of this agent was specific to the Shiga bacillus, but that the agent could be coerced to become antagonistic to other types of dysentery-causing bacteria by its maintenance with these insensitive bacteria in culture for a period of time. Therefore "the specificity of the antagonistic action is [...] not inherent to the very nature of the invisible microbe but is acquired by symbiotic culture with the pathogenic bacillus." He found, as had Twort, that this agent would not grow in any medium, and also showed that it was incapable of causing disease in experimental animals. The fact that the appearance of this agent was concomitant with the disappearance of the pathogenic bacteria and patient recovery suggested to d'Herelle that it was an "immunity microbe." d'Herelle called this agent "*Bacteriophagum intestinale*

9 Twort, 1915.

10 With all due respect to pathogenic prion proteins, which 'reproduce' and cause diseases such as Alzheimer's, but they do so only by converting other proteins of the same kind into a pathogenic state.

11 M.F. d'Herelle, 'On an invisible microbe antagonistic to dysentery bacilli', *Comptes Rendus Academie des Sciences* 165 (1917), 373–375. Translated from the French language by H.-W. Achermann via http://dx.doi.org/10.4161/bact.1.1.14941.

or *Bacteriophage*", a "living matter" that "develop(s) at the expense of" bacteria.[12]

d'Herelle did not cite any other work in his 1917 paper, not even Twort's, which had been published two years earlier. He might have been genuinely unaware of Twort's work, and there is precedent. Beijerinck did not know that the Russian botanist Dmitri Ivanowski had preceded him by six years in discovering a filtrable virus that causes tobacco mosaic disease; on being made aware of this, Beijerinck apologised and acknowledge Ivonowski's priority. Or d'Herelle might have been dishonest, ignoring Twort's work deliberately. We will never know. Either way, thanks to the obscurity of Twort's paper, d'Herelle "lived a life that every scientist must dream of."[13] Bacteriophage was hot. It held therapeutic potential against bacterial infections. And it was the "d'Herelle phenomenon." Four years after d'Herelle's report, in 1921, all hell broke loose.

A Japanese researcher, Tamezo Kabeshima, who had earlier worked in d'Herelle's laboratory, analysed a bacteriophage sample obtained from d'Herelle himself and concluded that the antimicrobial agent could not be a living being but was something he called a "pro-ferment",[14] which was merely a component of all microbes. Then the important Belgian immunologist and bacteriologist Jules Bordet, already a Nobel Prize winner, showed that a leucocyte exudate—fluid that leaks out of blood vessels during inflammation—could make bacteria initiate transmissible lysis. He and his Romanian collaborator Mihai Ciuca argued that this was an "autolytic phenomenon one that can start in perfectly normal microbes."[15] Another Belgian bacteriologist André Gratia supported this view.[16] This view appeared to be more in line with Twort's idea of an antimicrobial enzyme with the power of growth, rather than a virus.

Bordet and Ciuca further wrote that "the burden of an exact history makes it necessary for us to cite a previous work d'Herelle has not known [...] This remarkable work by F.W. Twort appeared in Lancet [...] two years before the research of d'Herelle."[17] d'Herelle should lose precedence, and be chastened for having advanced a bold hypothesis of a bacteria-killing virus prematurely. Bacteriophages, irrespective of their true nature, would no longer be the "d'Herelle phenomenon." For instance, Alexander Fleming, whilst contrasting the action of lysozyme with that of bacteriophage in 1922, would refer to the latter as "the lytic agent described by Twort, d'Herelle and later by many other authors."[18]

This would have been fine as far as precedence to scientific discoveries goes. But d'Herelle, in a reflection of his feisty personality, chose a combative response, seeking to deny precedence to Twort. In a sense this is understandable. In the years following his discovery of bacteriophage, d'Herelle had invested considerable effort into understanding

12 M.F. d'Herelle, *Bacteriophage, Its Role in Immunity* (Baltimore: Williams and Wilkins Company, 1922). https://archive.org/details/cu31924003218991
13 Duckworth, p. 797.
14 Phrase translation via Duckworth.
15 Duckworth, p. 797.
16 Ibid.
17 Ibid.
18 A. Fleming and V.D. Allison, 'Further Observations on a Bacteriolytic Element found in Tissues and Secretions', *Proceedings of the Royal Society B* 93 (1922), 142–151. https://doi.org/10.1098/rspb.1922.0051

the phenomenon, culminating in a 300-page volume, *The Bacteriophage, its Role in Immunity*, published in 1922. It is in the introduction to this book that d'Herelle discussed Twort's work and somehow came to the conclusion that "it is hardly probable that the cause is a bacteriophage."[19] Instead, he ascribed Twort's phenomenon—which in d'Herelle's view should be distinct from the d'Herelle phenomenon— "to a fragmentation of the bacteria."[20] He then made the point that any discovery should be counted as such only if it is recognised as such by the claimant to the discovery. Twort had been unsure about the nature of his antibacterial agent and ended his paper on an inconclusive note. d'Herelle claimed that many others had "accidentally encountered this strange phenomenon [...] but neglected [them] since their importance was not recognised." Then he went on to argue that the "fundamental experiment" leading to the discovery of the bacteriophage was his own work with the convalescent dysentery patients.

Things went back and forth between d'Herelle and the Belgian school in the early 1920s. Twort appears to have been largely silent, having stopped pursuing bacteriophage research and instead chasing "primitive viruses", work that was not looked upon favourably by funders.[21] In 1922, at a conference in Glasgow, in a rare appearance, Twort supported an interpretation that disagreed with both d'Herelle and the Belgian side. In the same meeting d'Herelle accused his naysayers of failing to consider the entire body of data in coming to their conclusions. This certainly would not have won him any favours and the debate quickly deteriorated to the realm of inanity and absurdity when Gratia pointed out how fire, though self-reinforcing, was non-living, and soda bubbles, though capable of growth, were not considered living.

This is all rather unfortunate considering that neither party was wholly wrong. The antibacterial agent in both cases was indeed bacteriophage, and in the Bordet and Ciuca study it was also an "autolytic phenomenon."[22] Bordet, Ciuca and Gratia, without knowing it, had discovered bacteriophage lysogeny, an important phenomenon in which a bacteriophage becomes a part of the host bacterium by integrating its own genetic material in the bacterial chromosome. The virus stays dormant, replicating with the host chromosome until certain conditions or random events allow it to suddenly and explosively develop into a full virus, lysing the host cell from within. This phenomenon plays important roles in bacterial evolution, and in many ways, in the last half a century and more, it has enabled us to understand several physiological properties of bacteria. The importance of this phenomenon would be recognised only later.

But through the 1920s, d'Herelle and the Belgian school continued to be at loggerheads with each other. In a book written in 1926,[23] d'Herelle complained that

---

19 D'Herelle, *Bacteriophage, role in immunity*, p. 17.

20 Ibid., p. 17.

21 G.H. Thomas, 'Frederick William Twort: Not Just Bacteriophage', *Microbiology Today*, 29 May 2014. https://microbiologysociety.org/publication/past-issues/world-war-i/article/frederick-william-twort-not-just-bacteriophage.html

22 Duckworth.

23 M.F. d'Herelle, *The bacteriophage and its behaviour* (Pennsylvania: Williams and Wilkins Company, 1926), p. 18. https://archive.org/details/bacteriophageits00dher

Twort would not be drawn into discussing the phenomenon with him and that this was proof that "the facts revealing the dissimilarity of the two phenomena are indisputable". He also called Bordet's work a "history of errors."[24] He did quickly move on to describing important work, including simple experiments that proved the "corpuscular" nature of the bacteriophage, contrasting it with the more solution-like characteristic that a chemical toxic to bacteria would display.[25] In describing this aspect of his work, d'Herelle mentioned how "in discussing this question with my colleague, Professor Einstein, he told me that, as a physicist, he would consider this experiment as demonstrating the discontinuity of the bacteriophage."[26] However, the great reputation of Bordet and colleagues meant that their view was quickly gaining ground. Not for long: d'Herelle always had his supporters. For example, and ironically, a Belgian researcher, Richard Bruynoghe, argued against Bordet at the Belgian Biological Society but rarely published his work in scientific journals. In 1934, the Hungarian biochemist Max Schlesinger showed that the bacteriophage comprised DNA and protein, and also measured its size to be 100 nm, larger than one would expect for an enzyme.[27] The Australian microbiologist Frank Burnet, having purchased and carefully read an English translation of d'Herelle's 1926 book,[28] believed in the conception of the bacteriophage as a virus and showed that it could be used in genetic studies.[29] These would dismiss the enzymatic explanation of the bacteriophage phenomenon.

The immediate consequence of Bordet, Ciuca and Gratia's efforts was that d'Herelle had to share credit for the discovery of bacteriophage with Twort. A more far-reaching consequence of the poor reputation that d'Herelle had built for himself by taking a combative position, despite his amazing research work, was that bacteriophage therapy for bacterial infections, of which d'Herelle was the primary votary, fell by the wayside in the West (and pretty much elsewhere except in the Stalinist Soviet Union).[30]

So, who discovered bacteriophage? All of the evidence presented above points to Twort first reporting its discovery, though he may not have ruled out alternative

24 Via A.V. Letarov, 'History of early bacteriophage research and emergence of key concepts in virology', *Biochemistry (Moscow)* 85 (2020), 1093–1112.

25 d'Herelle, *Bacteriophage behaviour*, p. 83. https://doi.org/10.1134/s0006297920090096

26 Ibid., p. 83, footnote.

27 Letarov, 2020 and G.S. Stent, 'Introduction: Waiting for the paradox', *Phage and the Origins of Molecular Biology* (Plainview, NY: Cold Spring Harbor Laboratory Press, 1966).

28 F.J. Fenner, 'Frank Macfarlane Burnet, 3 September 1899–31 August 1985', *Biographical Memoirs of the Fellows of the Royal Society* 33 (1987), 99–162. https://doi.org/10.1098/rsbm.1987.0005

29 Letarov, 2020 and Stent, 1966.

30 d'Herelle's colleague George Eliava set up an institute for bacteriophage therapy that is still a major research institute in Tbilisi, Georgia. Further discussion of bacteriophage therapy is beyond the scope of this account, but the following resources might interest the reader:

(1) Lawrence Osborne, 'A Stalinist Antibiotic Alternative', *The New York Times Magazine*, 6 February 2000;

(2) S. Hesse and S. Adhya, 'Phage Therapy in the Twenty-First Century: Facing the Decline of the Antibiotic Era; Is It Finally Time for the Age of the Phage?', *Annual Review of Microbiology* 73 (2019), 155–174. https://doi.org/10.1146/annurev-micro-090817-062535

(3) D.E. Fruciano, 'Phage as an antimicrobial agent: d'Herelle's heretical theories and their role in the decline of phage prophylaxis in the West', *Canadian Journal of Infectious Diseases and Medical Microbiology* 18 (2007), 19–26. https://doi.org/10.1155/2007/976850

hypotheses and therefore did not take a firm stand. If, to make a claim to a discovery, the claimant must recognise "all the ramifications"[31] of the original observation, then d'Herelle would take precedence, assuming of course that he made his discovery while unaware of Twort's work. Interestingly though, d'Herelle, much later in 1949, claimed to have discovered bacteriophage as early as in 1910 as agents afflicting bacteria that were killing locusts in Mexico; he had not reported this at that time,[32] though he mentioned this finding as initiating his curiosity and leading to his "definitive experiment" in his 1926 book. Perhaps he had not fully recognised its importance then, and did so only in retrospect, a failing he accused others of.

Whatever the truth may be, the discovery of bacteriophage, the recognition of its status as a virus, and its intensive study in the following decades would converge with other independent bacteriological studies in proving that DNA was the genetic material,[33] heralding the birth of all molecular biology.

2.3. The genetic material I: bacteria and the "transforming principle"

Bacteriology first. In January 1928, the *Journal of Hygiene* published a 46-page paper by a senior medical bacteriologist, Frederick Griffith.[34] The paper, considered today, is staid and makes for difficult reading. Griffith "has been described as a shy and reticent man."[35] His published works are few and he hardly ever participated in open scientific meetings, preferring to work alone. His interest in science was firmly restricted to the epidemiology of infectious diseases, primarily tuberculosis and the spectrum of maladies caused by pneumococci. He was highly regarded for his work on bacterial typing. Today he is famous for how his work on bacterial typing eventually helped identify DNA as the genetic material. The term typing in this context means classifying members of the same species of bacteria into different sub-groups, or types. Closely related bacteria, notably those causing infections, often differ from one another in the molecules they display on their cell surfaces. Such differences can be captured by the phenomenon by which these molecules interact with blood sera containing some kinds of immunity-determining molecules but not others. This would be akin to the different variants of SARS-CoV-2 spike protein reacting with some antibodies but not others. Depending on the immune sera that bacteria react to, the latter can be classified into distinct serological types.

---

31 Duckworth.

32 Ibid.

33 As predicted by Hermann Muller as early as 1922: (1) H.J. Muller, 'Variation due to change in the individual gene', *American Naturalist* 56 (1922), 32–50. https://doi.org/10.1086/279846 (2) J.E. Haber, '101 years ago: Hermann Muller's remarkable insight', *Genetics* (2023). https://doi.org/10.1093/genetics/iyad015

34 F. Griffith, 'The significance of pneumococcal types', *Journal of Hygiene* 27 (1928), 113–159. https://doi.org/10.1017/s0022172400031879

35 W. Hayes, 'Genetic transformation: a retrospective appreciation', *Journal of General Microbiology* 45 (1966), 385–397. https://doi.org/10.1099/00221287-45-3-385

In Griffith's time, serological typing was the preferred method of classifying members of any species of infectious bacteria. Such classifications were important for bacteriological diagnoses and, as expressed by Griffith, were of relevance to "issues of [...] greater importance [such as] the occurrence and remission of epidemics, the appearance of epidemic types in certain diseases and the attenuation of the infecting agent in others."[36] Griffith, at least in the context of the tubercle bacillus, appears to be "firmly convinced of the fixity of bacterial types",[37] i.e., a bacterial isolate would always maintain its serological type and not switch to another. His seminal work with pneumococcal typing, described below, would, however, persuade him to change his views on this matter.

In his 1928 paper, Griffith describes the prevalence of different pneumococcal types in the sputum of patients. It was known that virulence of any pneumococcus was correlated with the ability of the bacterium to produce a polysaccharide capsule. Those that produced a capsule were called S-form pneumococci and those that did not were of the R-form. Both R- and S-form pneumococci could belong to the same serological type. For example, Griffith showed that S pneumococci could be forced to lose their capsule and become attenuated R pneumococci of the same serological type on prolonged growth in certain culture media. Now, mice infected with these attenuated R pneumococci were usually fine. However, Griffith found that if a large dose of the R variant was inoculated under the skin of mice, the R form could revert back to the S-form of the same serological type, causing disease in the infected mice.

Griffith hypothesised that "it seemed possible that the mass of R pneumococci, disintegrating under the action of the animal tissues might furnish some substance which was utilisable by the survivor to build up their virulent structure."[38] To further develop and test this idea Griffith, "acting on the assumption that this material might be the S antigenic substance, which in varying amounts persists in the R form, [...] inoculated into the mouse [...] a very much smaller dose of living pneumococci together with a mass of killed virulent culture."[39] He found that the avirulent R form, though in small doses, would readily transform into the virulent S form "when the killed culture used is of the *same* serological type from which the R form was derived"[40] (italics mine).

Importantly he noticed that these transformations also occurred, albeit less frequently, when the live R and killed S pneumococci were of different serological types. In Griffith's words, "The inoculation into [...] mice of an attenuated R strain derived from one type, together with a large dose of virulent culture of another type killed by heating [...] resulted in the formation of a virulent S pneumococcus of the same type as that of the heated culture."[41] After ensuring that this had not occurred

36 Griffith, 1928.

37 M.R. Pollock, 'The discovery of DNA: an ironic tale of chance, prejudice and insight', *Journal of General Microbiology* 63 (1970), 1–20. https://doi.org/10.1099/00221287-63-1-1

38 Griffith, 1928, p. 150.

39 Ibid., p. 150.

40 Ibid.

41 Ibid., p. 159.

merely because some small fraction of the heat-killed S pneumococci had escaped death and developed to cause disease in the infected mice without really transforming the R pneumococci, Griffith concludes, "there seems to be no alternative to the hypothesis of transformation of type."[42] Despite his original belief in the permanence of bacterial types "his great care, perfectly planned controls and scrupulous honesty"[43] ensured that he deferred to his data and wistfully admitted to the falsity of his original belief.

Griffith did not speculate much on the mechanism by which the type transformation might have occurred in his paper. In fact, as noted by Hayes, "the most striking and important aspect of transformation as we see it now, namely, that it results in an inheritable change of character is neither mentioned nor implied."[44] The closest Griffith got to an explanation was in suggesting that the material responsible for type transformation might be "the specific protein structure of the virulent pneumococcus which enables it to manufacture a specific carbohydrate."[45] In his later work, Griffith appeared not to have pursued this line of research. Pollock, in a later review of Griffith's work, wrote, "I do not believe he could have been very much interested in the phenomenon of transformation which he himself had discovered."[46] Indeed at the only open scientific meeting Griffith spoke at, in 1936, he made no mention of type transformation and instead read out his paper on serological typing techniques. Nevertheless, without quite knowing it, he had triggered an avalanche.

Even if Griffith had no further interest in transformation, his paper was by no means ignored. Others managed to reproduce Griffith's findings in their own laboratories, and the scene of the action soon shifted across the ocean to America. Whereas Griffith was able to demonstrate transformation only during infection in mice, Richard Sia and Martin Dawson found ways to produce the phenomenon by mixing live R and dead S pneumococci in a test tube.[47] They argued that there are two ways in which transformation can be achieved: "either a latent attribute of the R cell may be stimulated by its association with the S vaccine,[48] or the R organism may acquire a new property from the vaccine." Either way, determining the "true nature and precise properties" of the agent responsible for transformation was of paramount importance. Towards this goal, Sia and Dawson made a beginning by showing that this agent was not the polysaccharide capsule itself. Lionel Alloway then showed that whole heat-killed S pneumococci were not strictly required for transformation but that an extract from lysed S pneumococci would suffice.[49] Alloway also asserted in his paper that the "exact

42 Ibid., p. 154.
43 Pollock, 1970, p. 8.
44 Hayes, 1966, p. 387.
45 Griffith, 1928, p. 151.
46 Pollock, 1970, p.10.
47 R.H.P. Sia and M.H. Dawson, 'In-vitro transformation of pneumococcal types', *Journal of Experimental Medicine* 54 (1931), 701–710, p. 708. https://doi.org/10.1084/jem.54.5.681
48 The term vaccine used to refer to killed *S pneumococci*.
49 J.L. Alloway, 'Further observations on the use of pneumococcal extracts in effecting transformation of type in vitro', *Journal of Experimental Medicine* 57 (1933), 265–278. https://doi.org/10.1084/jem.57.2.265

nature of the active material in these extracts still remains to be determined,"[50] setting the scene for Oswald Avery and his younger colleagues Colin MacLeod and Maclyn McCarty to do so over a decade later. Tragically, Griffith did not live to see where his work led: he was killed in a German bombing raid over London in 1941.

Oswald Avery was a medical bacteriologist working on the immunological properties of pneumococci in New York. According to René Dubos, who had worked in Avery's laboratory discovering pneumococcal capsule-degrading enzymes, Avery's most "original" contribution to bacteriology "for most workers [...] was the demonstration that the role played by complex carbohydrates in various immunological processes is at least as important as that classically attributed to proteins."[51] Though Avery had understood that immunological properties alone do not define bacterial virulence, his work was deemed to be the "masterplan [...] for the immunochemical study of infectious processes."[52] In fact, Avery's work had established the relationship between capsule production and virulence of pneumococcus that Griffith used to demonstrate type transformation. In addition, Avery had also shown that capsules promote virulence by protecting the bacteria from being consumed by immune cells.

When news of Griffith's work reached Avery's laboratory, it was as an exploding "bombshell".[53] As Griffith had been prior to showing type transformation, Avery was a staunch believer in the permanence of bacterial types. Initially, Avery was not convinced by Griffith's results and "was inclined to regard the finding as due to inadequate experimental controls."[54] However, he had no choice but to defer to the weight of mounting evidence and come around to the changed worldview when Griffith's results were reproduced by others, including Dawson in Avery's own laboratory. Avery quickly recognised the broader implications of type transformation. This is reflected in the opening lines of Avery, MacLeod and McCarty's famous paper of 1944:[55] "Biologists have long attempted by chemical means to induce in higher organisms predictable and specific changes which thereafter could be transmitted in series as hereditary characters. Among microorganisms the most striking example [...] is the transformation of specific types of pneumococcus." The holy grail going forward was the isolation and characterisation of the "active principle from crude bacterial extracts and identify [...] its chemical nature." Easier said than done. "It was some job, full of headaches and heartaches", as Avery told his brother Roy in a letter dated May 1943.[56]

From the crude extracts of killed S pneumococcus, Avery and his coworkers

---

50 Ibid., p. 277.

51 R. Dubos, 'Oswald Theodore Avery, 1877–1955', *Biographical Memoirs of the Fellows of the Royal Society* 2 (1956), 35–48, p. 39. https://doi.org/10.1098/rsbm.1956.0003

52 Ibid., p. 41.

53 Ibid., p. 40.

54 Ibid.

55 O.T. Avery, C.M. MacLeod and M. McCarty, 'Studies on the chemical nature of the substance inducing transformation of pneumococcal types', *Journal of Experimental Medicine* 79 (1944), 137–158. https://doi.org/10.1084/jem.83.2.89

56 Via R.D. Hotchkiss, 'Gene, transforming principle and DNA', *Phage and the Origins of Molecular Biology* (Plainview, NY: Cold Spring Harbor Laboratory Press, 1966).

isolated the active, transforming agent by removing proteins and polysaccharides and fractionating the resultant mixture. They describe that the active principle "separates out in the form of fibrous strands that wind themselves around the stirring rod."[57] This material was negative for protein, weakly positive for RNA and strongly positive for DNA in a suite of chemical tests. The ratio of nitrogen to phosphorus content in the active material was consistent with what one would expect for DNA. Treatment of the substance with enzymes that degrade proteins or RNA did not reduce its activity, whilst exposure to those that can depolymerise DNA abolished all activity. Finally, once transformation was complete and heritable, the transforming principle itself was also "reduplicated" during intergenerational transmission of the trait. Thus, the R pneumococcus acquired the capsule-producing trait from, or in response to, extraneous material that, most likely, was DNA, and the trait was heritable in the transformed cell concomitant with replication of the transforming active principle in the recipient.

In Avery's time, DNA was believed to exist in the form of short chains of only four nucleotides with a molecular weight of 1500. They existed in association with proteins and were believed to have purely structural roles. Until Avery et al's paper, DNA was not even known to be present in pneumococci. For example, in 1931, Charles White, while describing the contents of 'germs', or bacteria, listed enzymes (protein), pigments, sugars, fats and waxes, but not nucleic acid![58] Avery and colleagues estimated the molecular weight of the transforming principle to be around 500,000, calling it "a highly polymerised and viscous form of sodium desoxynucleate",[59] and, as he exclaimed to his brother, "who could have guessed it". With the benefit of hindsight, this estimate of molecular weight is much too low, and this error might have arisen from some degradation of the material during purification, something that would have made transformation notoriously difficult to achieve in the laboratory.[60]

In summary, an important implication of the whole of Avery, MacLeod and McCarty's work was the inevitable realisation that "nucleic acids are not merely structurally important but functionally active substances in determining the biochemical activities and specific characteristics of cells."[61]

Not everyone was convinced, however. Even if DNA were essential for transformation to occur, it was not wholly proven that its role was not merely in protecting a minute quantity of protein that might have escaped destruction in the preparation of the active material. Avery and colleagues themselves were being more than careful in not claiming that their preparation contained only DNA, in fact calling it "largely if not exclusively desoxyribonucleic acid."[62] Rollin Hotchkiss further purified the active principle until he could estimate that any protein contaminant did not constitute more than 0.02% of the total preparation. Work performed by many in the late 1950s showed that DNA in

57 Avery et al., 1944, p. 143.
58 C. White, 'What germs are made of', *Scientific Monthly* 32 (1931), 169–172.
59 An archaic term for deoxynucleic acid used by Avery et al, 1944.
60 Pollock, 1970.
61 Oswald Avery's letter to Roy Avery, May 1943.
62 Avery et al., 1944, p. 152.

these preparations was in fact being transported into the recipient bacteria and was then localised to the latter's own DNA. This was true not just in pneumococci but also in *Haemophilus influenzae*, in which transformation had also been demonstrated. One would not necessarily expect this if the DNA were playing a solely non-specific protective role. This also answered the question Sia and Dawson had posed, showing that the recipient cell was, in all probability, acquiring a trait from the donor cell via DNA. By this time, transformation in bacteria had been demonstrated not only for the trait of capsule production but many others. Reviewing the literature on transformation in 1961, Ravin wrote, "The burden of proof appears to be on those who would continue to claim that some substance other than DNA is the genetically specific substance in transformation reactions."[63] While all this was going on, other developments with bacteriophage had strengthened the case for DNA.

2.4. The genetic material II: the life cycle of bacteriophage

In May 1952, *The Journal of General Physiology* published a paper titled 'Independent functions of viral protein and nucleic acid in growth of bacteriophage' by Alfred Hershey and Martha Chase.[64] This work is often cited in textbooks as that which drove the final nail in the coffin of the candidature of proteins to be the genetic material. Indeed, the one-line summary of the findings of this paper reads thus: "The experiments reported in this paper show that one of the first steps in the growth of [a bacteriophage[65]] is the release from its protein coat of the nucleic acid of the virus particle after which the bulk of the [...] protein has no further function." This work, obviously, did not come out of the blue but was built on 15 years of research into bacteriophage growth and reproduction that followed the widespread recognition, in the 1930s, that the bacteriophage phenomenon was caused by a virus and not an autocatalytic enzyme.

This part of the bacteriophage story began with Emory Ellis, a biochemist at Caltech, USA. In the mid-1930s it was known that certain cancers are caused by filtrable viruses, a phenomenon that interested Ellis. However, studying viruses infecting animals was an expensive and time-consuming affair requiring the development and maintenance of animal colonies. Bacteriophage presented no such problem; growing bacteria and infecting them with bacteriophage were straightforward exercises. To some (including Ellis), it follows that if processes operating in animal viruses and bacteriophage "do indeed have common aspects, even though taking place in substrates as different as man and bacteria, the study of the process in the system lending itself to quantitative study seemed likely to be most rewarding [...] It made good sense [...] to try to learn all one could from this easily managed experimental subject before progressing to the more difficult viruses."[66]

63 A.W. Ravin, 'The genetics of transformation', *Advances in Genetics* 10 (1961), 61–163. https://doi.org/10.1016/s0065-2660(08)60116-9

64 A.D. Hershey and M. Chase, 'Independent functions of viral protein and nucleic acid in growth of bacteriophage', *Journal of General Physiology* 36 (1952), 39–56. https://doi.org/10.1085/jgp.36.1.39

65 The authors use the term T2 here, the name of the specific bacteriophage used in the study.

66 E. Ellis, 'Bacteriophage: One-step growth', *Phage and the Origins of Molecular Biology* (Plainview, NY: Cold Spring Harbor Laboratory Press, 1966), p. 55.

Ellis had read and was inspired by d'Herelle's 1926 book. In particular, he found d'Herelle's work on bacteriophage growth and infection process impressive. d'Herelle, through simple experimentation and brilliant reasoning, had indicated that bacteriophage action required direct contact between virus and bacteria and not just action at a distance, via, say, a diffusible toxin. d'Herelle had also argued, based on microscopic observations of the shape of the bacterial cell prior to death, that the bacteriophage lysed a bacterial cell from within the bacteria, implying that the virus, in one form or another, entered the cell and multiplied within. Most relevant to Ellis's work was d'Herelle's demonstration that the quantity of bacteriophage in a system was related to the number of lesions or plaques it could form on a plate growing bacteria. Later, in 1929, Burnet, keeping in mind that d'Herelle's work was quite neglected and was therefore missing "independent confirmation", substantiated the applicability of d'Herelle's plaque-counting method for enumerating bacteriophage.[67] Such measurements are important because being able to count bacteriophage would be an obvious prerequisite to any study of its growth, the term 'growth' in the context of microorganisms often referring to their multiplication or growth in number and not the size of individual cells or virus particles. It was while exploring the use of a "modification" of d'Herelle's plaque-counting method for developing a quantitative model of the bacteriophage life cycle that Ellis met Max Delbrück.

Max Delbrück (1906–1981) was a physicist interested in addressing fundamental problems in biology. Towards this end he was looking for simple, clearly definable biological systems and found bacteriophage, which Ellis introduced to him in 1937, attractive. Ellis and Delbrück wrote a paper together confirming d'Herelle and Burnet's view that bacteriophage growth during infection followed a step-like pattern. Infection of a bacterial cell by bacteriophage began with a latent phase in which there was no increase in virus titre for a while, followed by a sudden burst whence the bacteriophage quantity increased ~70-fold within 20 minutes. This was followed by another round of stasis or latency and explosive growth. Each sequence of latency and burst constituted an infection cycle. The latent phase would correspond to the period of attachment of bacteriophage to the bacterial cell surface and the part of the virus lifecycle inside the host cell. The burst would be the period of bacterial cell lysis and bacteriophage release. The released bacteriophage would then initiate a new cycle of infection. Ellis did not continue further with bacteriophage research, but Delbrück did, initiating, in collaboration with Alfred Hershey and Salvador Luria, an informal phage group which would go on to make seminal contributions to the origins of molecular biology. This includes proof that DNA is the genetic material; the discovery that protein synthesis based on information contained in genes requires an RNA intermediate; the establishment of principles underlying the generation of genetic variation; and early elucidation of mechanisms by which the expression of genes is regulated.

Of particular relevance to the story of DNA as the genetic material was the

67 F.M. Burnet, 'A method for the study of bacteriophage multiplication in broth', *The British Journal of Experimental Pathology* 2 (1929), 109–115.

phage group's contribution to what has been referred to as the eclipse[68] phase of the bacteriophage life cycle. This phenomenon had been noted a few times between 1939 and 1952, but the work of August Doermann,[69] published in 1952, was the most comprehensive and rigorous. Doermann used methods to lyse the bacterial cell prematurely, after infection by a bacteriophage but before cell death, to describe the eclipse phase thus: "During the first half of the latent period, no phage particles, not even those originally infecting the bacteria, are recovered."[70] Keeping in mind that bacteriophages are enumerated based on their ability to form plaques by killing host cells, Doermann's observation implies that soon after infection, or penetration, of the bacterial cell the bacteriophage assumes a form that is incapable of performing lysis. This phase is transient and soon a number of lysis-capable bacteriophage particles develop. Thus, it became clear that "a profound alteration of the infecting phage particle takes place before reproduction ensues."[71]

What could this profoundly altered form of the bacteriophage in eclipse be? This was an important question. As pointed out by Doermann, "on the thirtieth anniversary of the discovery of bacteriophages, one of the main difficulties in studying the life cycle was our complete ignorance about the state and nature of intracellular phage."[72] Some indications emerged between 1945 and 1947 from genetic studies. Such studies were well developed in eukaryotes, wherein the manner in which traits displayed by either parent recombine to produce emergent traits in the progeny had been explored for a century. It was a leap of faith to expect that such recombination would occur in bacteriophage as well, but works by Delbrück, Hershey and others showed that it did, and led to some important hypothesis development. Consider two traits of bacteriophage, A and B, that can be visually observed on microbiological plates growing bacteriophages and their bacterial prey. . Also assume that each trait is determined by its own dedicated component, called locus (plural loci), of the bacteriophage genetic material. If a bacteriophage showing trait A and not B were to infect a bacterium together with another showing B and not A, progeny viruses showing both traits or neither arise at some frequency. This would happen if the genetic material of the two bacteriophages combine and reassort in particular ways to form one that expresses some combination of properties of the two 'parental' bacteriophages.

Taking this further, Luria, in 1946–47, infected bacteria with multiple bacteriophages, all inactivated by exposure to ultraviolet radiation. Each such inactivated bacteriophage could singly attach to a bacterium but not proceed to lyse the host cell. However, infection of the same bacterium with multiple inactivated bacteriophages did, at

68 A.H. Doermann, 'The eclipse in the bacteriophage life cycle', *Phage and the Origins of Molecular Biology* (Plainview, NY: Cold Spring Harbor Laboratory Press, 1966).

69 A.H. Doermann, 'Liberation of intracellular bacteriophage T4 by premature lysis with another phage or with cyanide', *Journal of General Physiology* 35 (1952), 645–656, p. 655. https://doi.org/10.1085/jgp.35.4.645

70 Ibid.

71 Ibid.

72 Doermann, 1966.

some frequency, result in 'reactivation' producing fully functional viral particles. Luria showed that ultraviolet radiation introduced lethal lesions randomly on a subset of loci in the bacteriophage genetic material, and that recombination between bacteriophages with lesions at different sets of loci could produce virus progeny with all loci intact. This led Luria to make the inspired hypothesis that bacteriophage reproduction takes place "in an 'atomistic' way, by independent reproduction of a number of units and incorporation of these into phage[73] particles."[74] The suggestion arising from this hypothesis that the genetic material would disassemble into multiple pieces, each being replicated and then reassembled into a virus particle, is incorrect. It is nevertheless true that the phage is not intact within the host cell during eclipse and is re-formed before lysis. Luria also drew parallels "between the transfer phenomena here described and other phenomena in which a determinant of heredity is transferred into a new genetic complex as in the case of type transformation in pneumococci."[75]

Biochemistry would not be left behind. In an important work published in 1948,[76] Seymour Cohen showed that bacteriophage infection was immediately followed by DNA synthesis in the bacterium and that the amount of DNA synthesised before a certain time post-infection was equal to that contained in the viral particles liberated at that time. He also argued that the bacteriophage hijacked the host bacterium's resources for its own purposes, channelling the host's supply of phosphates and "enzymes for DNA and protein synthesis" towards the production of its own DNA and protein.

Taken together, around 1950, Hershey and Chase knew that during eclipse "genetic recombination takes place, apparently preceded by replication of genetic determinants of the phage."[77] It was also clear that "phage growth [...] consists of two stages: replication of some non-infectious form of the virus, followed by conversion of products of replication back into finished phage particles."[78] This also happens because "the bacterial metabolic system falls under the control of the nuclear apparatus of the phage, producing mainly phage specific materials."[79] To reiterate, this was a critical piece of information: deciphering the non-infectious form that the virus adopts during eclipse would help identify its genetic material. The active principle behind pneumococcal transformation had been shown to be DNA and tentative parallels between this phenomenon and recombination of bacteriophage genetic material had

73 Phage is a commonly used abbreviation of bacteriophage. I have used bacteriophage throughout this book, except while quoting others.

74 S.E. Luria, 'Reactivation of irradiated bacteriophage by transfer of self-reproducing units', *Proceedings of the National Academy of Sciences USA* 33 (1947), 253–264, p. 261. https://doi.org/10.1073/pnas.33.9.253

75 Ibid., p. 263.

76 S.S. Cohen, 'The synthesis of nucleic acid and protein in *Escherichia coli* B infected with T2r^{+} bacteriophage', *Journal of Biological Chemistry* 174 (1948), 281–293. https://doi.org/10.1016/S0021-9258(18)57397-X

77 A.D. Hershey, 'The injection of DNA into cells by phage,' *Phage and the Origins of Molecular Biology* (Plainview, NY: Cold Spring Harbor Laboratory Press, 1966).

78 Ibid.

79 Ibid.

been drawn by Luria. Northrop had been more explicit in 1951 that "the nucleic acid may be the essential, autocatalytic part [...] as in the case of the transforming principle of the pneumococcus, and the protein portion may be necessary only to allow entrance to the host cell."[80] This required proof, which Hershey and Chase provided.

The Hershey-Chase experiment was based on the idea that DNA and protein can be differentiated by their phosphorus and sulphur content. DNA is rich in phosphorus but contains no sulphur, whereas proteins contain sulphur and usually little or no phosphorus. After letting the phage attach to bacteria, Hershey and Chase separated the infected bacteria carrying bacteriophage in its eclipse state from any material the bacteriophage might have left behind outside the bacterial cell. By tracing the phosphorus and sulphur originally contained in the infecting bacteriophage, they showed the following: most bacteriophage DNA stayed in the infected bacterial cell. Most bacteriophage protein was left behind outside the cell. Infected bacteria separated from bacteriophage protein were perfectly capable of producing fully functioning virus particles. Therefore bacteriophage "protein probably has no function in the growth of intracellular phage. The DNA has some function."[81] The bacteriophage could be compared to a hypodermic needle made of protein, which injects its DNA cargo into the target bacterial cell. In its eclipse state, the bacteriophage is in the form of DNA that can replicate and produce new bacteriophages.

Within a year, the work of Raymond Gosling, Rosalind Franklin, Maurice Wilkins, James Watson and Francis Crick would demonstrate that DNA formed a double helical structure, which, together with the nucleotide composition studies of Erwin Chargaff (1905–2002), also suggested the base-pairing rule. A few years later, DNA polymerase, the enzyme that replicates DNA, would be discovered. In 1958, Mathew Meselson and Franklin Stahl showed, in one of the most famous experiments of all time, that DNA replication is semi-conservative, i.e., each newly synthesised daughter strand stays paired with the parental strand that had acted as the template for its synthesis.[82] Together, pneumococcal transformation, the genetics and biochemistry of bacteriophage, the elucidation of the structure of DNA, and the discovery of DNA polymerase and the semi-conservative mechanism of DNA replication made it reasonably clear that DNA was the genetic material and also suggested how replication might proceed. To quote Hershey, each of these lines of evidence:

> contributed in a different way to focus attention on the biological role of nucleic acids. It is useless to speculate, of course, what the impact of any one would have been without the support of the others. Surely though, the base-pairing rules plus one of the others could have generated very much the same molecular genetics that we pursue today.[83]

80 Via Ibid., p. 102.

81 Hershey and Chase, 1952, p. 56.

82 M. Meselson, and F.W. Stahl, 'The Replication of DNA in Escherichia coli', *Proceedings of the National Academy of Sciences USA* 44:7 (1958), 671–82. https://doi.org/10.1073/pnas.44.7.671

83 Hershey, 1966, p. 106.

2.5. The *annus mirabilis* for molecular genetics

The year 1961 has been called the "*annus mirabilis* for molecular genetics."[84] It was the year in which the isolation of messenger RNA, or mRNA, as the key intermediate between DNA and protein synthesis was first reported. It was also the year in which synthesis of a short polymer of amino acids in a test tube was described, a work that would lead to the genetic code being deciphered a few years later. The story of mRNA discovery is complex, involving a large number of researchers and an intricate body of scientific experiments and speculation. So intricate that sufficient priority to its discovery could not be assigned to anyone in particular, and nobody received the Nobel Prize for working out this central piece in the puzzle of life. Again, bacteriophage and bacteria played central roles in this story.

The biggest obstacle to, first, believing that DNA or nucleic acids in general could be the genetic material was their very conception in the minds of the scientists of the day. Early work on nucleic acids was limited to the study of their base composition in a subset of eukaryotic cells, all of which had nearly equal content of the four bases in their chromosomes. Any minor deviation from a perfectly balanced base composition was attributed to experimental error. This limited variation, or for that matter the complete absence of variation, in base composition did no favours to the idea that nucleic acids could determine the diversity of attributes across organisms. The prevalent notion that DNA was an aggregate of tetranucleotides comprising all the four bases also meant that the vast array of nucleotide sequences possible for long strands, even of the same overall base composition, was not appreciated. The genetic material was thought of as a molecule directly performing functions determining traits rather than carrying information that is used to produce functional molecules.

Bacteria to the rescue! Erwin Chargaff showed that base composition varied widely across bacteria. He also observed an important pattern: the content of G always equaled that of C and likewise for the pair of A and T.[85] This suggested the base-pairing rule to Watson and Crick in 1953, leading them to write one of the most famous lines in scientific literature: "It has not escaped our notice that the specific pairing we have postulated immediately suggests a possible copying mechanism for the genetic material."[86] Yet, through the 1950s, the idea that DNA was the genetic material remained a working hypothesis.[87] The problem, of course, was that nobody knew how DNA could direct the synthesis of proteins, which, at the end of the day, are the workhorses of the cell.

---

84 Cold Spring Harbor Symposia on Quantitative Biology, '1961: Cellular Regulatory Mechanisms, Vol. XXVI'. http://symposium.cshlp.org/site/misc/topic26.xhtml

85 Chargaff's body of work has been summarised in N. Kresge, R.D. Simoni and R.L. Hill, 'Chargaff's rules: the work of Erwin Chargaff', *Journal of Biological Chemistry* 280 (2005), 172–174. https://doi.org/10.1016/s0021-9258(20)61522-8

86 J.D. Watson and F.H.C. Crick, 'Molecular Structure of Nucleic Acids: A Structure for Deoxyribose Nucleic Acid', *Nature* 171 (1953), 737–738. https://doi.org/10.1038/248765a0

87 M. Cobb, 'Who discovered messenger RNA?', *Current Biology* 25 (2015), R525–R532. https://doi.org/10.1016/j.cub.2015.05.032

In 1953, with the structure of DNA solved and the base-pairing rule proposed, the physicist George Gamow wrote to Watson and Crick suggesting that proteins were synthesised on the DNA, with "free amino-acids from the surrounding medium get[ting] caught into the 'holes'"[88] on the DNA and getting linked together. Crick refuted this: it was already known that in eukaryotic cells, DNA was housed in a nuclear compartment separated from the surrounding cytoplasm in which protein synthesis occurred. This would necessitate the presence of an intermediate. As early as 1947, André Boivin (1895–1949), a strong supporter of the conclusions of Avery and colleagues' 1944 paper, had suggested that "the macromolecular desoxyribonucleic acids govern the building of macromolecular ribonucleic acids, and, in turn, these control the production of cytoplasmic enzymes."[89] Much later, in 1957—by which time the ribosome, the protein synthesising machine, had been discovered by George Palade—Crick proposed the Central Dogma of molecular biology, stating that proteins were produced from DNA via RNA and that the reverse was not possible. However, in contrast to what is commonly believed, Crick's view of the RNA intermediate bore no resemblance to the mRNA as we know them today, or as we would know them only a few years after Crick had made his proposal.[90]

The bulk of RNA produced by cells form part of the ribosome. These RNAs are stable. On the other hand, mRNAs are turned over rapidly. Early experimental evidence for the presence of transitory RNAs came throughout the 1950s. Jacques Monod's (1910–1976) and Arthur Pardee's (1921–2019) groups, in 1952 and 1954, showed that protein synthesis in bacteria, in response to the availability of certain nutrients, required RNA synthesis. In 1953, Hershey demonstrated that bacteria produced large quantities of unstable RNA shortly following infection by a bacteriophage. Between 1956 and 1958, Elliot Volkin and Lazarus Astrachan confirmed this, but also showed that the base composition of the RNA produced following bacteriophage infection was different from that of total RNA normally present in the host bacteria. One interpretation of this finding, offered by the authors though by no means proven, was that "such RNA molecules may be [...] possibly related to phage growth."[91] Volkin and Astrachan were somehow led to believe, wrongly, that their work had suggested that RNA was a precursor to DNA synthesis, an error that "obscured its true significance."[92]

In an important paper written in 1960, Sol Spiegelman and Benjamin Hall took the resolution at which nucleic acids were being studied one notch upwards. They demonstrated that RNA synthesised following bacteriophage infection showed sequence complementarity to bacteriophage DNA. At this point in time, we were still far from being able to sequence

88 G. Gamow, 'Possible Relation between Deoxyribonucleic Acid and Protein Structures', *Nature* 173 (1954), 318. https://doi.org/10.1038/173318a0

89 The original work is in French and this quote is part of an English-language summary of Boivin's article accessed via Cobb.

90 Cobb.

91 E. Volkin and L. Astrachan, 'Phosphorus incorporation in Escherichia coli ribo-nucleic acid after infection with bacteriophage T2', *Virology* 2 (1956), 149–161. https://doi.org/10.1016/0042-6822(56)90016-2

92 Cobb.

DNA, i.e., determine the sequence of bases that make up a stretch of DNA (or RNA), an extremely challenging problem that has now become highly accessible thanks to technique and technology development over half a century! However, if two strands of nucleic acids could stably bind to each other, in a process called hybridisation or annealing, then these strands must show high, if not perfect, sequence complementarity. This is the principle Spiegelman and Hall had exploited to demonstrate:

> that RNA molecules synthesised in bacteriophage-infected cells have the ability to form a well-defined complex with denatured DNA[93] of the virus [...] [and] conclude that the most likely inter-relationship of the nucleotide sequences of [bacteriophage] DNA and RNA is one which is complementary in terms of the scheme of hydrogen bonding proposed by Watson and Crick.[94]

In a second follow-up paper, Spiegelman, Hall and Storck showed that such RNA-DNA hybrids were native to bacteriophage-infected bacterial cells, "consistent with the assumption that DNA serves as a template for the synthesis of complementary informational RNA."[95]

In 1961, Sydney Brenner, Francois Jacob and Mathew Meselson (1930–) published a paper definitively proving that "a new RNA with a relatively rapid turnover is synthesised after phage infection. This RNA, which has a base composition corresponding to that of phage DNA, is added to pre-existing ribosome."[96] This paper was published back-to-back with another work, by James Watson and co-workers, also proving the existence of an RNA intermediary between DNA and protein synthesis.[97] In a review published in the same month, Jacob and Monod used the term "messenger RNA" to refer to this short-lived intermediary.[98] Later in the year, Crick, Brenner and others proposed that a code mapping each nucleotide triplet to an amino acid might exist, suggesting how DNA sequence could ultimately determine protein sequence. It was clear that such a code would be degenerate, with the same amino acid encoded by multiple nucleotide triplets. Crick and colleagues appreciated this and recognised that this degeneracy in the genetic code could even resolve "the major dilemma [...] that while the base composition of the DNA can be very different in different microorganisms the amino acid composition of their proteins only changes by a moderate amount."[99]

---

93 Denatured DNA refers to DNA in which the two strands have been separated, for example by heating.

94 B.D. Hall and S. Spiegelman, 'Sequence Complementarity of T2-DNA and T2-specific RNA', *Proceedings of the National Academy of Sciences USA* 47 (1961), 137–146, p. 143. https://doi.org/10.1073/pnas.47.2.137

95 S. Spiegelman, B.D. Hall and R. Storck, 'The occurrence of natural DNA-RNA complexes in E. coli infected with T2', *Proceedings of the National Academy of Sciences USA* 47 (1961), 1135–1142. https://doi.org/10.1073/pnas.47.8.1135

96 S. Brenner, F. Jacob and M. Meselson, 'An Unstable Intermediate Carrying Information from Genes to Ribosomes for Protein Synthesis', *Nature* 190 (1961), 576–581. https://doi.org/10.1038/190576a0

97 F. Gros, H. Hiatt, W. Gilbert, C.G. Kurland, R.W. Risebrough and J.D. Watson, 'Unstable ribonucleic acid revealed by pulse labelling of Escherichia coli', *Nature* 190 (1961), 581–585. https://doi.org/10.1038/190581a0

98 F. Jacob and J. Monod, 'Genetic regulatory mechanisms in the synthesis of proteins', *Journal of Molecular Biology* 3 (1961), 318–356. https://doi.org/10.1016/s0022-2836(61)80072-7

99 F.H.C. Crick, L. Barnett, S. Brenner, R.J. Watts-Tobin, 'General nature of the genetic code for proteins',

While prominent researchers like Watson, Crick and Monod were thinking about mRNA and the regulation of its synthesis, an "obscure researcher",[100] Marshall Nirenberg, was doing the same thing elsewhere. He had in fact used the term messenger RNA in his diaries before it was first used in a published work by Jacob and Monod. Some consider Nirenberg the first to isolate mRNA but Cobb, writing about the history of mRNA discovery, argued that "neither the discussion nor the data [presented in the paper concerned] justify this claim."[101] Nevertheless, Nirenberg set up cell-free systems to perform protein synthesis in a test tube. He showed that adding viral RNA to this system resulted in rapid and abundant protein production. It was only a matter of time before he would use it towards cracking the genetic code.[102] Heinrich Matthaei and Nirenberg fed the cell-free system synthetic poly-U RNA to produce a chain of several phenylalanines, an amino acid. This was again published in 1961 and would inspire the chemist Har Gobind Khorana to use cell-free protein synthesis and a panel of synthetic RNA to fully draw out the genetic code a few years later. The Central Dogma was complete: the sequence of DNA and base-pairing rules determine the sequence of messenger RNA, which, together with the genetic code, accurately predicts the amino acid sequence of the corresponding protein.

What remained to be done was determine, precisely, the sequence of any given DNA molecule. Frederick Sanger would achieve this,[103] adding to his earlier work on protein sequencing, and in the process becoming one of the very few individuals to have won two Nobel Prizes. Technology development building upon Sanger's DNA sequencing method, and growth in computation and computing power, would ensure that DNA sequencing, over the next few decades, became easier, cheaper and scalable. It has now produced complete catalogues of whole genome sequences of many tens of thousands of organisms from all known kingdoms of life. But before all this could happen, even before Sanger published his most developed DNA sequencing method in the late 1970s, Carl Woese had used a more rudimentary form of sequencing very short bits of ribosomal RNA to tell us something path-breaking about the tree of life and the place of bacteria and archaea on this tree.

2.6. The tree of life and the place of bacteria in it

It was, coincidentally, in the year 1961 that Roger Stanier wrote a draft of an article 'The Concept of a Bacterium' and sent it to his senior collaborator Cornelis van Niel. Stanier and van Niel together published the final version of their paper the year after in the journal *Archiv für Mikrobiologie*. At this time, there was much interest among bacteriologists in the role the field was playing in expanding our understanding of

Nature 192 (1961), 1227–1232, p. 1231. https://doi.org/10.1038/1921227a0

100 Cobb.

101 Ibid.

102 As defined in the primer that opens this chapter.

103 As would others, such as Allan Maxam and Walter Gilbert.

DNA as the genetic material and the manner of its functioning. It was not so much in the diversity of microorganisms and their relationships. The scientific lineage of Stanier and van Niel goes back to Martinus Beijerinck and his special interest in the ecological approach to microbiology in which relationships among microorganisms take centre stage. Thus, it is no surprise that they opened their paper with a very microbiological, and not molecular genetic, question for bacteriology: "Since the earliest days of microbiology [...] the abiding intellectual scandal of bacteriology has been the absence of a clear concept of a bacterium."[104]

How do we define a bacterium? This problem of definition is as much a question of classification. Large life forms, rich in easily identified distinguishing morphological features, have been classified into groups at different levels of granularity at least since the times of the ancients. For example, Aristotle and his student Theophrastus systematically classified animals and plants respectively.[105] In the mid-18th century, Carl Linné (later Carl Linnaeus after his ennoblement) developed a systematic, hierarchical two-part naming convention, which, at its highest level, grouped natural objects and living things into vegetable, animal and mineral. However, such classification schemes were not informed by evolutionary theory. The idea that one life form arose from a different, ancestral life form over aeons of evolution did not quite exist. As such, these classifications were not unlike how books are catalogued: Tolkien's *The Lord of the Rings* is epic fantasy, listed under fantasy and in turn under 20th-century fiction and so on and so forth, just as most of Stephen King's *The Dark Tower* series would be. No relationship such as one being ancestral to the other is assumed, even though some degree of inspiration may not be denied.

In the mid-19th century, Charles Darwin, in developing the theory of evolution, also recognised that for all life forms, "all true classification is genealogical; that community of descent is the hidden bond which naturalists have been unconsciously seeking." He hoped, in a letter to Thomas Huxley, that "the time will come [...] though I shall not live to see it, when we shall have very fairly true genealogical trees of each great kingdom of nature."[106] The scientist and master draughtsman Ernst Haeckel was listening, for he, with the help of his friend August Schleicher, drew a series of trees of life, some resembling gnarled great trees and others schematic. Though Haeckel's view of how inheritance of traits happen over generations was Lamarckian—the mostly erroneous 18th-century view that an organism would transmit acquired traits, such as an injury scar, to its progeny—Haeckel's contributions to biology and art remain profound. He conceived a tree of life as one with three branches: plant, animal and a more primaeval branch comprising all microorganisms including bacteria. The idea behind these trees is that the origin of life is at the bottom, on the tree trunk. As one moves up the tree,

104 R. Stanier and C.B. van Niel, 'The concept of a bacterium', Archiv *für* Mikrobiologie 42 (1962), 17–35. https://doi.org/10.1007/bf00425185

105 M. Morange, *A History of Biology* (Princeton, NJ: Princeton University Press, 2021) (originally published in French in 2016). https://doi.org/10.2307/j.ctv1bhg2m4

106 J. Sapp and G.E. Fox, 'The singular quest for a universal tree of life', *Microbiology and Molecular Biology Reviews* 77 (2013), 541–550. https://doi.org/10.1128/mmbr.00038-13

the stem branches and each branching event indicate a diversification of an ancestral life form at the base of the branch into two or more descendant forms. This conception largely holds true for the modern-day tree of life, though the nature of the stem and the branches have undergone innumerable revisions and probably will continue to do so. Haeckel also coined the term phylogeny ("phylogenese") to refer to genealogical relationships among organisms.[107]

In a genealogical or phylogenetic classification scheme, organisms are grouped together not just because they share some common traits, as *The Lord of the Rings* and *The Dark Tower* do, but because they share a common ancestor, as I do with my first cousins or as all primates do at a deeper point in evolutionary history. This is easier said than done considering the vast timescales over which large evolutionary changes happen. Even deep genealogical relationships have to be built using extant traits supplemented with fossil evidence where available. The choice of traits for making useful phylogenetic relationships is not straightforward either. However, in "higher" organisms, the ability of two individuals—one male and the other female—to come together to produce offspring allows them to be classified naturally into the same species. In other words, members of the same species can exchange genetic material with one another. Beyond that, it would again be observable morphological characteristics—even at the level of the embryo as Darwin favoured—that help build good hypotheses of phylogenetic relationships among species.

Bacteria[108] were (and, at a deep philosophical level, still are) a problem. They are observable only under a microscope and distinguishing morphological features are few. Thanks to the ability of a single bacterium to divide on its own to produce two daughter cells, the species definition that works for higher organisms does not apply to bacteria. Further, in the 19th century, microorganisms were believed to arise by spontaneous generation, leaving evolution and genealogies moot, until Pasteur showed otherwise. In addition, it was believed that all bacteria were of the same kind and any morphological variants noted were merely different life stages of the same organism. We owe the first systematic classification of bacteria into 'species' to the pioneering 19th-century microbiologist Ferdinand Cohn.[109] Cohn went against the grain by insisting that bacteria comprised multiple species each with its own defining characteristics, while also recognising that the idea of a species for bacteria would be different from that for higher organisms. Cohn grouped bacteria into four *genera* (singular *genus*) based on their shape, and each genus into multiple species based on traits such as pigmentation and ability to cause infection. In a sense I believe that his classification scheme, developed well after the publication of Darwin's *On The Origin of Species*, was genealogical in intention.[110] This is

107 R. Willmann and J. Voss, *The Art and Science of Ernst Haeckel* (Cologne: TASCHEN GmbH, 2021).

108 And microbes more broadly.

109 G. Drews, 'The roots of microbiology and the influence of Ferdinand Cohn on microbiology of the 19th century', *FEMS Microbiology Reviews* 24 (2000), 225–249. https://doi.org/10.1016/s0168-6445(00)00026-7

110 Being dependent on recent secondary literature on Cohn, I am not sure if Cohn himself saw his classification as genealogical.

because Cohn believed that within a species, new variants would emerge and that these would propagate their unique traits to future generations.

By the time Stanier and van Niel wrote about their concept of a bacterium, nearly a century after Cohn's work, they had come to believe that attempting a phylogenetic classification of bacteria was a "waste of time".[111] They had not always believed so however. In 1941, they had proposed a detailed genealogical classification of bacteria placing the spherical cocci at the most basal branch.[112] By 1961, however, they no longer "care[d] to defend"[113] their scheme. Instead, they focused their energies on identifying parameters by which bacteria could be clearly defined and distinguished from other microorganisms. If this definition managed to maintain, incidentally, a phylogenetic character, so be it. Is such a systematic, scientific demarcation of the bacteria even possible? The fact that microbiologists, in practice through experience and intuition, rarely argued over the placement of any given microbe into bacteria or otherwise told Stanier and van Niel that there exists "solid, pragmatic basis for the belief that a scientific definition of the bacteria constitutes an attainable goal."[114]

As with any such exercise in classification, problems arose for extreme cases, for example intracellular parasites such as *Rickettsia* that are barely resolvable under the light microscope. Can they be differentiated from large viruses? It was thought that such parasites represented some ill-defined intermediate state between viruses and cellular forms of life. This idea could be maintained only as long as the nature of viruses remained a mystery. Work on bacteriophage and other animal and plant viruses had ensured that this was no longer the case. As summarised by André Lwoff in 1957,[115] viruses clearly differed from cellular microorganisms in many ways. Cells differed from viruses in:

1. Containing both DNA and RNA;
2. In containing enzymes performing reactions leading to energy production or the synthesis of macromolecules;
3. In reproducing autonomously by division.

Rickettsia and similar intracellular parasites are cells and not viruses or any intermediate life forms; in fact, such intermediate forms are not known to exist, making the boundary between viruses and cells a great discontinuity.

At the opposite extreme, can one distinguish among cellular microorganisms, differentiating large bacteria from fungi, protozoa and algae? For example, what we now know as a type of cyanobacteria were then called blue-green algae for their

111 Via Sapp and Fox, 2013.

112 R. Stanier and C.B. van Niel, 'The main outlines of bacterial classification', *Journal of Bacteriology* 42 (1941), 437–466. https://doi.org/10.1128/jb.42.4.437-466.1941

113 Stanier and van Niel, 1962.

114 Ibid., p. 18.

115 A. Lwoff, 'The concept of virus', *Journal of General Microbiology* 17 (1957), 239–253. https://doi.org/10.1099/00221287-17-2-239

photosynthetic ability. It is in this context that Stanier and van Niel reintroduced the terms prokaryote and eukaryote,[116] referring to their original usage by Édouard Chatton in 1937. In their view "our only hope of more clearly formulating a concept of the bacterium" rests on the "essential differences between" prokaryotic and eukaryotic cells, which find a prominent place in any modern textbook of cell biology. In their own words,

> the principal distinguishing features of the prokaryotic cell are: 1. absence of internal membranes which separate the resting nucleus from the cytoplasm, and isolate the enzymatic machinery of photosynthesis and of respiration in specific organelles;[117] 2. Nuclear division by fission, not by mitosis[118] [...] [and] the presence of a single structure which carries all the genetic information of the cell;[119] and 3. the presence of a cell wall which contains a specific muco-peptide[120] as its strengthening element.[121]

Based on these criteria, blue-green algae would be prokaryotic, as all bacteria are.

The classification system of Chatton, Stanier and van Niel is entirely not without its evolutionary implications, at least at first glance. If prokaryotes and eukaryotes each form a separate clade with its own common ancestor, then the classification becomes phylogenetic. Evolutionary diversification following the branching of prokaryotes and eukaryotes could have then produced a diversity of forms within each. Stanier and van Niel recognised this by concluding, "if we look at the microbial world in its entirety, we can now see that evolutionary diversification through time has taken place on two distinct levels of cellular organisation [...]"[122]

At the end of the day, evolutionary change primarily acts on the genetic material, and any observable trait, morphological or biochemical, is a function, sometimes distant, of the action of one or more genetic loci and their interaction with the environment. Therefore, any phylogenetic classification would be best served by being developed based on detailed comparisons of DNA sequences, coupled with models of how the results of such comparisons can be mapped to a long-term course of events.

The inspiration to use molecular information to build phylogenetic trees came from a paper published in 1965 by Emile Zuckerlandl and Linus Pauling. Pauling had made major breakthroughs in the study of chemical bonding and applied these to the three-dimensional structure of proteins. He had won the Nobel Prize for Chemistry in 1954 and the Nobel Peace Prize in 1962. In the century or more since the publication of Charles Darwin's great work, our appreciation of the nature of biological inheritance

116 Stanier and van Niel, 1962.

117 Chloroplast and mitochondria respectively.

118 An elaborate process by which eukaryotic cells divide, ensuring that each daughter cell receives a copy of every chromosome.

119 The single bacterial chromosome, based on the paradigm of the model bacterium *Escherichia coli*.

120 Called peptidoglycan.

121 Stanier and van Niel, 1962, pp. 32–33.

122 Ibid. This view was criticised by C.R. Woese and N. Goldenfield, 'How the microbial world saved evolution from the Scylla of molecular biology and the Charybdis of the modern synthesis', *Microbiology and Molecular Biology Reviews* 73 (2009), 14–21. https://doi.org/10.1128/mmbr.00002-09

and the role of molecules therein, most notably of nucleic acids, saw quantum leaps. Since evolutionary change initiates in molecular variation in cells, which may or may not manifest in visible or otherwise easily observable traits, the time was now ripe to revisit the manner in which phylogenetic relationships were built. The important question that still needed answering was "where in the now living systems the greatest amount of their past history has survived and how it can be extracted."[123] The short answer to this is that the "the relevance of molecules to evolutionary history" is greatest for what Zuckerlandl and Pauling called "*semantophoretic molecules* or *semantides*." Semantides are "molecules that carry the information of the genes or a transcript thereof." Nucleic acids and proteins are semantides.

By the time Zuckerlandl and Pauling wrote their paper, the amino acid sequences of about 20 proteins, each with 100 or more monomeric units, had been determined.[124] In contrast, Sanger's paper on an early method of sequencing very short fragments of nucleic acids was published only the same year,[125] and his more transformative work on sequencing long stretches of nucleic acids would not be published for more than a decade. No wonder that the bulk of the article by Zuckerlandl and Pauling centred around amino acid sequences of protein and not base sequences of nucleic acids.

Just as with morphological traits, building phylogenetic relationships with semantide sequences must use comparisons of data from extant organisms and not ancestral and extinct beings.[126] Zuckerlandl and Pauling, obviously, were well aware of this and recognised what could be achieved by comparisons of extant semantide sequences. The extent of variation between equivalent semantide sequences from two extant organisms should contain information on the time elapsed since their divergence from a common "molecular ancestor." Analysis of monomeric units present at individual sites on the semantides being compared can predict "the probable amino acid sequence of the ancestral protein" and also help develop hypotheses on the "lines of descent along which given changes in amino acid sequence occurred."[127] Though nucleic acid sequences were not available then, the degeneracy of the genetic code was known. There can be isosemantic changes in the DNA, which will not change the amino acid sequence of the protein encoded, and therefore "there probably is more evolutionary history inscribed in the base sequence of nucleic acids than in the amino acid sequence of the corresponding protein."[128]

In 1964 Sol Spiegelman invited Carl Woese, a biophysicist by training, to pursue his research work at the University of Illinois. Woese was interested in protein synthesis and the evolutionary underpinnings of the genetic code. Woese was inspired by the

123 E. Zuckerlandl and L. Pauling, 'Molecules as documents of evolutionary history', *Journal of Theoretical Biology* 8 (1965), 357–366. https://doi.org/10.1016/0022-5193(65)90083-4

124 B. Foltmann, 'Protein sequencing: past and present', *Biochemical Education* 9 (1981), 2–7.

125 F. Sanger, G.G. Brownlee and B.G. Barrell, 'A two-dimensional fractionation procedure for radioactive nucleotides', *Journal of Molecular Biology* 13 (1965), 373–398.

126 Though, more recently, fragmented DNA isolated from fossils and burial sites have been sequenced.

127 Zuckerlandl and Pauling.

128 Ibid.

article by Zuckerlandl and Pauling, which, in Woese's words, "launched biology into the world of molecular chronometers."[129]

What is a molecular chronometer? A molecule that contains information that can be used to measure times of evolutionary divergence between two or more organisms. This molecule has to be semantophoretic and preferably a nucleic acid. Which part of a genome, comprising a large number of genes and genetic loci, would be best suited to be a molecular chronometer? Not all traits, and therefore not all genes or loci, are found across the entire diversity of life forms, but a molecular chronometer has to be, for it to be used as a common point of comparison. Furthermore, this gene must be one for which the degree of sequence divergence is proportional to the time over which the organisms being compared have diverged from their last common ancestor. Ideally such a chronometer should undergo changes randomly. This can happen in genes or loci on which no selective constraints operate. In other words, these should be loci at which changes will not affect the survival of the organism, or more precisely its ability to produce offspring—or that natural selection as an external force does not make changes at the loci more, or less, frequent than one would expect by random chance, which operates equally on all loci. Intuitively, sites of isosemantic changes might fit the bill. However, Woese had concluded that such sites undergo changes more rapidly than desired. They act as "*evolutionary stopwatches*", which measure "short-term evolutionary events", and not chronometers that can be used to build deep phylogenetic relationships.[130]

Woese argued, perhaps paradoxically, that the "most useful molecules for phylogenetic measurement all represent highly constrained functions." An example would be genes involved in non-negotiable processes, essential to all life, such as protein synthesis. Because they are under a strong pressure to be maintained as they are, they change slowly. Needless to say, a version of these genes will be present in all forms of cellular life, enabling the construction of universal phylogenies. The flipside is that these constraints can make changes far from random, making a true molecular chronometer hard, and probably impossible, to find.

Finally, if the molecule to be used as the best-possible-under-the-circumstances chronometer has enough domains, or distinct segments, each undergoing changes at different rates, it could be used to measure phylogenetic relationships over a broad range of evolutionary distances: a molecule used to measure the time of divergence of two closely related bacteria may also be used to measure that between bacteria and fungi. It turned out, rather conveniently, that ribosomal RNA (rRNA), which best satisfied criteria for being a molecular chronometer, is also highly abundant in all cells, making it amenable to purification and sequence determination using the laborious and rudimentary methods available to Woese in the early and mid-1970s.

In 1977, when Woese and George Fox published their seminal paper on the

129 C.R. Woese, 'Bacterial evolution', *Microbiology and Molecular Biology Reviews* 51 (1987), 221. https://doi.org/10.1128/mmbr.51.2.221-271.1987

130 Ibid.

phylogenetic structure of the prokaryotic domain,[131] the sequence of a type of rRNA called the 16S rRNA (18S rRNA in eukaryotes) was available for a diverse selection of prokaryotes and, for a yeast, a plant and an animal cell among eukaryotes. By comparing these sequences, Woese and Fox concluded that prokaryotes and eukaryotes "do not constitute a dichotomy" from an evolutionary standpoint. A class of organisms called methanogens, which were capable of reducing carbon dioxide to methane, were considered bacteria at that time. Woese and Fox showed that these "appear to be no more related to typical bacteria than they are to eukaryotic cytoplasms." Because their metabolism seemed suited to conditions believed to be prevalent in a young Earth 3–4 billion years ago, Woese and Fox called these organisms archaebacteria (now just archaea). In a review published the year after, Woese, Magrum and Fox described how archaebacteria do not include just methanogens but also a variety of cells of the prokaryotic type (but for differences in the composition of their cell walls) encompassing a range of shapes, metabolisms and DNA base composition. "In short, the archaebacteria exhibit a degree of diversity roughly comparable to that seen among the eubacteria."[132; 133]

Woese's work over a period of nearly a decade had established a framework for constructing phylogenetic relationships among organisms from nucleic acid sequence data, in the process discovering that the "phylogenetic structure of the living world [...] is not [...] bipartite [...] along the lines of the organisationally dissimilar prokaryote and eukaryote. Rather it is [...] tripartite comprising (i) the typical bacteria, (ii) the line of descent manifested in eukaryotic cytoplasms, and (iii) a little explored grouping",[134] the archaebacteria. The first complete bacterial genome would be published nearly 20 years later. We now have complete, or nearly complete genomes of tens of thousands of bacteria, archaea and eukaryotes. The number of molecular chronometers available for deep phylogenetic analysis has increased, and methods to perform such analyses have become more powerful, concomitant with the astonishing rise in computing power available to us. Yet the tripartite phylogenetic structure of the tree of life more or less still holds at a certain level, with adjustments therein in light of a flood of new data clearly maintaining that bacteria and archaea are distinct (Fig. 2.1). One major alternative hypothesis is the eocyte model, according to which eukarya emerges from an already diversified archaeal branch, a model that would thoroughly demolish any aspirations that the prokaryote-eukaryote dichotomy might have to being phylogenetic in nature.[135] This revises the three-branch structure of Woese's phylogeny back to a

131 C.R. Woese and G.E. Fox, 'Phylogenetic structure of the prokaryotic domain: the primary kingdoms', *Proceedings of the National Academy of Sciences USA* 74 (1977), 5088–5090. https://doi.org/10.1073/pnas.74.11.5088

132 Typical bacteria.

133 C.R. Woese, L.J. Magrum and G.E. Fox, 'Archaebacteria', *Journal of Molecular Evolution* 11 (1978), 245–252. https://doi.org/10.1007/bf01734485

134 Woese and Fox, 1977, p. 5090.

135 C.J. Cox, P.G. Foster, R.P. Hirt, S.R. Harris, and T.M. Embley, 'The archaebacterial origin of eukaryotes', *Proceedings of the National Academy of Sciences USA* 105 (2008), 20356–20361. https://doi.org/10.1073/pnas.0810647105

two-branch system, but with the important difference that the eukarya branch out of archaea, thus placing the archaea, considered prokaryotes, closer to eukarya than to bacteria, the other major group of prokaryotes.

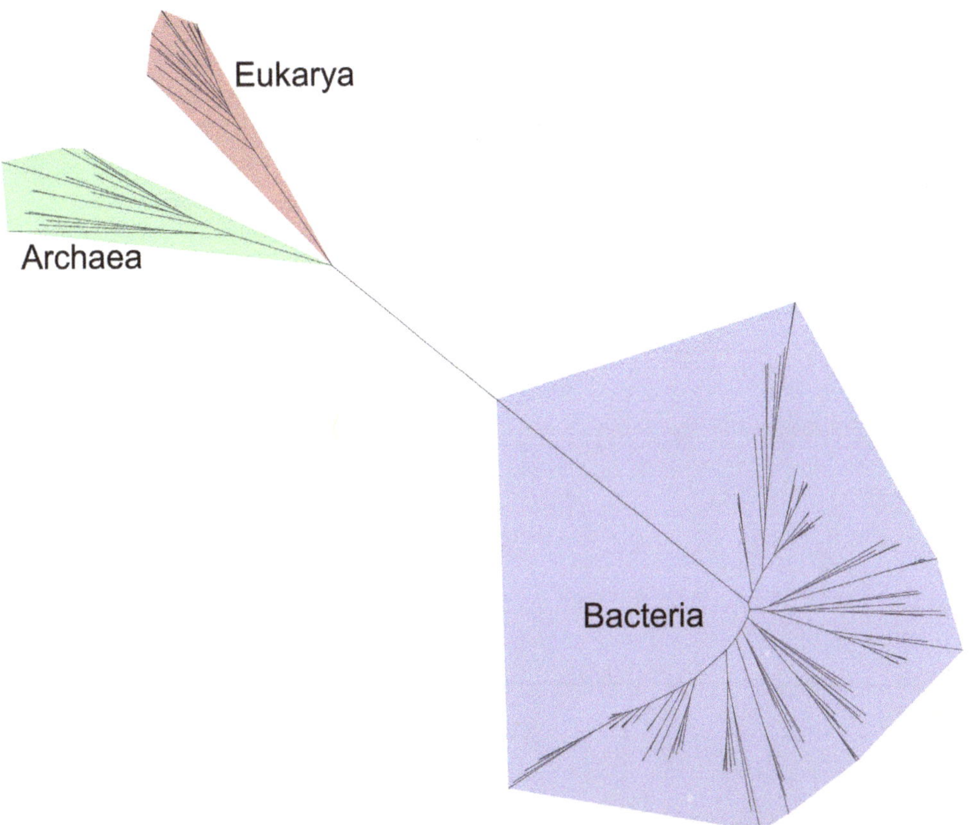

Fig. 2.1. A tree of life. A tripartite tree of life from iTOL (https://itol.embl.de) showing bacteria, archaea, and eukarya. Latest iTol paper: I. Letunic and P. Bork, 'Interactive Tree of Life (iTOL) v6: recent updates to the phylogenetic tree display and annotation tool', *Nucleic Acids Research* 52 (2024), W78–W82.

In the first three quarters of the 20th century, our understanding of bacteria increased exponentially, not just through research on bacteria for bacteria's sake, but also through the use of bacteria, especially *Escherichia coli*, and bacteriophage as models for studying universal properties of life. This is seen, in particular, in the manner in which Griffith's work on pneumococcal typing, performed to catalogue the diversity within this species of bacteria, led Avery and co-workers to use the phenomenon of type transformation to demonstrate the role of nucleic acids in biological inheritance. Similarly, the discovery of bacteriophage and the acceptance of their nature as viral led to their use as models to study—in principle—virus-induced cancers, which eventually converged with Griffith and Avery's bacteriology in establishing the role of DNA as the genetic material. The use of bacteria and bacteriophage as models for life did not stop there, and continued in showing how the information contained in the DNA is read to produce proteins via messenger RNA, something that applies to all life. The recognition that DNA is an informational molecule and that DNA sequence is the source of an organism's traits meant that efforts would begin to try and obtain nucleic acid sequences. This then fed back into the study of bacteria for bacteria's sake, thanks to Woese's work showing the place of bacteria and the newly recognised archaea in the vast tree of life. The 1980s, in our story, saw a lot of technological development, which we will skip. These developments resulted in a flood of complete genome sequences of cellular life forms starting in the mid-1990s. And this is where we will jump to in the next chapter.

3. The genome: how much DNA?

3.1. The first genome sequences

The structure of DNA, the discovery of the base-pairing rule by Watson and Crick, and the recognition that the genetic material was informational rather than merely functional implied that the secrets of life lay in the *sequence* of bases and not merely the base composition of the genome. The first genome sequence to be unveiled was that of a bacteriophage called MS2. Ironically, the genetic material of this bacteriophage is not DNA but RNA. Only viruses, and of course not all viruses, are known to use RNA as the genetic material. The small ~4,000-base sequence of the MS2 RNA was fully determined by Fiers and colleagues in Ghent, Belgium in 1976[1]. This landmark effort has been hailed by Eugene Koonin and Michael Galperin as "a truly heroic feat of direct determination of an RNA sequence."[2]

The year after the publication of the sequence of the MS2 RNA, Frederick Sanger and coworkers published the first complete sequence of a DNA genome, that of the bacteriophage ΦX174 (*phi-X174*).[3] This short sequence of ~5,400 bases (per strand) was in fact printed in whole as part of the paper describing it. The sequence allowed the researchers to identify and describe the sequence of all the nine protein-coding genes of this virus. The genome also taught us that some pairs of genes can be found in overlapping positions on the genome, something that we now know is seen in many viruses but is rarer in the genomes of prokaryotes, and where present, involves shorter overlaps of only a few bases.

The protein sequences of a subset of the nine genes whose DNA sequences were described by Sanger and colleagues had been determined earlier. This made these gene sequences easy to find. But Sanger et al. also proposed the sequences of genes they had known nothing about. How was this possible? The answer, in principle, lies in basic probabilities. Protein synthesis on an mRNA begins at a trinucleotide codon called

1 W. Fiers, R. Contreras, F. Duerinck, G. Haegeman, D. Iserentant, et al., 'Complete nucleotide sequence of bacteriophage MS2 RNA: primary and secondary structure of the replicase gene', *Nature* 260 (1976), 500–507. https://doi.org/10.1038/260500a0 . This work describes the sequencing of the third and final gene in this bacteriophage genome.

2 E.V. Koonin and M.Y. Galperin, *Sequence-Evolution-Function: Computational Approaches in Comparative Genomics* (Boston: Kluwer Academic Press, 2003).

3 F. Sanger, G.M. Air, B.G. Barrell, N.L. Brown, A.R. Coulson, C.A. Fiddes, C.A. Hutchison, P.M. Slocombe, and M. Smith, 'Nucleotide sequence of bacteriophage ΦX174 DNA', *Nature* 265 (1977), 687–695. https://doi.org/10.1038/265687a0

https://doi.org/10.11647/OBP.0446.03

the START codon, which is commonly AUG, encoding the amino acid methionine. In a genome, a three-base sequence would just be too frequent for each instance of AUG to act as a bona fide START codon. So, which among the many AUG trinucleotides are likely to mark the start of protein-coding genes? There are three STOP codons at which protein synthesis terminates. Just by random chance, and assuming a balanced base composition, one in every ~21 trinucleotides would match a STOP codon sequence. A 21-amino acid sequence is too short for most proteins. But real protein sequences are non-random and those rare instances where an AUG codon is followed by a *sufficiently long* sequence of non-STOP codons would be candidates for being protein-coding genes. Here is an example: "the dog and the cat are not big" may not be a particularly intelligent sentence, but is readable. But a part of this sentence with word breaks positioned by shifting them by one character—"...hed oga ndt hec ata ren otb..."—sounds like garbage. The former is equivalent to a protein-coding gene whereas the latter is not, though both are derived from the same sequence of characters.

Predicting protein-coding sequences this way would require highly accurate genome sequences, for even a single error can throw a spanner in the works. Imagine a single character missing in the sentence referred to above. If the 'd' in 'dog' were to go missing as a result of a typing error, and when faced with the condition that every word has three characters, we will get "the oga ndt hec ata ren otb ig...", quite similar to the garbled version described above. This was a concern for Sanger and coworkers. Alternatively, a longer sequence motif called the Shine-Dalgarno (SD) sequence is far less likely to be present in a long sequence by random chance. It was known that the SD sequence was a site to which the ribosome binds to initiate protein synthesis. Therefore, an AUG codon adjacent to an SD site would be a likely START codon. This principle is something that Sanger and colleagues exploited to find previously unknown genes in the ΦX174 genome. Modern methods for gene finding are more sophisticated than this, but long stretches of STOP-free sequences in highly-accurate genome sequences are often used to train these modern, automated gene-finding methods.

Just prior to his retirement, Sanger published the sequence of the relatively large genome of the bacteriophage λ[4] (*lambda*; a bacteriophage that we will revisit later in this book). The genome, at over 48,000 base pairs, was the longest to have its sequence determined when it was published in 1982. It was still short enough to be printed in full as "Figure 2", spanning over 20 pages of the paper describing it. Again, the genome sequence allowed accurate determination of the sequence and the location of the ~60 protein-coding genes in this bacteriophage. Many features of this genome would turn out to be similar to that of prokaryotic genomes. Many of the researchers involved in the λ genome sequencing project had been working on this bacteriophage for years. They could therefore channel their expertise into the curation of the genome and gene assignments, making these largely correct, though it was then still very

4 F. Sanger, A.R. Coulson, G.F. Hong, D.F. Hill, and G.B. Petersen, 'Nucleotide sequence of bacteriophage λ DNA', *Journal of molecular biology* 162 (1982), 729–773. https://doi.org/10.1016/0022-2836(82)90546-0

early in the history of whole genome sequencing.[5] Today, however, many prokaryotic genomes being sequenced are of organisms about which little, if anything, is known. There may be a high chance of large errors in such genomes and their annotations, yet the speed and cost-effectiveness of modern sequencing methods together with the power of modern, automated gene finding and annotation methods allow us to *know* an organism by knowing just its genome sequence! But we are jumping the gun.

In the mid-1980s, no cellular genome had been sequenced, but a moonshot project was announced, which kickstarted hectic activity in this direction. In March 1986, the Department of Energy of the US Government organised a meeting to discuss the feasibility of sequencing the entire 3 billion base pairs of the human genome. From 50,000 base pairs of λ to 3,000,000,000 base pairs of *Homo sapiens*! Some would call this madness, others ambition and vision! Whatever it might have been, Hamilton Smith (b. 1931)—microbiologist and Nobel Prize winner, who was at the meeting—reports that there was "unanimous enthusiasm for the project even though sequencing was not advanced enough at the time."[6] At around the same time, Renato Dulbecco, who had won the Nobel Prize a decade earlier for his work on cancer-causing viruses, wrote an article advocating the need for a human genome sequence in advancing our understanding of cancer.[7] Within a few months, the US Department of Energy apportioned funds for a pilot project. And five years later, on 1 October 1990, the full $1 billion human genome project was officially launched with a bulk of the funding coming from the US National Institutes of Health. Sequencing the smaller genomes of model organisms including the bacterium *Escherichia coli*, whose many deaths at the hands of a plethora of bacteriophages had launched molecular biology a few decades earlier, was also envisioned as part of the project.

Frederick Blattner had been working for years on *E. coli* and λ at the University of Wisconsin at this time. He would eventually sequence the ~4.6 million base pair (Mbp) genome of *E. coli* in 1997. But he had mulled over the prospect of doing so well before the human genome project was formally proposed. In an article published in 1983 in the journal *Science*, Blattner presciently wondered if the complete *E. coli* genome, crunched by the most powerful computers, would allow us to "reconstruct what the organism looks like, what it lives on, what it could be poisoned by and how it behaves? Or will a biological '*uncertainty principle*' be discovered that would preclude such a development?"[8] He started preparing to sequence the *E. coli* genome in 1988. In 1990, he received a grant from the human genome project to go ahead and get it done. It would take him seven years from that point, and a nudge from the funders, to complete the sequence. He would, in the process, lose the chance to attain the distinction of sequencing the first genome of a cellular organism.

---

5 Koonin and Galperin, 2003.

6 H.O. Smith, 'History of microbial genomics', in C.M. Fraser, T.R. Read, and K.E. Nelson (ed.) *Microbial Genomes* (Totowa: Humana Press, 2004), p. 6.

7 R. Dulbecco, 'A turning point in cancer research: sequencing the human genome', *Science* 231 (1986), 1055–1056. https://doi.org/10.1126/science.3945817

8 F.R. Blattner, 'Biological frontiers', *Science* 222 (1983), 719–720, p. 720. https://doi.org/10.1126/science.222.4625.719

3.2. The first bacterial genomes: minimal genomes for cellular life

The first cellular genome sequence to be fully determined was that of the bacterium *Haemophilus influenzae* in 1995.[9] *H. influenzae* was by no means the favourite in the race to this title, despite a failed attempt to secure funding for its sequencing by Sal Goodall in 1988, but managed to get there and by some distance! Despite its name, *H. influenzae* does not cause influenza (or flu), a viral disease; the misnomer is a historic contingency arising from its discovery and erroneous identification as the causative agent of a flu epidemic in the 1890s. It does, however, cause a variety of diseases ranging from mild ear infections to life-threatening infections of the blood stream. Richard Moxon, who had been working on this bacterium for years at the University of Oxford, was interested in laying his hands on its complete genome sequence. He discussed its feasibility with Hamilton Smith in 1993. At this time, genome sequencing was too challenging to be pursued by individual laboratories; it was for large, dedicated sequencing centres employing specialist personnel to attempt. Recognising this, Smith seeded the idea of sequencing the *H. influenzae* genome in J. Craig Venter (b. 1946), the founder of The Institute for Genome Research (TIGR),[10] which at the time possessed some 30 automated DNA sequencers. Venter, a polarising figure in the history of genome sequencing, was in. Robert Fleischmann at TIGR was designated the project lead in 1994 and the complete genome sequence of *H. influenzae* was published the following year in a paper with 40 authors![11]

DNA sequencers do not and cannot, at least in the near future, sequence entire bacterial genomes end to end in one go.[12] Automated Sanger method-based DNA sequencers, used for sequencing genomes in the 1990s and much of the 2000s, could generate sequences of ~1,000 bases in one go. For example, the average length of a sequencing *read*—a contiguous stretch of DNA sequence produced in one go—in the *H. influenzae* genome project was less than 500 bases. This is >3,000-fold smaller than the length of the *H. influenzae* genome. Such short fragments must be stitched together to assemble a complete genome. In a sense, this is a two-dimensional equivalent of building a complex structure out of hundreds of LEGO™ blocks.

The earliest approaches to creating DNA fragments that, after sequencing, can be stitched together, was laborious and time-consuming. This was why the *E. coli* genome, though first off the blocks, took nearly a decade to complete. In contrast, the *H. influenzae* genome project elected to harness the power of randomness and probabilities to obtain the bulk of the genome sequence: multiple copies of the genomic DNA were fragmented randomly by mere physical shearing, the ends of fragments

9 The brief history of the *H. influenzae* genome sequencing effort described in this paragraph is based on the recollections of Hamilton Smith, described in Smith, 2004.

10 Now called the J. Craig Venter Institute.

11 R.D. Fleischmann, M.D. Adams, O. White, R.A. Clayton, E.F. Kirkness, et al., 'Whole-genome random sequencing and assembly of *Haemophilus influenzae* Rd', *Science* 269 (1995), 496–498, 507–512. https://doi.org/10.1126/science.7542800

12 Figuratively speaking for circular genomes.

thus generated sequenced and then assembled, all in a matter of weeks. If I were to randomly tear two identical strips of paper in two separate trials (not, for example, by aligning them end to end and tearing them together), the fragments of paper I produce from one strip will not be identical to that from the other. In a similar manner, two identical molecules of genomic DNA from, say, two different cells of bacteria from the same colony, when randomly sheared would each be cut at their own set of sites, the two sets being non-identical. However, the ends of fragments derived from one DNA molecule would *overlap* with those from the other molecule. These overlapping ends allow two short fragments to be stitched together into a longer fragment.

The greater the number of DNA molecules that are fragmented and sequenced, out of a much larger population of DNA molecules fed into the sequencer, the greater the chance that any given base is sequenced. A theoretical model developed by Eric Lander and Michael Waterman in the late 1980s showed that as *coverage*—or, in effect, the number of DNA molecules successfully sequenced—increases, the chance that any position is left unsequenced decreases exponentially. Systematic features of the genome, such as stretches of sequence that are repeated multiple times in the same DNA molecule, would be exceptions but can be handled by procedures specifically targeted at addressing them. This is how the first cellular genome, and most other genomes since, were completely sequenced. Thus, the *H. influenzae* genome sequencing effort was historically significant, not just because it was the first past the chequered flag but also because it proved that simple random DNA shearing was enough for full genome sequencing, precluding the need for highly complex preparatory procedures.

The nearly 1.8 Mbp genome of *H. influenzae* encodes ~1,750 protein-coding genes. Though this was the first cellular genome to be fully sequenced, many viral genome sequences had been previously determined, as had the sequences of individual genes from a variety of organisms. Much of this sequence data was (and is) publicly available. The most popular database of DNA sequences, GENBANK,[13] maintained by the National Centre for Biotechnology Information (NCBI) at the National Institutes of Health (NIH), USA, held nearly 270,000 entries in 1995, having grown its collection ~500-fold since 1982. Genes with similar sequences often share functions. Given a database of gene sequences, of which a subset is of known function, it is possible to predict the function of a newly sequenced gene by comparing its sequence with those deposited in such a database. This is now an essential cornerstone of genome annotation. Using this principle, Fleishmann and colleagues annotated the *H. influenzae* genome and showed that it encodes proteins that can import and process sugars that the bacterium was known to utilise for nutrition. The phenomenon of transformation, first described by Griffith and later interpreted by Avery and co-workers in pneumococcus, had been, in time, shown to occur in *H. influenzae* as well. The genome sequence showed how genes involved in this process are organised as a cluster on the chromosome. The strain of *H. influenzae* that Fleishmann et al. sequenced is not pathogenic. The genome helped

13 'GenBank Overview', *National Library of Medicine*. https://www.ncbi.nlm.nih.gov/genbank/

to rationalise this knowledge by showing that a segment of the DNA, carrying genes responsible for the attachment of a pathogenic variety of *H. influenzae* to host cells, had, over the course of evolution, been cut out of the chromosome of its benign counterpart. Thus, the *H. influenzae* genome project served as proof of principle for the hopes that a genome sequence can help predict the metabolic capabilities of an organism, as well as help find the molecular basis for its other traits such as virulence.

Much insight into how genomes function and evolve derives from comparisons between the genome sequences of multiple organisms. Fleischmann et al. set the tone for this by comparing the genes of *H. influenzae* with the 1,000-odd genes of *E. coli* that had been sequenced at the time. However, the field of comparative genomics was launched in earnest by the second complete cellular genome sequence, that of the parasitic bacterium *Mycoplasma genitalium*—determined, again, at TIGR, by a team led by Claire Fraser.

Even while the *H. influenzae* genome was nearing closure, Venter was thinking about the next genome that should be sequenced.[14] *M. genitalium* is a parasitic bacterium that infects the human urogenital tract. Its genome, at less than 0.6 Mbp, was the smallest cellular genome known at the time. It was therefore "an important system for exploring a minimal functional gene set."[15] As noted by Arcady Mushegian and Koonin, "however small, a *cellular gene set* has to be self-sufficient in the sense that *cells* generally import metabolites but not functional proteins; therefore they have to rely on their own gene products to provide housekeeping functions."[16] Viral genomes, though small, do not count in this respect because viral replication is achieved by the direct use of host proteins. This made the sequencing of the genome of *M. genitalium* a momentous milestone in cataloguing and understanding the minimal requirements of cellular life.

Clyde Hutchison at the University of North Carolina had been studying the biology of *M. genitalium* for years, and agreed to collaborate with the TIGR group in sequencing the bacterium's genome. The genome sequencing project was completed in "less than four weeks."[17] The team predicted 470 protein-coding genes in the genome. A remarkably high proportion of these (~32%) are involved in protein synthesis, a non-negotiable component of the central dogma, essential to cellular life. In contrast, only 14% of the genes encoded by the ~four-times larger genome of *H. influenzae* are involved in translation-associated processes. This constituted early evidence that the minimal gene set of *M. genitalium* is particularly rich, as a proportion of the total gene complement, in functions *a priori* considered essential to life. This was more firmly

14 Smith, 2004.

15 C.M. Fraser, J.D. Gocayne, O. White, M.D. Adams, R.A. Clayton, et al., 'The minimal gene complement of *Mycoplasma genitalium*', *Science* 270 (1995), 397–403, p. 397. https://doi.org/10.1126/science.270.5235.397

16 A.R. Mushegian and E.V. Koonin, 'A minimal gene set for cellular life derived by comparison of complete bacterial genomes', *Proceedings of the National Academy of Sciences USA* 93 (1996), 10268–10273, p. 10268. https://doi.org/10.1073/pnas.93.19.10268

17 Smith, 2004.

established a few years later when the TIGR team, again in association with Hutchison, used genetic techniques to render as many genes as possible non-functional in *M. genitalium*, one at a time.[18] This effort showed that between 50–70% of all the genes in this genome are essential for the growth of this bacterium under a standard laboratory growth condition. Compare this with similar experiments done on the nearly 10-fold larger genomes of cosmopolitan bacteria such as *E. coli*, from which we know that well under 10% of its genes would be essential under nutrient-rich laboratory growth conditions.[19]

Within a few months of the publication of the *M. genitalium* genome, Mushegian and Koonin described a systematic comparison of the genes encoded by both the genomes sequenced thus far. Their goal was to "attempt to define the minimal gene set that is necessary and sufficient for supporting cellular life."[20] The idea behind such a comparative genomic approach to delineate a minimal gene set is the following: the more disparate genomes a gene is found in, the greater its centrality to life. This, in many ways, contrasts with the genetic approach of disrupting individual genes with a view to testing their essentiality to growth under defined laboratory conditions. By necessarily limiting its scope to conditions definable in a laboratory, the genetic approach may, at best, only tangentially point to the essentiality of a gene in an organism's ecological niche. Further, the essentiality of a gene may be dependent on the presence of another gene, something that is often neglected in large-scale genetic studies.[21] For example, there might be redundancy in gene function, i.e., two genes may be capable of performing the same essential function in the same cell and one may take over the extra workload in the absence of the other. On the other hand, the comparative genomic approach "is more applicable to deriving a minimal set of genes that are sufficient to sustain a robust evolutionary trajectory ... rather than just to support a cell under artificially favourable conditions."[22] That said, genes identified as constituting a minimal gene set by comparative genomic approaches are likely to be discovered as essential by genetic means, though the overlap between the two will be far from perfect.

18 C.A. Hutchison, S.N. Peterson, S.R. Gill, R.T. Cline, O. White, C.M. Fraser, H.O. Smith, and J.C. Venter, 'Global transposon mutagenesis and a minimal mycoplasma genome', *Science* 286 (1999), 2165–2169. https://doi.org/10.1126/science.286.5447.2165

19 https://www.genome.wisc.edu/resources/essential.htm. This webpage, last updated in 2004, provides links to various lists of essential genes in *E. coli*. For a more recent study, see E.C.A. Goodall, 'The Essential Genome of *Escherichia coli* K-12', *mBio* 9 (2018), e02096–17. https://doi.org/10.1128/mbio.02096-17

20 Mushegian and Koonin, 'Minimal gene set', p. 10268.

21 It is possible to stage multiple gene disruptions in the same cell. However, doing this comprehensively for all genes in *E. coli* would require well over 15 million disruptions of pairs of genes. With every step increase in the number of genes to be disrupted together, the total number of variant strains of *E. coli* that will have to be constructed increases 4,000-fold! Such screens, though limited to subsets of genes, have been carried out, and cases in which the presence of one gene affects the essentiality of another have been identified. Similarly, large-scale screens have also attempted to identify single genes that are essential to the growth of *E. coli* in hundreds of laboratory growth conditions.

22 E.V. Koonin, 'How many genes can make a cell: the minimal gene set concept', *Annual Review of Genomics and Human Genetics* 1 (2000), 99–116, p. 107. https://doi.org/10.1146/annurev.genom.1.1.99

Rosario Gil, and colleagues[23] point out that "the notion of a minimal cell" or that of a minimal gene set "cannot be sharply defined." For example, different environmental conditions or ecological niches might recommend different sets of minimally required functions and genes. Despite this caveat, the search for a minimal gene set is a valuable effort to "define which functions would be performed in any living cell and to list the genes that would be necessary to maintain such functions." Through this, we may even be able to define the fundamental genetic basis of all cellular life.

A comparative genomic approach is not, of course, without its complexities and limitations. It depends on our ability to predict genes of equivalent functions in multiple genomes, which—in the absence of experimental data on the function of these genes—is not always straightforward. This arises in part because of the myriad ways in which a group of genes, similar in sequence, are related in evolutionary terms. Evolution of life forms is characterised by divergence.[24] Starting from an ancestor, multiple descendants who are different from one another arise. This is true not only at the gross level of the organism but also at the level of individual genes.[25] A gene during replication may undergo a change in its sequence, which may be transmitted to one daughter cell but not the other, causing divergence between the two sisters. At this point, the gene sequence in the second daughter cell may be identical to that of the ancestor, but over time will undergo changes—some of which may be of some consequence and others not. The two descendants of the ancestral gene, one in each sister lineage, are related by divergence from a single last common ancestor. Such pairs (or groups) of genes "originating from a single ancestral gene in the last common ancestor of the compared genomes" are called *orthologs*.[26] In one or each sister lineage, molecular processes that operate on DNA may end up creating a second copy of a gene in a process called *duplication*. The two copies of the gene, though initially identical, will diverge over time and eventually look just about recognisable as similar. Genes which are "related by gene duplication",[27] irrespective of whether they are in the same genome or not, and the time of duplication notwithstanding, are called *paralogs*. *Homologs* encompass genes "sharing a common origin"[28], regardless of whether their relationship is through diversification from a common ancestor or by duplication. In general, orthologs are considered to perform the same function in different organisms. Paralogs may have diverged to perform different but mechanistically related functions. Differentiating between orthologs and paralogs is a matter of inference from sequence data from extant organisms. It becomes more and more arduous as the evolutionary divergence

---

23 R. Gil, F.J. Silva, J. Pereto, and A. Moya, 'Determination of the core of a minimal bacterial gene set', *Microbiology and Molecular Biology Reviews* 68 (2004), 518–537. https://doi.org/10.1128/mmbr.68.3.518-537.2004

24 This is not to neglect convergent evolution in which multiple lineages might acquire the same trait independently, or the phenomenon of horizontal gene transfer (see Chapter 4).

25 In fact, every base in an extant genome has its own evolutionary history.

26 As defined in E.V. Koonin, 'Orthologs, paralogs and evolutionary genomics', *Annual Review of Genetics* 39 (2005), 309–338, p. 311 (side box). https://doi.org/10.1146/annurev.genet.39.073003.114725

27 Ibid., p. 311 (side box).

28 Ibid.

increases between the genomes compared. This task is often a compromise between using the best available computational methods and the logistics of performing these computations within a reasonable timeframe.[29]

M. genitalium and *H. influenzae* are very different types of bacteria. They diverged from their last common ancestor as early as 1.5 billion years ago. Both genomes have undergone much gene elimination since their divergence event. This suggests that both genomes have shed considerable flab and evolved towards their own minimal state over aeons. Therefore, Mushegian and Koonin hypothesised that genes common to the two species—even if a set of two organisms represents but a paltry sample of the great diversity of life—should approximate the minimal gene set applicable to two very distant segments of the tree of life (Fig. 3.1 below). They found that 240 genes shared an orthologous relationship between the two organisms. This represents about half of the total gene complement of the smaller genome. Inspecting this set of genes in light of our general understanding of essential cellular functions—metabolic balance, reproduction, and evolution—Mushegian and Koonin concluded that it covered a necessary but not sufficient gene set. Not sufficient because the researchers found small gaps in essential functions that are otherwise fully represented in both genomes. Such gaps, if real and not a mere artefact of the sequence comparison methods, would render these critical processes entirely non-functional. This revealed the importance of what is called *non-orthologous gene displacement*, the phenomenon by which "the same function is performed by unrelated or very distantly related and non-orthologous proteins",[30] in evolution. This might have had a role to play in the evolution of ~10% of the minimal gene repertoire inferred by Mushegian and Koonin.[31] A few years later, when Koonin synthesised the scientific literature on the minimal genome concept and expanded these analyses by comparing 21 genomes, including those of a few eukaryotes and archaea, he found that only ~30% of the original minimal gene complement was ubiquitous. A comprehensive review of bacterial genomes,[32] published by Koonin and Wolf many years later in 2008, showed that only ~70 orthologous groups of genes were universal across hundreds of prokaryotic genomes. The rest are likely to have seen some degree of non-orthologous gene displacement in some clades.

The take home message from several attempts at finding a minimal gene set for cellular life,[33] mostly based on a synthesis of widely-present orthologous gene clusters and inferences of non-orthologous gene displacements, is that it consists of ~200 genes give or take a few. What functions do these represent (Table 3.1)?

---

29 There are other fine-grained classes of relationships with names like co-orthologs, in-paralogs, out-paralogs, etc., but these will not concern us here.

30 Koonin, 2000, p. 102.

31 The 265 genes (both orthologous and non-orthologous relationships) inferred by Mushegian and Koonin as minimally required for cell function showed a large, but imperfect, overlap with the list of essential genes identified by genetic means in *M. genitalium*. This is expected, as described earlier in this chapter.

32 E.V. Koonin and Y.I. Wolf, 'Genomics of bacteria and archaea: the emerging dynamic view of the prokaryotic world', *Nucleic Acids Research* 36 (2008), 6688–6719. https://doi.org/10.1093/nar/gkn668

33 Gil et al., 2004.

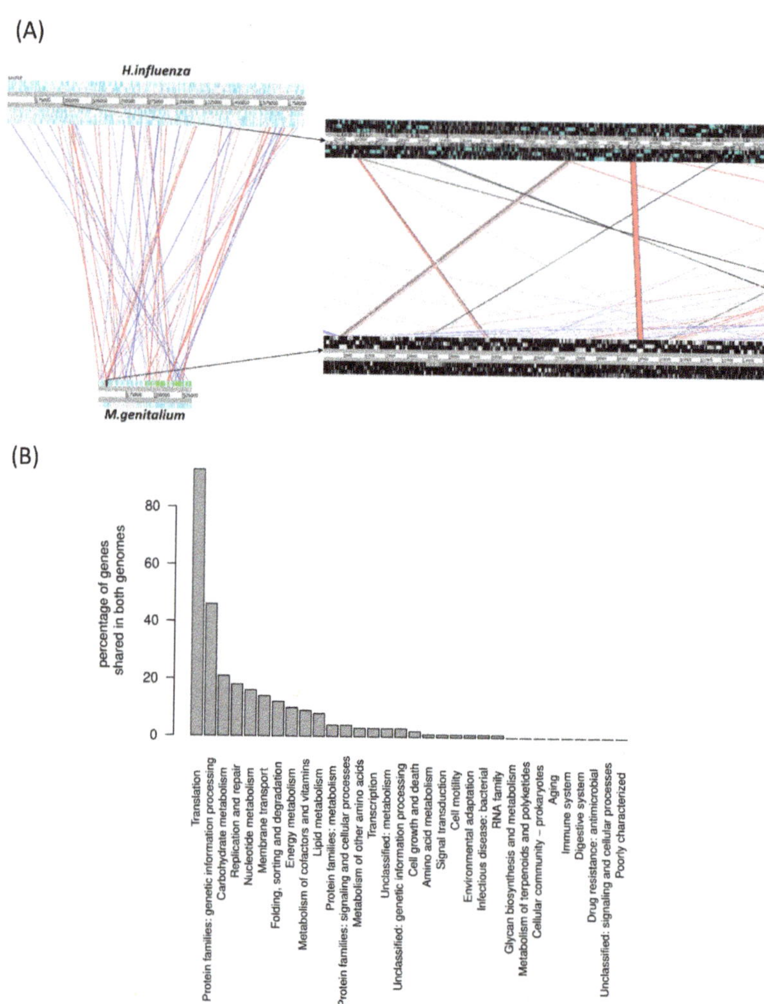

Fig. 3.1. Comparison of *H. influenzae* and *M. genitalium* genomes. (A) This figure shows linear representations of the *H. influenzae* and *M. genitalium* genomes, with lines marking genes sharing homology between the two genomes. On the right is a zoom-in of a small region of the two genomes. Figure generated using the stand-alone Artemis Comparison Tool (http://sanger-pathogens.github.io/Artemis/ACT/) by Ganesh Muthu. (B) This figure shows the percentage of genes from different gene functions, as defined by the KEGG database, that are conserved in the genomes of *H. influenzae* and *M. genitalium*. Note that most genes involved in translation or protein synthesis are present in both genomes. In general, genes involved in genetic information processing are more conserved than those involved in metabolism. Figure generated based on data available in the KEGG database, by Nitish Malhotra.

The processes that constitute the central dogma of molecular biology—replication, transcription and translation—are indispensable to any cellular life form. As such, genes encoding proteins involved in these processes form part of the minimal gene repertoire. Replication requires proteins for DNA unwinding, stabilisation of single-stranded DNA, initiation and continuation of DNA synthesis, and, where required, stitching together fragments of newly-synthesised DNA. From the perspective of cellular reproduction, the segregation of the two DNA molecules post replication will be followed by cell division, and proteins that form the septum for division are also crucial. Transcription cannot occur without an enzyme for RNA synthesis, proteins that ensure its initiation at precise sites upstream of a gene start site, and those that ensure its proper execution and completion. Translation requires several tens of proteins and RNA molecules that form or help assemble the ribosome, factors that ensure its efficient operation, transfer RNAs (tRNA) that bring the correct amino acid to the extending protein chain, as well as proteins that process tRNAs as required. A certain degree of quality control must attend to these processes as well. Errors during replication as well as chemical transformations of DNA bases lead to lesions that compromise genome integrity. Most genomes encode a variety of mechanisms to address such lesions, and a basal set of genes for DNA repair is part of the minimal gene set. Many long chains of amino acids cannot fold, on their own, into defined shapes that are competent to perform biochemical reactions. Improper folding of proteins, and the consequent accumulation of poorly folded proteins, is often pathological, as amply shown by the variety of human neurodegenerative diseases it causes. Cells minimise the chance of protein misfolding by carrying chaperone proteins that provide a safe space for newly synthesised protein chains to fold. Further, proteins, when misfolded or when past their sell-by date, need to be degraded. Together, genes encoding systems for protein quality control and clearance turn out to be essential; as are proteins that degrade RNA, thereby recycling nucleotides.

Whereas these constitute the minimal gene set for the maintenance and operation of the informational or the genetic component of the cell, the rest ensure the supply of energy and other molecular precursors for building the cell, its machines and the informational polymers. It is in this metabolic component that the impact of an organism's ecological niche on what would be essential to its growth and survival becomes particularly apparent. For example, what a cell can use for nutrition and how it would be converted into a supply of energy would depend on what is available in its immediate environment, as would the decision on whether to make or import precursors for, say, protein synthesis. Nevertheless, some basic machinery to import nutrients from the environment would be needed. For example, the minimal gene set built by Gil and colleagues includes a set up for the import of glucose and phosphate. Nutrient molecules, once imported, must be processed towards energy generation in the form of the nucleotide molecule ATP (adenosine triphosphate), and proteins that catalyse the relevant transformations would be essential. Monomeric components of macromolecules like DNA, RNA, and proteins must be either synthesised or imported from the environment. For example, enzymes that synthesise the ribose and the base component of nucleotides find a place in the minimal gene set. However, enzymes that synthesise amino acids do not, presumably because amino acids are often available to be imported from the environment. Synthesis of lipids (fats) that form the cell membrane is also essential. Overall, however, it is hard—and in all likelihood futile—

to define a *universal* minimal metabolic gene set, and as Gil et al. argue, "future studies will highlight a diversity of minimal *ecologically dependent* metabolic charts supporting a universal genetic machinery."[34]

Table 3.1. Gene functions in the minimal cellular genome. This table lists the set of gene functions and some example gene names present in the hypothetical/theoretical minimal genome for cellular life. Based on the work described in Gil et al. (2004), which presents a full, detailed table.

Gene function	**No of genes**	**Example genes**
Replication	13	dnaB, dnaE, dnaG, dnaN, dnaQ, dnaX, gyrA, gyrB, holA
Transcription	8	deaD, greA, nusA, nusG, rpoA, rpoB, rpoC, rpoD
Translation	96	alaS, valS, serS, pth, rnpA, rplA, rplB, rpsS, rpsT, obg, tsf, infA
DNA repair, restriction and modification	3	nth, polA, ung
Protein processing	15	map, pepA, dnaJ, grpE, ffh, ftsY, secE, secY, gcp
Cellular processes	5	ftsZ, pitA, ptsG, ptsI, ptsH
Energy and intermediary metabolism	56	eno, gapA, ldh, pgi, pgk, tpiA, atpA, atpB, atpC, rpe, cdsA, pssA, adk, upp, coaA, ribF

3.3. Minimal genomes in action

A fundamental question arising from these efforts towards defining a minimal set of genes or gene functions essential to cellular life is whether organisms with truly minimal genomes exist, or can survive if artificially constructed. Nearly every gene in such an organism should be essential to its survival under highly benevolent conditions. If these genomes are to approximate the theoretical minimal genome, they should be ~0.2—0.3 Mbp long. Catalogues of individual genes essential in one organism or another, even when determined experimentally using genetic means, do not address this question. What these catalogues tell us is whether an organism can survive or not when one gene among the many it encodes is removed from its repertoire. It is reasonable to expect that

34 Gil et al., 2004, p. 535.

simultaneous deletions of large chunks of dispensable genes from a genome, achieved by any tools at our disposal, could eventually enable us to get close to the answer. Several such attempts have been made on the large genomes of model bacteria such as *E. coli*, often with a goal of engineering a cell factory for biotechnological applications.[35] Such endeavours have successfully removed as much as 25% of the genome, still producing a viable organism that can thrive in some laboratory media. However, the resultant genomes, at over 10 times the size of the theoretical minimal genome described above, are still a far cry from our goal. Further, bacteria like *E. coli* are highly flexible and capable of thriving in a range of different environments. This complicates any attempt at defining (nearly-) universal minimal genomes from genetic analyses of such cosmopolitan bacteria. For now, let us put attempts at engineering a minimal genome on the backburner and instead search nature for any cellular life form encoding a genome approximating the minimal genome.

An important development in our quest for a naturally existing (nearly) minimal genome came with the sequencing of the genomes of *insect endosymbionts*. These are bacteria that live in an obligate, mutually beneficial relationship with their insect hosts, and in fact live inside specialised insect cells called bacteriocytes. When the insect reproduces, the endosymbionts are also transmitted down generations just as organelles such as mitochondria, which are integral parts of eukaryotic cells, are. This contrasts with the transmission of pathogens or symbionts such as the nitrogen fixing rhizobia in plant roots, which jump from one host to another *horizontally*, uncoupled from host reproduction. In return for providing a safe environment for their survival and propagation, the bacteria supply their hosts with useful, essential nutrients that may be inadequately available in the latter's diet.

The genomes of insect endosymbionts are small and code for a fairly limited set of biochemical functions. Their sizes are comparable to and at times much smaller than the genome of *M. genitalium*. A classic example of an insect symbiont with a small genome is that belonging to members of the bacterial genus *Buchnera. Buchnera* began an endosymbiotic relationship with insects called aphids some 200–250 million years ago. Since then, the genomes of the host and symbiont have evolved together in parallel such that different species of aphids harbour distinct varieties of *Buchnera*. The first *Buchnera* genome sequenced, at the turn of this century by Shuji Shigenobu and colleagues in Tokyo, was nearly 0.65 Mbp long,[36] and was second only to *M. genitalium* on the list of the smallest known bacterial genomes.

Buchnera are distantly related to *E. coli* and *H. influenzae* despite the great disparity among their genome sizes. The *Buchnera* genome appears to have undergone considerable minimisation by gene loss (a process referred to as *genome reduction*) since divergence

35 H. Mizoguchi, H. Mori, and T. Fujio, '*Escherichia coli* minimum genome factory', *Biotechnology and Applied Biochemistry* 46 (2007), 157–167. https://doi.org/10.1042/ba20060107

36 S. Shigenobu, H. Watanabe, M. Hattori, Y. Sakaki, and H. Ishikawa, 'Genome sequence of the endocellular bacterial symbiont of aphids Buchnera sp. APS', *Nature* 407 (2000), 81–86. https://doi.org/10.1038/35024074

from its last common ancestor with *E. coli* and *H. influenzae*. Along the way, *Buchnera* has lost genes encoding enzymes for the synthesis of components of the cell membrane called phospholipids. This suggests that this *Buchnera* either imports phospholipid molecules from the host cell, or hijacks the host enzymatic machinery to synthesise its cell membrane but in a manner that is not particularly inimical to the host's wellbeing. On the other hand, it has retained all the genes necessary for the synthesis of essential amino acids. Aphids feed on phloem, the plant sap that is poor in protein content, and hence require their permanent bacterial residents to produce and supply them with these critical nutrients. Recollect that genes for the synthesis of amino acids are not part of the theoretical, universal minimal genome for cellular life described earlier. Yet, these are encoded by the small genome of *Buchnera*,[37] because its amino acid-poor habitat dictates that it does so and supplies itself and its host with these essential molecules. Thus, the example of the *Buchnera*-aphid symbiosis emphatically illustrates the importance of recognising that there could be multiple ecologically-conditioned minimal genomes, applicable in particular to their metabolic components.

Analysis of the genome sizes of multiple *Buchnera* species, reported by Jennifer Wernegreen and colleagues around the same time as the first *Buchnera* genome, suggested that these would fall within a narrow range just under 0.65 Mbp.[38] A couple of years later, sampling of a more diverse variety of aphids and their resident *Buchnera* species by Gil et al. showed that this was not the case.[39] Instead, Buchnera genomes span a much broader spectrum of sizes extending downwards to under 0.45 Mbp, ~20% smaller than the genome of *M. genitalium*.

A few years later, in 2006, the small 0.42 Mbp genome of a *Buchnera* species was finally sequenced by Vincente Perez-Brocal and co-workers.[40] Curiously, this endosymbiont had undergone extensive genome reduction, losing even the ability to synthesise the essential amino acid tryptophan; the provision of this molecule to the host was achieved by a coexisting secondary bacterial symbiont. This pointed to an exquisite multi-way symbiosis between a large eukaryotic host cell and more than one bacterial endosymbiont. It even led to the suggestion that the 250-million-year-old association between *Buchnera* and aphids might be ending with newer endosymbionts functionally replacing *Buchnera*.

The xylem sap of plants is an even poorer source of organic nutrients than the phloem. The xylem fluid transports water and inorganic minerals from the root to

37 As we will shortly see, so do other small insect symbiont genomes.

38 J.J. Wernegreen, H. Ochman, I.B. Jones, and N. Moran, 'Decoupling of genome size and sequence divergence in a symbiotic bacterium', *Journal of Bacteriology* 182 (2000), 3867–3869. https://doi.org/10.1128/jb.182.13.3867-3869.2000. This study did not sequence the genomes, but just measured genome sizes using a technique called pulsed-field gel electrophoresis.

39 R. Gil, B. Sabater-Muñoz, A. Latorre, F.J. Silva, and A. Moya, 'Extreme genome reduction in Buchnera spp.: toward the minimal genome needed for symbiotic life', *Proceedings of the National Academy of Sciences USA* 99 (2002), 4454–4458. https://doi.org/10.1073/pnas.062067299

40 V. Pérez-Brocal, R. Gil, S. Ramos, A. Lamelas, M. Postigo, J.M. Michelena, F.J. Silva, A. Moya, and A. Latorre, 'A small microbial genome: the end of a long symbiotic relationship?', *Science* 314 (2006), 312–313. https://doi.org/10.1126/science.1130441

the rest of the plant and contains very little organic nutrients. Incredibly however, a group of insects called sharpshooters feed on xylem. As in the case of the phloem-feeding aphids, bacterial endosymbionts provide these sharpshooters with essential organic nutrients. In 2006, Dongying Wu and colleagues characterised the genome of a species of the bacterial genus *Baumannia*, which lives in an endosymbiotic relationship with the glassy-winged sharpshooter insect.[41] The *Baumannia* genome, at under 0.7 Mbp, is similar in size to that of the first *Buchnera* genomes to be studied. Wu et al. describe *Baumannia* as a "machine" for the synthesis of vitamins, which it provides its host with. However, unlike *Buchnera, Baumannia* lacks the ability to synthesise most essential amino acids but for the odd exception. The *Baumannia* genome encodes a general amino acid transporter however, suggesting that the bacterium imports these molecules from its local environment. But where do these molecules come from, considering that the host cannot source them from its regular diet?

The sharpshooter insect harbours other bacterial endosymbionts besides *Baumannia*, for example of the genus *Sulcia*. Though Wu and coworkers did not initially set out to sequence the genome of *Sulcia*, their efforts to isolate and sequence the *Baumannia* genome incidentally produced partial sequence data for the *Sulcia* genome as well. The data obtained for *Sulcia* clearly showed that this bacterium has the ability to synthesise most amino acids, which are lacking in the host diet and cannot be supplied by *Baumannia* either. In fact, the complementarity in the metabolic abilities of the two endosymbionts is nearly perfect (Fig. 3.2). For example, while *Baumannia* is the vitamin machine, *Sulcia* lacks the ability to produce these vital molecules. The *Sulcia* genome does not encode enzymes for the synthesis of the amino acid histidine, which turns out to be the rare amino acid produced by *Baumannia*. Additionally, precursors for the synthesis of some important molecules in one organism are supplied by the other. Wu and colleagues' microscopy-based observation that *Baumannia* and *Sulcia* are physically attached to each other, and not merely sharing the same habitat, further underlines the complementarity and cooperation between the two endosymbionts.

Yet another example of complementarity in metabolic abilities between two endosymbionts and their host involves the bacterial genera *Tremblaya and Moranella*, which live in a mealybug.[42] Many metabolic systems for the synthesis of essential amino acids are incomplete in both endosymbionts. In other words, neither endosymbiont can, on its own and given the necessary substrates, synthesise essential amino acids. But the genes that one bacterium lacks are present in the other such that the two gene sets assemble together to form more or less complete systems for essential amino acid synthesis. Any gaps that persist may be filled in by the host. These are examples of how close association between two microorganisms sharing limited space and resources

41 D. Wu, S.C. Daugherty, S.E. Van Aken, G.H. Pai, K.L. Watkins, H. Khouri, L.J. Tallon, J.M. Zaborsky, H.E. Dunbar, P.L. Tran, N.A. Moran, and J.A. Eisen, 'Metabolic complementarity and genomics of the dual bacterial symbiosis of sharpshooters', *PLoS Biology* 4 (2006), e188. https://doi.org/10.1371/journal.pbio.0040188

42 J.P. McCutcheon and C.D. von Dohlen, 'An interdependent metabolic patchwork in the nested symbiosis of mealybugs', *Current Biology* 21 (2011), 1366–1372. https://doi.org/10.1016/j.cub.2011.06.051

could, over a long process of evolution, define each organism's metabolic abilities such that redundancy is kept to a minimum, and to the extent that even basic capabilities may be achieved only by the sum of the two partners.

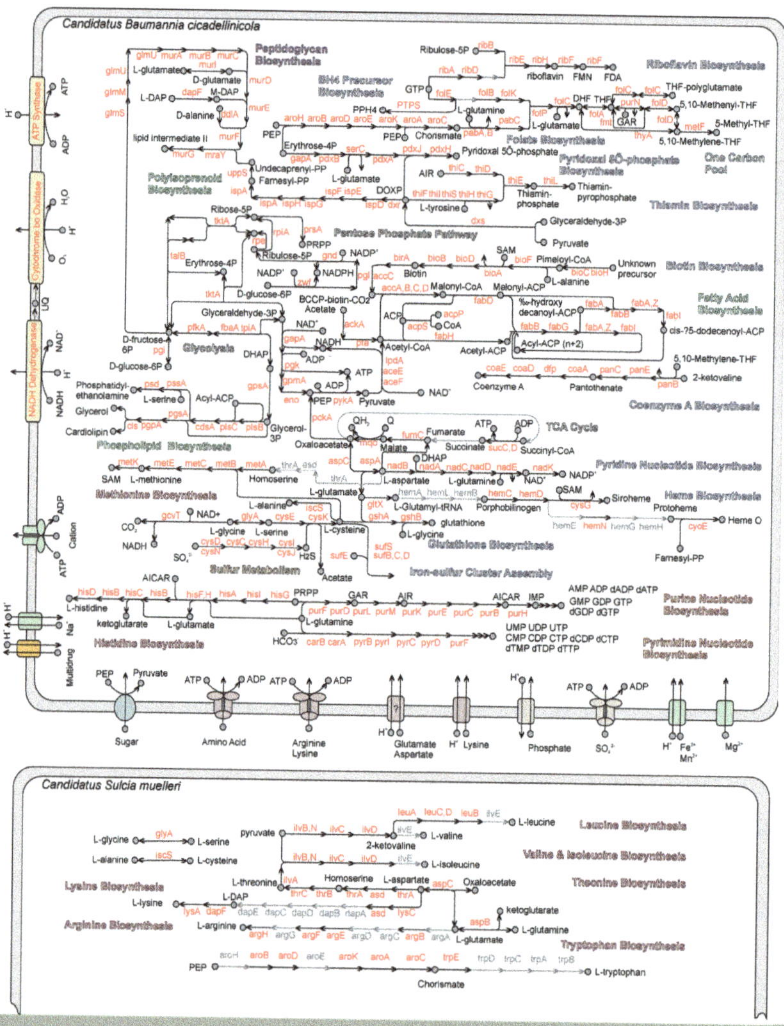

Fig. 3.2. Metabolic complementarity between endosymbiont genomes. This figure shows that the metabolic production in *Baumannia* is focused on vitamins and a couple of amino acids, whereas *Sulcia* is an amino acid producing machine. Originally published as Figure 4 in D. Wu, S.C. Daugherty, S.E. Van Aken, G.H. Pai, K.L. Watkins, H. Khouri, L.J. Tallon, J.M. Zaborsky, H.E. Dunbar, P.L. Tran, N.A. Moran, and J.A. Eisen, 'Metabolic complementarity and genomics of the dual bacterial symbiosis of sharpshooters', *PLoS Biology* 4 (2006), e188, Creative Commons Attribution License.

Sulcia and *Tremblaya*, the endosymbionts of the sharpshooter insect and the mealybug respectively, have exceptionally small genomes. The *Sulcia* genome data acquired by Wu and colleagues covered only ~0.15 Mbp, but it was incomplete. However, the genomes of other *Sulcia* species, sequenced fully in the years following the publication of Wu et al.'s work, showed that these genomes range in size from less than 0.2 Mbp to under 0.3 Mbp. It may be the case that *Sulcia* who live in insects that feed on the nutrient-poor xylem need the relatively large 0.3 Mbp genome, whereas those that can access the relatively more nutritious phloem can get by with a 0.2 Mbp genome, close in size to the theoretical universal minimal genome. The *Tremblaya* genome is even smaller at 0.14 Mb. This tiny genome lacks a whole suite of essential genes involved in protein synthesis,[43] raising the question of whether it can directly use the proteins encoded by the four-times larger genome of the co-resident *Moranella* to perform protein synthesis. The exceptional complementarity between *Tremblaya* and *Moranella*, likely encompassing the sharing of both metabolites and proteins, is probably achieved by the fact that *Moranella* lives *inside Tremblaya*, thus forming a nested endosymbiotic structure.[44]

Both *Tremblaya*, in the mealybug, and *Sulcia*, in the sharpshooter, have a bacterium with a much larger genome as a partner in symbiosis. Together, the two endosymbionts in each case create a composite, multi-species genetic repertoire that is 0.7–0.9 Mbp in size. The question then arises whether bacteria with such small and potentially sub-minimal genomes can establish endosymbiotic relationships in the absence of a partnership with a bacterium containing a three- or four-folds larger genome. In other words, can a pair of bacteria, both with minimal or sub-minimal sized genomes, complement each other and build symbiosis with a larger host?

The phloem-feeding leafhopper insect *Macrosteles* harbours two primary endosymbionts: a *Sulcia* and a bacterium of the genus *Nasuia*. Similar to other endosymbiotic systems involving a *Sulcia* (the sharpshooter insect for example),[45] it is the *Sulcia* that provides most essential amino acids to the system, with the co-symbiont (*Baumannia* in the sharpshooter and *Nasuia* in the *Macrosteles*) producing and supplying the rest.[46] Yet another example of complementarity between *Sulcia* and *Nasuia* is in the pathway for respiration, which reduces oxygen, resulting in the release of energy. Both endosymbionts encode only part of the respiration machinery, but together assemble a complete system. Successful operation of this system cannot be with the mere exchange of small-molecule metabolic products, but would require each endosymbiont to provide its partner with the proteins that the latter lacks. This would be similar to the situation in the mealybug where *Moranella* probably provides *Tremblaya* with a subset of proteins necessary for protein synthesis. In contrast to *Baumannia* or *Moranella*, *Nasuia* itself harbours a tiny genome clocking at a mere 0.11 Mbp. Smaller than the already small genome of *Sulcia*, and that of various other endosymbiont genomes sequenced over the years and not discussed here.[47] And just

43 The Tremblaya genome lacks genes encoding amino acid tRNA synthetases, which charge tRNAs with amino acids. The tRNAs are essential for the ribosome to read the genetic code on the mRNA, and it is the amino acid attached to a tRNA that is added to a growing protein chain at every step in protein synthesis.

44 McCutcheon and von Dohlen.

45 Yet another example is the spittlebug *Clastoptera*, which harbours a *Sulcia* and a *Zinderia* as endosymbionts. Here again *Sulcia* produces most amino acids and *Zinderia* supplies the remaining.

46 G.M. Bennett and N.A. Moran, 'Small, smaller, smallest: the origins and evolution of ancient dual symbioses in a phloem-feeding insect', *Genome Biology and Evolution* 5 (2013), 1675–1688. https://doi.org/10.1093/gbe/evt118

47 For example, *Zinderia* (see footnote 41 above), also carries a small ~0.21 Mbp genome. See J.P.

around 50% of the size of the theoretical minimal genome for cellular life.

As one might expect from the dramatic difference in genome size between *Nasuia* and *Baumannia*, the former codes for hardly any complete large metabolic pathways except for those for the synthesis of two essential amino acids. The phloem, though lacking in nitrogenous nutrients such as amino acids, carries ~1,000 times as much carbohydrates as the xylem. Thus, carbohydrates would form the bulk of the nutrition that the phloem-feeding leafhopper supplies to its endosymbionts. Both *Sulcia* and *Nasuia* encode but fragments of pathways for carbohydrate utilisation. It is likely that the host provides carbohydrates not only in their raw form, but also in partially digested states.

Like in *Tremblaya*, the remarkable lack of genes in *Sulcia* and *Nasuia* extends beyond the ecology-sensitive metabolism to even the core genetic component. The 0.19 Mbp *Sulcia* genome lacks a protein that, in other bacteria, ensures that the machinery for chromosome replication does not drop off the DNA frequently. Neither *Sulcia* nor *Nasuia* encodes proteins that help maintain the DNA in a state amenable to the unwinding of the two strands necessary for replication and transcription. These proteins are known to be essential to the survival of other bacteria and are even targets of antibiotics. Their absence in these severely reduced genomes suggests an ability to compensate for and adjust to highly inefficient implementation of parts of the processes that constitute the indispensable Central Dogma of molecular biology.

In the leafhopper insect, the two tiny genomes of the bacterial endosymbionts together assemble a gene set that approaches the minimal requirements for cellular life, though they might compromise the efficiency of operation of genetic processes. *Carsonella ruddii* is a bacterium that lives in an endosymbiotic relationship in psyllid insects which form galls on the hackberry tree. Unlike *Sulcia* or *Tremblaya* or *Nasuia,* this *Carsonella* is the sole bacterial endosymbiont of its phloem-feeding psyllid host. *Carsonella* carries a small 0.16 Mbp genome.[48] Its complement of genes for the synthesis of essential amino acids falls short of what it needs to supply its host with these nutrients. It also lacks many essential genes involved in replication, transcription, and translation, and has no bacterial partner to depend on for compensation. Is the host providing *Carsonella* with these proteins? How could a eukaryotic host supply its endosymbiont with proteins for bacteria-specific processes? Have these genes been transferred from a *Carsonella* ancestor to a psyllid ancestor such that these genes of bacterial origin are transcribed and translated by the eukaryotic host?[49]

McCutcheon and N.A. Moran, 'Functional convergence in reduced genomes of bacterial symbionts spanning 200 My of evolution', *Genome Biology and Evolution* 2 (2010), 708–718. https://doi.org/10.1093/gbe/evq055

48 A. Nakabachi, A. Yamashita, H. Toh, H. Ishikawa, H.E. Dunbar, N.A. Moran, and M. Hattori, 'The 160-kilobase genome of the bacterial endosymbiont Carsonella', *Science* 314 (2006), 267. https://doi.org/10.1126/science.1134196

49 For example, transfer of genes from bacterial endosymbionts called *Wolbachia* to their insect and nematode hosts has been documented. See J.C. Hotopp, M.E. Clark, D.C. Oliveira, J.M. Foster, P. Fischer, et al., 'Widespread lateral gene transfer from intracellular bacteria to multicellular eukaryotes', *Science* 317 (2007), 1753–1756. https://doi.org/10.1126/science.1142490

Organelles such as mitochondria and chloroplasts, which are integral parts providing energy-generating and photosynthetic abilities to their hosts, were once bona fide bacteria. A switch to an endosymbiotic lifestyle and the consequent minimisation of their genomes to an extreme has culminated in their present state. Is *Carsonella* on the path to becoming an organelle like the mitochondria or chloroplast, which do not satisfy the conditions for being living organisms in their own right, despite their ancient origins as bacterial cells? There is evidence now that the host compensates for the extreme gene losses in *Carsonella*.[50] This is in part due to genes of the host, as well as genes that have been transferred from the *Carsonella* genome to the DNA of the host, similar to what we often see described for mitochondria. As argued by Tamames and colleagues in a report describing a detailed analysis of the *Carsonella* genome, "the extreme degradation of the genome is not compatible with its consideration as a mutualistic endosymbiont and, even more, as a living organism."[51]

In summary, some insect endosymbiont genomes get close to the minimal genome in size and can even be sub-minimal. These arise through a process of gene loss. Their cozy ecological niche, which includes a eukaryotic host cell as well as one or more bacterial partners, protects them from any lethal consequence of such losses. In many cases, a combination of multiple sub-minimal genomes together assembles the minimal requirements for life. But in others, gene loss has proceeded to such an extent that the 'bacteria' may not even satisfy various conditions for being living organisms. Let us now move on to efforts at engineering minimal genomes in the lab.

Can a genome be engineered in the laboratory such that it supports free-living cellular life while being as small as it can get? Doing so would require knowledge and use of a bacterium which, in spite of being host to a small genome, can be conveniently grown in laboratory conditions. None of the insect endosymbionts would fit the bill. But, *Mycoplasmas* such as *M. genitalium*, despite their parasitic lifestyle in the wild, can be maintained in and grown on laboratory media. Given a natural, small genome from such an organism, can we reduce it by removing all or most non-essential genes and still produce a viable cell? Deleting a few hundred genes one after another from a population of bacteria, testing each intermediate mutant for survival and growth, always wondering which of the innumerable paths towards a final, minimal genome would be viable seems rather forbidding, and it is. The alternative is to create 'synthetic' genomes chemically from scratch in a test tube using existing genome sequence data as a template. This is easier said than done, but Craig Venter and Clyde Hutchison's team in the 2000s put in the heroic effort necessary to develop methods to achieve this formidable goal and also to transplant it into a recipient cell such that the newly-

---

50 D.B. Sloan, A. Nakabachi, S. Richards, J. Qu, S.C. Murali, R.A. Gibbs, and N.A. Moran, 'Parallel histories of horizontal gene transfer facilitated extreme reduction of endosymbiont genomes in sap-feeding insects', *Molecular Biology and Evolution* 31 (2014), 857–871. https://doi.org/10.1093/molbev/msu004

51 J. Tamames, R. Gil, A. Latorre, J. Peretó, F.J. Silva, and A. Moya, 'The frontier between cell and organelle: genome analysis of *Candidatus Carsonella ruddii*', *BMC Evolutionary Biology* 7 (2007), 181. https://doi.org/10.1186/1471-2148-7-181

introduced genome would take control over keeping its host cell running. This suite of techniques is what Hutchison and others used to create a synthetic minimal cellular genome.

Hutchison and colleagues chose the genome of the bacterium *Mycoplasma mycoides* as the starting point of their genome minimisation exercise.[52] *M. mycoides* has a nearly 1.1 Mbp genome and had been previously used by the group as a testing ground for chemical genome synthesis in the laboratory. The genome is nearly twice the size of the *M. genitalium* genome but is still on the smaller end of the spectrum of bacterial genome sizes.[53] However, *M. mycoides* grows over 15 times faster than *M. genitalium* in the laboratory, making it that much more convenient an organism to work with. Hutchison and coworkers used genetic screens and their knowledge of biochemistry to identify essential and quasi-essential genes encoded by the *M. mycoides* genome. Quasi-essential genes are genes whose loss would not be lethal to the cell but would cause serious growth impairments. Both essential and quasi-essential genes would be part of the final minimised genome. However, these would prove not to be enough to produce vital cells. Redundancy in gene function, even in a relatively small genome, meant that not all non-essential genes were truly dispensable. A redundant gene was dispensable in single-gene disruption experiments only because there was another gene, also singly-dispensable, that compensated for its loss. Using a clever trick, which again appears to be practical only in a genome synthesis paradigm, the researchers identified a couple of dozen such important but singly-dispensable genes. These were added to the minimal genome's repertoire. After ensuring that extragenic sequence elements that are required for transcription and translation were not inadvertently disrupted during their genome engineering and synthesis process, the group produced a minimal, 0.53 Mbp *M. mycoides* genome encoding ~470 genes capable of supporting the growth of its host cell.

As one would expect, nearly all genes involved in core genetic processes like replication, transcription, and translation in the original *M. mycoides* genome were included in its minimal form. In contrast, a variety of *selfish* genetic elements (discussed later in this chapter) whose intrinsic value, if any, to the host organism is unclear could be safely deleted. The metabolic component of the natural *M. mycoides* genome was minimised such that genes involved in the utilisation of glucose were retained whereas those responsible for the metabolism of non-glucose sugars were deleted. This is consistent with the organism's 'ecological niche' in the laboratory in which glucose was the main carbon source. A few genes of unknown function had to be retained. Whether these perform fundamentally novel yet essential processes or whether they play important accessory roles in well-known processes remains unknown. In summary, this synthetic or engineered organism carries the smallest genome for any

52 C.A. Hutchison III, R.-Y. Chuang, V.N. Noskov, N. Assad-Garcia, T.J. Deerinck, et al., 'Design and synthesis of a minimal bacterial genome', *Science* 351 (2016), aad6253. https://doi.org/10.1126/science.aad6253

53 We will discuss the distribution of bacterial genome sizes shortly.

organism capable of survival and multiplication in the laboratory. The minimal cell grows slower than the parental *M. mycoides*[54] but multiplies several times faster than *M. genitalium* with its slightly larger genome, which may be the smallest known for any natural organism that can be grown under laboratory conditions.

3.4. Genome size and information content

Most discussions of theoretical minimal cellular genomes would, for obvious reasons, centre around small bacterial genomes of the kind described above. These clock at less than 1 Mbp in length. However, fully sequenced bacterial genomes, of which tens of thousands are available to us through public databases today, encompass a two order of magnitude range of sizes. The distribution of bacterial genome sizes shows two peaks, one at around 2 Mbp and the other at around 5 Mbp. The smallest known bacterial genomes, as seen earlier, are less than 0.2 Mbp, whereas the longest genome, belonging to the genus *Sorangium*, extends to over 13 Mbp. In general, cosmopolitan bacteria, which can thrive in multiple niches and handle extremes of feast and famine plus a whole gamut of toxic agents, tend to have larger genomes. Such bacteria include members of the versatile actinomycete clade, which includes bacteria from which the antibiotic streptomycin was first isolated. Genomes of eukaryotes span a much broader range. The smallest among these are comparable in size to the larger bacterial genomes. At the other extreme, some plants and fishes have extremely large 100,000 Mbp genomes, two orders of magnitude longer than the human genome.

Besides size, bacterial and eukaryotic genomes differ more importantly in gene density, i.e., the percentage of genome sequence coding for genes. Bacterial genomes are gene rich: genes, mostly encoding proteins, cover ~90% of most bacterial genomes. The bacterial genome is effectively a necklace densely strung with beads (genes) with little exposed thread (non-genic DNA). The longer the genome, the more genes it encodes, the relationship between genome size and the number of genes being linear. This does not apply to the genomes of many kinds of eukaryotes, most of which encode 10^4 genes irrespective of genome size (Fig. 3.3). The genome of the baker's yeast, which is not much larger than the largest bacterial genomes, encodes under 10,000 genes, whereas the several hundred times longer human genome encodes only three to four times as many genes. This is not because genes encoded by higher eukaryotic genomes are orders of magnitude longer than bacterial and yeast genes, but is due to the presence of vast tracts of DNA devoid of genes in the genomes of many higher eukaryotes.[55] So, why are the genomes of many higher eukaryotes so large and, despite supporting complex life styles, so gene poor? What 'purpose', if any, do these gene-poor regions serve? The history of thought on genome size

---

54 The following very recent work showed, however, that this minimal cell, when cultured over ~2,000 generations, can fully recover from this growth deficit. R.Z. Moger-Reischer, J.I. Glass, K.S. Wise, L. Sun, M.C. Bittencourt, et al. 2023. 'Evolution of a minimal cell', *Nature* (2023). https://doi.org/10.1038/s41586-023-06288-x.

55 Even some genomes of amoeba—a microbial eukaryote—are large and gene poor.

is rich for eukaryotes and that is what we will consider to try and explain the differences in genome size and gene density between eukaryotic and bacterial genomes.

We go back in time to 1951, before the publication of the experiment of Alfred Hershey and Martha Chase, and well before that of the double helical structure of DNA by James Watson and Francis Crick. Though Oswald Avery's team had demonstrated the nature of Frederick Griffith's transforming principle by then, that DNA is the genetic material was still far from being universally accepted. It was at this time that A.E. Mirsky and Hans Ris, in a paper[56] published in the Journal of General Physiology, reported their measurement, using chemical means, of the amount of DNA[57] contained in various eukaryotic cells, in particular those of vertebrates. Today, we take for granted that different cells in the same organism have more or less[58] the same DNA content. However, this was not a given in 1950. In fact, cytologists, who studied DNA in cells using dyes that stain this material, believed otherwise.[59] By bringing together previously published results and their own data, Mirsky and Ris first concluded that there is "considerable support for the rule"[60] that all cells in any given organism carry the same amount of DNA. However, the DNA content of cells varies enormously across vertebrate species and, most remarkably, in highly unpredictable ways!

It is not unreasonable to predict *a priori* that a more *complex*[61] organism, performing a vast variety of actions, would require more genes to support its complexity. To accommodate a larger number of genes, a more complex organism would need to carry more DNA than a simpler one. Again, it would be intuitive to argue that phylogenetically closely-related species would share many traits and would therefore carry similar amounts of DNA. Mirsky and Ris showed that neither expectation was borne out by facts! For example, cells of the freshwater vertebrate lungfish carry 40 times as much DNA as birds and, even more strikingly, 20 times as much DNA as the cells of the considerably more complex human. Even the enormous amount of DNA carried by lungfish cells pales in comparison to that carried by cells of humble aquatic worm-like creatures called amphiuma. That phylogenetically related species need not carry similar amounts of DNA was illustrated by a comparison of the DNA content of the ocean-dwelling triggerfish with that of other organisms. Triggerfish cells carry 170-fold less DNA than the somewhat related amphiuma, but only nine times as much DNA as the cells of early-branching multicellular eukaryotes called sponges, which are not even vertebrates! Thus, the DNA content of many multicellular eukaryotes, most notably vertebrates for which Mirsky and Ris had collected data,

---

56 A.E. Mirsky and H. Ris, 'The desoxyribonucleic acid content of animal cells and its evolutionary significance', *Journal of General Physiology* 34 (1951), 451–462. https://doi.org/10.1085/jgp.34.4.451

57 We refer here to *haploid* DNA content—i.e., one copy of each chromosome.

58 Subject to somatic variations and DNA modifications that may be specific to one cell type but not another in the same organism.

59 Mirsky and Ris note that this erroneous belief likely arose from differences in the volume of the nuclear compartment in which DNA of eukaryotic cells are housed, and not from differences in DNA content itself.

60 Mirsky and Ris, p. 453.

61 However vaguely the term may be defined.

does not correlate with phylogenetic relatedness, nor with any notion of organism complexity. A take home message from the work of Mirsky and Ris is: if an organism's complexity is proportional to the information content (reflected in, say, the number of genes) of its DNA, then the DNA content of many eukaryotes may not predict how much *useful* information it encodes. This once again brought to the fore the question of what function DNA might serve, adding ammunition to the cause of researchers who were not entirely convinced of DNA's central role in heredity.

Time passed and more data along similar lines accumulated. Two decades after the publication of Mirsky and Ris's paper, C.A. Thomas coined the term *c-value paradox*, *c-value* being used to refer to the DNA content of a cell, to formulate the puzzle.[62] In particular, the "wide variations of c-value within single genera of insects and plants"[63], besides expanding Mirsky and Ris's finding to well beyond vertebrates, intrigued Thomas. A particular example quoted by Thomas is the 2000-fold variation in c-value among species of algae. It had also become clear that many eukaryotic cells carried more DNA than they would need. Thus, the puzzle was not about how a complex organism would fit as many genes as it needed within a small piece of DNA. For example, using evidence of gene activity obtained from nucleic acid staining experiments, Thomas estimated that the DNA of the fruitfly *Drosophila melanogaster* carries 50 times more DNA than what it would need to accommodate its genes. What might all the excess DNA be for? Thomas writes: "We could solve the problems by saying that 98% of the DNA is irrelevant to the organism. This is a currently popular way of dealing with problems and, in one restricted sense, the informational sense, it must be true ... It might be argued that this non-functional DNA is being held in reserve by nature for future evolutionary experimentation. However most evolutionary biologists argue that selection[64] is applied to the organism as it is, not as it *might be*. If suggestions of this kind are reflected, how can we explain an animal replicating and maintaining most of its DNA if it is no advantage to it? It probably isn't true: it offends the principle of parsimony"[65].

Mammalian genomes are typically of the order of 1000 Mbp long. Let us first assess the feasibility of such a large genome being dense in useful information. Let us, for the purpose of this discussion, assume that all useful information in a genome is represented by protein-coding genes. The human genome is 3,000 Mbp long. Assuming that an average protein-coding gene is about 1,000 bp long, an information-rich human genome should carry about 3 million protein-coding genes. From the discussion above we know that this is not true. However, is it *even possible* that the human genome *can* code for over a million genes and successfully sustain the species, given what we know about how genomes function and how they evolve?

62 C.A. Thomas Jr, 'The genetic organization of chromosomes', Annual Review of Genetics 5 (1971), 237–256. https://doi.org/10.1146/annurev.ge.05.120171.001321

63 Ibid., p. 247.

64 Darwinian selection, according to which any trait or piece of DNA would be maintained if it confers a benefit to the population of organisms.

65 Thomas, 1971, p. 251.

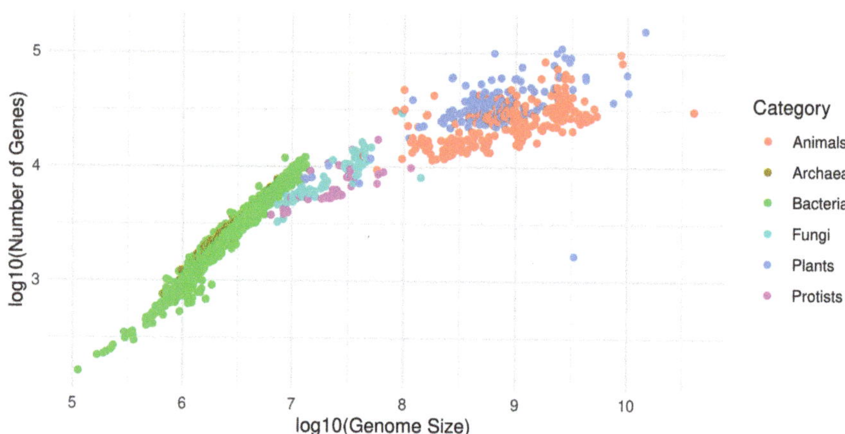

Fig. 3.3. c-value paradox. Figure showing the relationship between gene count and genome size for different categories of organisms on the $\log_{10}$ scale. Note that the variation in gene count fails to match variation in genome size for higher eukaryotes such as plants and animals. Figure produced by Ganesh Muthu using data from NCBI Genomes database.

Genomes change over time, and that is a given. DNA bases are chemical entities subject to slowly reacting with other molecules in their neighbourhood or with radiation. These molecules include reactive species of oxygen such as peroxides and superoxides, some of which are products of cellular metabolism. Reactions of DNA bases with such molecules or their contact with radiation can change, or mutate, them. DNA replication, though highly accurate, is not entirely perfect and can introduce errors in the copied DNA. Though mechanisms to correct such mutations exist, they are again not foolproof. So, changes are inevitable! These can be *substitutions*, i.e., a base is replaced by another, or they can be *insertions* or *deletions* of one or more bases. In a DNA sequence coding for a protein, a single-base substitution can be one of the following types. (1) *Synonymous* mutations cause no change in protein sequence because the genetic code, which maps every three-base codon on an mRNA sequence to an amino acid, is redundant, with 61 codons mapping to 20 amino acids. (2) *Non-synonymous* mutations change the amino acid sequence. (3) *Nonsense* mutations introduce a translation STOP signal, prematurely producing a truncated protein. Only about 20% of all possible substitutions are synonymous. The rest change the protein sequence in some way. Even synonymous mutations are not always without consequence[66] and can, for example, affect how well a protein is expressed or even folded into a functional form. A large majority of insertions and deletions would garble a protein sequence and introduce a STOP signal prematurely.

66 See, for example, D. Agashe, M. Sane, K. Phalnikar, G.D. Diwan, A. Habibullah, N.C. Martinez-Gomez, V. Sahasrabuddhe, W. Polachek, J. Wang, L.M. Chubiz, and C.J. Marx, 'Large-Effect Beneficial Synonymous Mutations Mediate Rapid and Parallel Adaptation in a Bacterium', *Molecular Biology and Evolution* 33 (2016), 1542–1553. https://doi.org/10.1093/molbev/msw035

Mutations that change a protein sequence to the detriment of the organism would be lost from the population as a consequence of natural selection,[67] which favours the fittest. How detrimental a non-synonymous mutation is to the activity of a protein and subsequently to the survival of the species depends on the protein, the precise nature of the amino acid change and where it is located on the protein. Some proteins are essential to the organism's survival, are often old and have had their amino-acid sequences highly fine-tuned during the course of evolution. Mutations in such proteins often affect function adversely, leading to morbidity or mortality of the host organism. Amino acids fall into a few physicochemical categories.[68] Some carry a negative charge, others are positively charged and some others are nonpolar and water-repelling. Some contain sulphur and others contain a hydroxyl group. Some are very flexible, others rigid and a few others large and aromatic. A mutation that changes an amino acid into another of the same physicochemical type can be expected to be more conservative than one that changes the physicochemical type of the amino acid. A large majority of amino acid changes would be of the latter type, and, depending on the protein and the location of the amino acid, likely to affect function. Finally, a protein, like a machine, has many parts. Some of these are immediately relevant to its function and highly sensitive to any mutation. Other parts may play structural roles or play passive roles in which the amino acid sequence does not matter much; these are more likely to take mutations without hugely impacting function.

While we can come up with these general rules for whether mutations may or may not affect protein function, what do the data on protein sequences tell us about the likelihood of mutations in protein coding genes? The rate at which a protein sequence has changed over the course of evolution can serve as a measure of how well mutations are tolerated by this protein. This has parallels with the comparative genomics approach for identifying a minimal gene complement, in which the presence of a gene in most organisms is considered as evidence for its centrality to cellular life. Here however, assuming that a gene is present in some form across a set of organisms, the extent to which it differs in its sequence among these organisms would be a measure of its tolerance to change. For an everyday example, the human nose is not identical in shape and size across our populations, but always retains its central ability to allow air to pass through. In other words, there is some flexibility in what the human nose looks like, but none in its ability to allow air flow.

As early as in the 1960s, around the time Zuckerlandl and Pauling wrote about using protein and nucleic acid sequences to construct phylogenetic trees,[69] Margaret Dayhoff[70]—a computational chemist and a pioneer in the use of computers to analyse

67 However, we will see shortly that things are not all that simple and quantitative considerations are involved.

68 For example, C. Pommié, S. Levadoux, R. Sabatier, G. Lefranc, M.P. Lefranc, 'IMGT standardized criteria for statistical analysis of immunoglobulin V-REGION amino acid properties', Journal of Molecular Recognition 17 (2004), 17–32. https://doi.org/10.1002/jmr.647

69 See Chapter 2.

70 B.J. Strasser, 'Collecting, comparing, and computing sequences: the making of Margaret O. Dayhoff's

biological data—had developed an "atlas" of protein sequences that were then available and published it in the form of a book. She, with her colleague Richard Eck, showed how comparing sequences of the same protein from different organisms can help trace the protein's evolutionary history.[71] Along the way, Dayhoff and Eck developed a model that showed how large proteins evolve from combinations of much shorter sequences.[72] Recall that Carl Woese, much later, demonstrated that by choosing the right gene for such analyses one can infer not just the evolutionary history of the gene but that of the species itself.[73] Data in Dayhoff and colleagues' Atlas showed that the *histone* protein, which is absolutely essential for DNA compaction in eukaryotes, has a mutation rate that is 1,500 times less than that of protein molecules called *fibrinogen peptides A and B* whose sequences are not consequential to the biological process they are involved in.[74] This showed that mutations that are deemed detrimental to protein function and organism survival would be rare (Fig. 3.4).

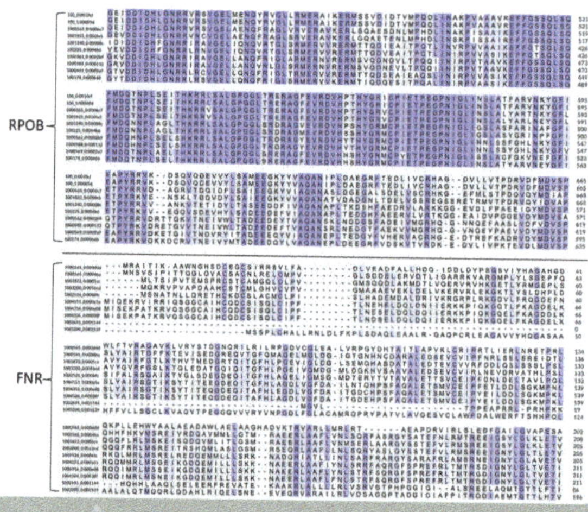

Fig. 3.4. Variation in protein sequences. This figure shows a sequence alignment of fragments of (A) the catalytic subunit of RNA polymerase (RPOB), an essential protein, and (B) FNR, a non-essential regulator of gene expression (see chapter 5). Note the higher number of residues conserved across the various examples of the catalytic subunit of the RNA polymerase than in FNR. Figure produced using Uniprot (https://www.uniprot.org) and OrthoDB (https://www.orthodb.org).

Atlas of Protein Sequence and Structure, 1954-1965', *Journal of the History of Biology* 43 (2010), 623–660. https://doi.org/10.1007/s10739-009-9221-0. This article uses the example of the Atlas to discuss attitudes towards computational and theoretical aspects of biology in the 1960s. It also points out that all of Dayhoff's collaborators, except Richard Eck, were women and wonders whether this might have had any bearing on how the Atlas was initially received.

71 R.V. Eck and M.O. Dayhoff, 'Evolution of the structure of Ferredoxin based on living relics of primitive amino acid sequences', *Science* 152 (1966), 363–366.

72 For a retrospective view of Dayhoff and Eck's work, see M.L.R. Romero, A. Rabin, and D. Tawfik, 'Functional Proteins from Short Peptides: Dayhoff's Hypothesis Turns 50', *Angewandte Chemie International Edition* 55 (2016), 15966–15971. https://doi.org/10.1002/anie.201609977

73 See Chapter 2.

74 A classic example quoted often in papers from the late 1960s and the 1970s.

In the late 1960s and early 1970s, Motoo Kimura, an influential thinker in the field of molecular evolution, used data from Dayhoff's Atlas to estimate that the mutation rate for a small number of proteins performing important cellular functions is around ~10^{-9} per amino-acid site per year.[75] The actual rate at which mutations occur is likely to be higher than this because mutations in important proteins, being often disadvantageous, would be regularly lost from the population. For the fibrinogen peptides briefly noted above, in which mutations are of little consequence to function, Kimura estimated the mutation rate to be ~10^{-8} per amino-acid site per year.[76] Kimura argued that the latter number, which is less affected by natural selection that purges highly detrimental mutations from the population, would be closer to the true mutation rate.[77] These mutation rates may appear to be very low, being 1 to 10 in a billion, but for a genome containing 3 billion base pairs, all encoding genes, amounting to a billion amino-acid sites, the number of mutations would be ~10 (all in information-rich genes) per genome per year. Of these, ~1 could be highly deleterious. A much more recent study of sequence data for ~300 genes from across ~100 individuals also showed that the fraction of mutations that are highly deleterious in humans would be <15%, while a majority of mutations would have mild negative effects.[78] Assuming a generation time of 20 years, we can expect ~20 seriously damaging mutations every generation, which is an awful lot.

Susumu Ohno (1928–2000), yet another pioneer in molecular evolution, arrived at similar results in 1972.[79] Based on a conventional estimate that the rate of deleterious mutation per locus—read protein coding gene—is ~10^{-5} per generation, Ohno calculated that an organism with 3 million genes would accumulate ~30 detrimental mutations per generation. Ohno, further by quoting other studies, pointed out that to sustain a population with such a high rate of deleterious mutations, every mated pair of parents will have to produce an astronomical 10^{78} zygotes to ensure that a few survive into the next generation and maintain a stable population size! Quoting earlier estimates that the actual number of deleterious mutations is only ~0.5 per generation, much lower than the 20–30 one would expect had the genome been filled with information-rich genes, Ohno argued that most mutations—and by inference, most of the genome—is inconsequential to species survival and propagation. The human genome, at 3 billion base pairs, may code for only about 40,000 genes, accounting for a little more than 10%

75 M. Kimura, 'Evolutionary rate at the molecular level', *Nature* 217 (1968), 624–625. https://doi.org/10.1038/217624a0

76 A recent study estimating mutation rate with large whole genome sequences arrived at a ball-park similar estimate. See L.A. Bergeron, S. Besenbacher, J. Zheng, P. Li, M.F. Bertelsen, et al., 'Evolution of the germline mutation rate across vertebrates', *Nature* 615 (2023), 285–291. https://doi.org/10.1038/s41586-023-05752-y

77 T. Ohta and M. Kimura, 'Functional organization of genetic material as a product of molecular evolution', Nature 233 (1971), 118–119. https://doi.org/10.1038/233118a0

78 A. Eyre-Walker, M. Woolfit, and T. Phelps, 'The distribution of fitness effects of new deleterious amino acid mutations in humans', 173 (2006), 891–890. https://doi.org/10.1534/genetics.106.057570

79 S. Ohno, 'An argument for the genetic simplicity of man and other mammals', *Journal of Human Evolution* 1 (1972), 651–662. https://doi.org/10.1016/0047-2484(72)90011-5

of the genome. Most of the rest of the DNA, being non-functional, can tolerate changes and not count towards the number of deleterious mutations per generation. Ohno concludes that "90% of our genomic DNA is 'junk' or 'garbage' of various sorts."[80] In contrast to mammalian genomes that run into billions of base pairs, a gene-dense bacterial genome encoding only a few thousand genes would accumulate less than one in ten thousand deleterious mutations per genome per generation,[81] making its high gene density that much more manageable.

These arguments, however simple and unaware of the complexities of eukaryotic genome evolution, had been made some three decades before the human genome was published. Incredibly though, the landmark event did not qualitatively change anything significant in this regard and instead served to validate Ohno's stand. It showed that the human genome codes for 30,000–40,000 genes,[82] not very different from the estimates made by Ohno. The number of genes encoded by the human genome has only been revised somewhat downwards since. Studies measuring mutation rates from genome sequence data across humans and mammals showed that only 10–15% of the human genome is likely to be functional,[83] consistent with Ohno's back-of-the-envelope calculations.

In 2012, a post-human-genome project called ENCODE, which aims to experimentally identify regions of the human genome that undergo transcription—or are bound by a set of DNA-binding proteins, or undergo chemical changes called *epigenetic* modifications—came to a stunning conclusion that at least 80% of the human genome is functional[84] and that it was time to sing a requiem for the concept of junk DNA! However, this conclusion, which has been severely criticised since its publication,[85] ignores decades of well-supported arguments from evolutionary biology arising from the c-value paradox, some of which we have described here or will do so shortly; it does not quite explain why

---

80 Ibid., p. 656.

For a classic break down of the concept of junk and garbage, see S. Brenner, 'Refuge of spandrels', *Current Biology* 8 (1998), R669.

81 Assuming a generation time of less than one day and a mutation rate of 10^{-8} per site per year.

82 The International Human Genome Sequencing Consortium, 'Initial sequencing and analysis of the human genome', *Nature* 409 (2001), 860–921. https://doi.org/10.1038/35057062

83 C.P. Ponting and R.C. Hardison, 'What fraction of the human genome is functional?' *Genome Research* 21 (2011), 1769–1776. https://doi.org/10.1101/gr.116814.110

84 The ENCODE Project Consortium, 'An integrated encyclopedia of DNA elements in the human genome', *Nature* 489 (2012), 57–74. https://doi.org/10.1038/nature11247

85 For a very balanced critique of the ENCODE project, see F.W. Doolittle, 'Is junk DNA bunk? A critique of ENCODE', *Proceedings of the National Academy of Sciences USA* 110 (2013), 5294–5300. https://doi.org/10.1073/pnas.1221376110

See also S.R. Eddy, 'The ENCODE Project: missteps overshadowing a success', *Current Biology* 23 (2013), R259–261. https://doi.org/10.1016/j.cub.2013.03.023

For a perspective that introduces the c-value paradox, leading to the ENCODE project, see S.R. Eddy, 'The C-value paradox, junk DNA and ENCODE', *Current Biology* 22 (2012), R898–899. https://doi.org/10.1016/j.cub.2012.10.002

And for an entertaining, if aggressive critique, see D. Graur, Y. Zheng, N. Price, R.B.R. Azevedo, R.A. Zufall, and E. Elhaik, 'On the Immortality of Television Sets: "Function" in the Human Genome According to the Evolution-Free Gospel of ENCODE', *Genome Biology and Evolution* 5 (2013), 578–590. https://doi.org/10.1093/gbe/evt028

this conclusion—if broadly applied to the genomes of other multicellular eukaryotes—would not imply that a fish needs 100 times as much functional DNA as a human; and plays "fast and loose"[86] with the definition of the term 'function'. While the ENCODE project, a great success in many ways, has provided an invaluable resource for the study of human molecular biology, we can safely ignore its ill-fated conclusion on what fraction of the human genome is functional.

Now back to where we were following Thomas' definition of the c-value paradox. If only ~15% of the genome is functional, how come the remaining 85% is still hanging around? Can the human species afford to carry such a large excess of baggage? How come the bacterial genome is not saddled with as much junk?

3.5. The fate of a DNA sequence

What determines whether a DNA sequence, or for that matter a single base substitution mutation, is found and conserved in the genomes of a population of cells or organisms? Let us start with two intuitive parameters that immediately come to mind. First is the probability or the rate at which this piece of DNA sequence or mutation forms on a genome. Second is the selective advantage or disadvantage this DNA sequence or mutation, once formed, confers on its host. The latter is something we have casually alluded to several times as the prediction of Darwin's theory of natural selection and survival of the fittest, but is one that needs to be addressed with a bit more nuance here. One can expect that the fate of the DNA or mutation will be proportional to a version of the product of these two parameters. Let us delve a bit more deeply into them.

Some in the field of evolutionary biology consider changes in the genome to occur purely randomly; therefore mutation rate variation should be of little consequence to how evolution proceeds.[87] However, there is no good reason to assume this at a molecular level. One common cause of mutation is chemical reaction of bases with environmental factors. For example, singlet oxygen is a reactive form of dioxygen gas, as are peroxide and superoxide which form under certain kinds of cellular metabolisms. These reactive oxygen species can mutate DNA, but do not cause all possible changes equally.[88] For example, these reactive oxygen molecules often modify the guanine base into 8-oxo guanine. This change in molecular structure of guanine changes the base pairing ability of guanine. 8-oxo guanine can base pair with adenine, in contrast to the highly specific pairing of normal guanine with cytosine. Such an erroneous base-paired structure, as a consequence of replication that matches adenine with thymine, would

86 Graur et al., 2013, p. 579.

87 For a systematic criticism of this view, see A. Stoltzfus, *Mutation, Randomness and Evolution* (Oxford: Oxford University Press, 2021). https://doi.org/10.1093/oso/9780198844457.001.0001

88 There are several reviews of this field. For example, J. Lunec, K.N. Holloway, M.S. Cooke, F. Faux, H.R. Griffiths, and M.D. Evans, 'Urinary 8-oxo-2'-deoxyguanosine: redox regulation of DNA repair in vivo?', *Free Radical Biology and Medicine* 33 (2002), 875–885. https://doi.org/10.1016/s0891-5849(02)00882-1; A. Poetsch, 'The genomics of oxidative DNA damage, repair, and resulting mutagenesis', *Computational and Structural Biotechnology Journal* 18 (2020), 207–219. https://doi.org/10.1016/j.csbj.2019.12.013

result in a guanine to thymine mutation. Though cells carry systems that can repair such errors, these are not 100% efficient. Thus, the genome of a population of cells routinely exposed to DNA damage specifically from reactive oxygen molecules can be expected to display more guanine to thymine substitutions than other types of single-base changes. Similarly, another commonly cited source of mutation, UV radiation, leaves its own signature, namely elevated rates of cytosine to thymine mutations and replacements of tandem cytosine-cytosine with thymine-thymine.[89]

Life forms encode a plethora of mechanisms for repairing damaged DNA and thus limit mutation rates. However, each DNA repair mechanism has its own unique ability to limit some kinds of damage but not others. Therefore, the types of DNA repair mechanisms that an organism encodes can further determine which mutations occur in the genome post-repair, and at what rates. In fact, some recent evidence from Mrudula Sane et al. even suggests that the repertoire of DNA repair machinery encoded by an organism, by shifting the nature of mutations the organism experiences, can determine how well it can adapt to environmental pressures.[90] Further, mutation rates at some positions on a genome may be dependent on the sequence of bases surrounding them. For example, a very recent study showed that a run of five or more guanines preceding a thymine increases the rate of mutation of the thymine to guanine a few hundred times in the bacteria *Salmonella*![91] Finally, experiments measuring mutation rates in bacteria have shown that different types of mutations occur at different rates, and also that these rates differ across bacterial species.[92] Similarly, work in the single-celled eukaryote yeast[93] and the bacterium *E. coli*[94] has shown that environmental conditions can dictate the rates at which different types of base changes occur.

Even if all base changes were to occur at more or less equal rates, the genetic code is such that all amino acid changes that these base changes cause will not occur with equal probabilities. First, because each codon comprises three bases, some codon changes require only a single base change, whereas others would need three. Next, some amino acids such as methionine are encoded by a single codon, whereas others like the positively-charged arginine are coded for by as many as six codons. In general, it is more likely to get an arginine-encoding codon by random chance than the one

89 H. Ikehata and T. Ono, 'The mechanisms of UV mutagenesis', *Journal of Radiation Research* 52 (2011), 115–125. https://doi.org/10.1269/jrr.10175

90 M. Sane, G.D. Diwan, B.A. Bhat, and D. Agashe, 'Shifts in mutation spectra enhance access to beneficial mutations', *Proceedings of the National Academy of Sciences USA* 120 (2023), e2207355120. https://doi.org/10.1073/pnas.2207355120

91 J.L. Cherry, 'T residues preceded by runs of G are hotspots of T→G mutation in bacteria', *Genome Biology and Evolution* 15 (2023), evad087. https://doi.org/10.1093/gbe/evad087

92 See, for example, S. Kucukyildirim, H. Long, W. Sung, S.F. Miller, T.G. Doak, and M. Lynch, 'The rate and spectrum of spontaneous mutations in Mycobacterium smegmatis, a bacterium naturally devoid of the post replicative mismatch repair pathway', *Genes Genomes Genetics* 6 (2016), 2157–2163. https://doi.org/10.1534/g3.116.030130

93 H. Liu and J. Zhang, 'Yeast spontaneous mutation rate and spectrum are environment-dependent', *Current Biology* 29 (2019), 1584–1591. https://doi.org/10.1073/pnas.1323011111

94 W. Wei, W.-C. Ho, M.G. Behringer, S.F. Miller, G. Bcharah, and M. Lynch, 'Rapid evolution of mutation rate and spectrum in response to environmental and population-genetic challenges', *Nature Communications* 13 (2022), 4752. https://doi.org/10.1038/s41467-022-32353-6

that codes for methionine. All the three possible changes in the third base of the AUG codon that encodes methionine will convert this amino acid to isoleucine. Therefore, any environmental change that requires a methionine to isoleucine switch in a protein for adaptation would be relatively more accessible to a bacterial population than, say, a methionine to a threonine change which would be possible only with a second base change from uracil to cytosine. Similarly, any single base mutation in the third position of the only codon for the amino acid tryptophan will either convert it into a translation STOP signal or into the amino acid cysteine. Again, third base mutations in either of the two codons for the negatively charged aspartate will either be synonymous or switch it to glutamate, another negatively-charged amino acid which preserves the physicochemical properties of that amino acid position. Thus, even if all mutations are possible at the same rate at the DNA level, the changes to the amino acid sequence will be somewhat skewed, thus constraining the direction in which protein function can change.

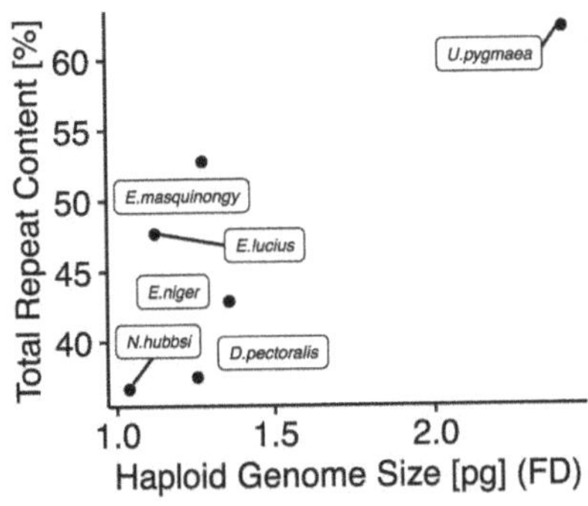

Fig. 3.5. Repetitive DNA in eukaryotic genome expansion. This figure shows the correlation between genome size and its repetitive DNA content for fishes called mudminnows. Originally published as Figure 3B in R. Lehmann, A. Kovařík, K. Ocalewicz, L. Kirtiklis, A. Zuccolo, J.N. Tegner, J. Wanzenböck, L. Bernatchez, D.K. Lamatsch, and R. Symonová, 'DNA Transposon Expansion is Associated with Genome Size Increase in Mudminnows', *Genome Biology and Evolution* 13 (2021), evab228, CC BY 4.0.

While non-uniformities in mutation rates and their consequences on the evolution of protein function and adaptation are important in evolution, they do not immediately help us understand variations in genome size and the c-value paradox. The c-value paradox is driven by a large excess of non-functional DNA. Thus, it is not merely a result of any variation in mutation rate, but one of how some DNA sequences, despite being non-functional in terms of their contribution to the organism's ability to survive and reproduce, happen to expand within a genome. To help understand this, one will have to first recognise that the ability of a piece of DNA to confer a *benefit to its host* organism is not the only parameter that determines its success. If a sequence of DNA can multiply and propagate *itself* in some manner within a genome such that it does not compromise its own survival by being very expensive for its host to maintain, then there is no need to invoke any concept of benefit for the host to explain its success.

This underlies the concept of *selfish* DNA, popularised by Richard Dawkins in the mid-1970s,[95] and articulated only a few years later by Doolittle and Sapienza,[96] and by Orgel and Crick,[97] as the source of much non-functional DNA in mammalian genomes (Fig. 3.5).

A particular class of selfish DNA, called *transposons*, has the insidious ability to multiply itself rapidly once incorporated into a genome, thus increasing the size of the genome. These DNA copies mostly offer little benefit to their host cells, except when, in the course of evolution, portions of some of these may be fortuitously utilised (or *exapted*) to perform roles in, for example, the regulation of gene transcription. A large proportion of many mammalian and plant genomes can have their origins traced back to transposons. However, multiple copies of the same DNA can, instead of being beneficial to their host, be detrimental to it by promoting inappropriate rearrangements or juggling of the genome sequence. Such rearrangements can at times be severely problematic to the host, compromising the very survival of the host and the transposon. Yet, there are always ways around such potential troubles. Some copies of such DNA, through mutational processes that change their sequence, end up changing so much that they are no longer repetitive and are therefore unable to precipitate problematic genome rearrangements. They become *inactive*, i.e., unable to make additional copies of themselves in their host genome, but quietly remain successful in being propagated down generations along with the rest of the genome during normal cellular and organismal reproduction. Thus, some types of DNA are able to multiply themselves at a high rate within a genome despite being mostly useless to their host cells, which contributes to the excess DNA in many eukaryotic genomes.

Now that we are clear that some mutations can occur more frequently than others and that some pieces of DNA can propagate themselves successfully in a selfish manner, we can ask what it is that ensures that a mutation or a piece of DNA (selfish or otherwise) is maintained in or lost from a population of cells or organisms. The principle of natural selection is fairly straightforward to understand. If a mutation or a newly-acquired DNA sequence, which we call X, enables its bearer to reproduce much more successfully than do cells without X, thus producing more offspring carrying X, then the proportion of cells with X will increase over the course of generations. If X causes its bearer to be very poor at reproduction, it will eventually be lost from the population. Very intuitive. But curiously enough, this is *not* a given! It is possible that a highly beneficial mutation may not stick and be lost, just as it is possible that a totally neutral, or even slightly deleterious (but not lethal) mutation or DNA sequence can hang around and even take over the population. This once again is a consequence of probabilities.

---

95 R. Dawkins, *The Selfish Gene* (Oxford: Oxford University Press, 1976).

96 W.F. Doolittle and C. Sapienza, 'Selfish genes, the phenotype paradigm and genome evolution', *Nature* 284 (1980), 601–603. https://doi.org/10.1038/284601a0

97 L.E. Orgel and F.H.C. Crick, 'Selfish DNA: the ultimate parasite', *Nature* 284 (1980), 604–607. https://doi.org/10.1038/284604a0

If all cells could reproduce forever at a precise rate prescribed by their genetic material, there would be no question of probabilities. However, resources are finite and there are limitations to an organism's ability to utilise even excess resources. Any given habitat has a carrying capacity for a population of organisms. This brings us to the idea of a population size for a species, which is the total number of organisms of that species. A second term that is often used in evolutionary biology is *effective population size* which, being representative of the number of breeding individuals of a species, will often, but not always, be less than the population size.[98] We will revisit the latter term shortly.

Now let us consider a simplified toy situation in which a species maintains a stable population size, as a result of, say, a balance between birth and death rates. Let us consider a situation where two genetic variants, *A* and *B*, of the same species exist in a population in equal proportions. Let us assume that the difference in the DNA between the two variants is neutral, i.e., both variants are equally adept at survival and reproduction. They start at a population size equal to the carrying capacity of their habitat and after a generation, during which they reproduce at the same rate, they maintain their population size through, say, a balance between birth and death. From a probabilistic standpoint, this boils down to a *sampling* of a proportion of individuals post-reproduction such that the total population size remains the same as in the previous generation. Let the total population be N, the number of individuals of type A after sampling be n_A, and that of B be n_B. The assumption of the neutral nature of the variants implies that the probability that any one individual from N will be A is 0.5 and that it will be B is also 0.5. Assume that N is 1,000. Let us first ask what is the probability of the likely event that the two variants maintain their proportion at 50-50 after this sampling exercise, i.e, what is the probability that $n_A = 500$ and $n_B = 500$. We call this $P(n_A = n_B = 500)$. This can be answered by the binomial probability distribution.

$$P(n_A = n_B = 500, N = 1000) = \binom{1000}{500} \times 0.5^{500} \times 0.5^{500}$$

This comes to about 0.025. This may appear to be small, but given that for a total population of 1,000 n_A and n_B can take any of 1,000 values, this is fairly high. But this already tells us that the moment we involve probabilities, things get uncertain. Now, let us ask what is the chance that this sampling exercise would produce $n_A=0$, i.e., the variant A is lost from the population despite A not being in any way more detrimental than B.

$$P(n_A = 0, N = 1000) = \binom{1000}{0} \times 0.5^{0} \times 0.5^{1000}$$

The answer is a very small number, $\sim 10^{-300}$, very close to zero even if not quite zero. When N is 1,000, the chance that either of A or B will be lost from the population is almost non-existent. Now let us change the population size—after all, different species

98 M. Husemann, F.E. Zachos, R.J. Paxton, and J.C. Habel, 'Effective population size in ecology and evolution', *Heredity* 117 (2016), 191–192. https://doi.org/10.1038/hdy.2016.75

have different population sizes. If $N = 100$, then the probability that $n_A = 0$ is

$$P(n_A = 0, N = 100) = \binom{100}{0} \times 0.5^0 \times 0.5^{100}$$

Again, a small number, $\sim 10^{-30}$, but a whole lot bigger than when $N = 1000$. Now, let us consider a species with a very small population size, $N = 10$. Then the probability that $n_A = 0$ will be

$$P(n_A = 0, N = 10) = \binom{10}{0} \times 0.5^0 \times 0.5^{10}$$

This is close to 10^{-3}. In other words, one every 1,000 such samplings could result in the loss of the variant *A* from the population when $N = 10$. Of course, if the population suddenly finds itself in a serious bottleneck and there is one reproducing individual ($N = 1$), then it is 50-50 whether it is *A* or *B*.

Those who do not trust the binomial distribution but are comfortable writing a computer program and trust the results that a computer would produce can try to run a simulation that attempts to recreate this sampling exercise and introduce a *time* component to it. We can set up an initial population of some size *N* that is 50-50 for *A* and *B*, let both variants reproduce at equal probabilities and then sample down from this population randomly such that the population size is again *N* and count the number of *A* and *B* type members in the resultant sample. Let us call these $n_{A,1}$ and $n_{B,1}$ for the numbers obtained after one generation. Repeat the sampling exercise but instead of starting with a 50-50 distribution of *A* and *B*, start with $n_{A,1}$ and $n_{B,1}$ to obtain $n_{A,2}$ and $n_{B,2}$ after two generations. Repeat this for as many generations as possible using, for every generation i, $n_{A,i-1}$ and $n_{B,i-1}$ as the starting population distribution to get $n_{A,i}$ and $n_{B,i}$ as the resultant distribution. With a computer, running this for a million generations or more is no big deal! Plot the number of generations (i) on the x-axis and $n_{A,i}$ or $n_{B,i}$ on the y-axis. Repeat this multiple times to get a sample of possible trajectories. And as we had done with the binomial calculation, repeat this exercise for multiple *N* and compare the results.

Thus, when the population sizes are small, there is a good chance that a neutral mutation or DNA sequence, initially present in 50% of the population, would either be totally lost from the population or fully dominate it, making population size a third determinant of the fate of a DNA sequence. Motoo Kimura derived a series of equations for the probability that any mutation or DNA sequence will be fixed in a population and the number of generations it would take to get there (we will not discuss the latter here). These equations require the term *effective population size* to be defined. We agree that random processes, or more technically for evolutionary biology *neutral drift*, can change the frequencies of occurrence of a mutation or a piece of DNA in a population at a certain rate. The size of an ideal population that approximates the rate at which the variation in the occurrence of a DNA sequence changes for a real population is called *effective population size* or N_e.[99] N_e is more important than the total

99 Husemann et al., 2016; D. Graur and W.-H. Li, 'Fundamentals of Molecular Evolution', *Sinauer Associates Inc* (2000).

census population in studies of evolutionary processes because not all individuals of a species participate in reproduction and it is these individuals who contribute to genetic transmission and variation. N_e tends to be less than the census population, but apparently not always.[100] Some effects, beyond the simple fact that the number of individuals participating in reproduction is often less than the total population, exacerbate the difference between census population and N_e. For example, difference in the numbers of males and females results in a large reduction in N_e. The difference between census population and N_e is especially stark when there have been large long-term changes in census population size, for example through catastrophic natural disasters, even if the numbers have recovered since. In such cases, N_e is closer to the size of the smallest known population than to the largest. For example, though the census population size for humans is around 8 billion, the estimated N_e is only around 10,000.[101] As one might guess, calculating N_e is not straightforward[102] and is a discipline in itself, and here we will take a few published estimates with support from multiple efforts for granted.

Kimura showed that the probability that a mutation or DNA sequence will fully take over the population, also known as *fixation probability P*, is given by the following equation:[103]

$$P = \frac{1 - e^{-2s}}{1 - e^{-4N_e 2s}}$$

In the above equation, N_e is the effective population size and s is a measure of the selective advantage or disadvantage the mutation or DNA sequence gives its host. The assumption here also is that the organism is diploid, i.e., has one paternal and one maternal copy of its chromosomes, which in effect doubles N_e. s is measured as the selective advantage or disadvantage conferred by a single copy of the mutation or DNA sequence.

Let us again consider a neutral mutation or DNA sequence and ask what its fixation probability would be for different N_e values. When s is very small and negative (say -10^{-10}), or neutral, the numerator in the above equation simplifies to $2s$. For $N_e = 1000$, P will be 0.0005. As N_e decreases to 10, P increases to 0.05. Thus, the probability that a neutral mutation will be fixed in a population increases as N_e decreases. One can see that even a deleterious mutation ($s < 0$) can be fixed in a population with some probability, which again increases dramatically as N_e decreases. For a DNA sequence with $s = -0.01$, the fixation probability can be as high as 0.04 at $N_e = 10$ but drops rapidly to 10^{-19} when N_e is 1,000.

Similarly, there is no guarantee that an advantageous mutation ($s = +0.01$) will

100 Husemann et al., 2016.

101 Graur and Li, 2000.

102 M. Kimura and J.F. Crow, 'The measurement of effective population number', *Evolution* 17 (1963), 279–288. https://doi.org/10.2307/2406157; J. Wang, E. Santiago, and A. Caballero, 'Prediction and estimation of effective population size', *Heredity* 117 (2016), 193–206. https://doi.org/10.1038/hdy.2016.43

103 Via Graur and Li, 2000.

be fixed in the population. Kimura's equation shows that even for a population size of 1,000, there is a 98% chance that it will not be fixed! However, compared to the 10^{-19} probability that a piece of DNA with an equivalent selective disadvantage will be fixed, the fixation probability of the beneficial DNA is high when N_e is 1,000; although the same piece of beneficial DNA at $N_e = 10$ has only a slightly higher probability of fixation than one with an equal detrimental effect. Note here that N_e, once it crosses very low levels at which noise plays a big role, has very little impact on the fixation probability for an advantageous piece of DNA, very unlike the massive effect it has on the fate of detrimental DNA pieces. In fact, when *s* is positive and N_e large, the probability of fixation approximates to 2*s*, independent of N_e. Thus, the effect of population size on the fixation probability of an advantageous DNA piece is much smaller than that of a detrimental piece of DNA.

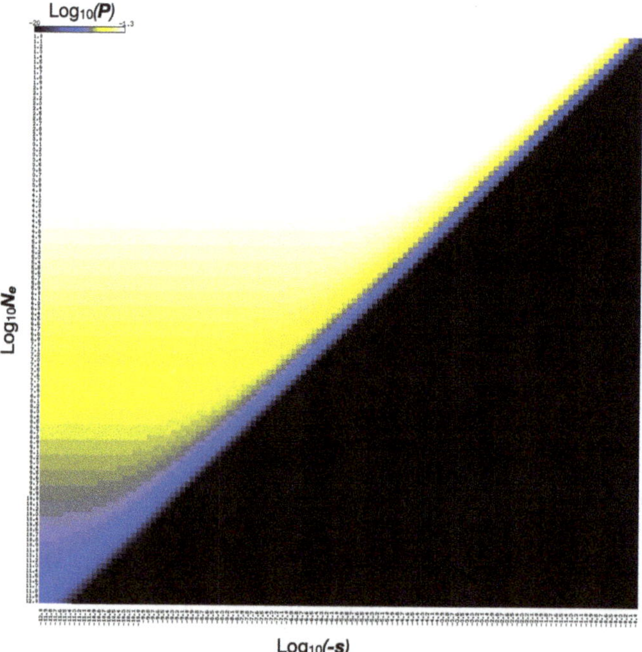

Fig. 3.6. Relationship between N_e and *s*. This figure shows the relationship between N_e and *s* on the $\log_{10}$ scale for negative values of *s*. Note that the fixation probability of a deleterious mutation drops sharply when $|s| > 1/N_e$.

Now, assuming that a mutation or a new DNA sequence has appeared in a population, its success depends on the selective advantage or disadvantage it confers on its host as well as the population size of the host species. A three-way plot of the two parameters, N_e and *s*, against *P* will show that the fixation probability of a deleterious mutation drops sharply when $|s| > 1/N_e$.[104] When the selective disadvantage offered by a piece

104 M. Lynch and M.S. Conery, 'The origins of genome complexity', *Science* 302 (2003), 1401–1404. https://doi.org/10.1126/science.1089370

of DNA is not strong enough to overcome a limitation presented by the population size of a species, random processes or neutral drift dominates (Fig. 3.6). When the strength of selection is high enough to overcome this population size barrier, adaptive evolution by which a disadvantageous mutation is lost from a population becomes more likely despite it still being subject to the vagaries of chance.

A prediction of this theory is that a neutral piece of selfish DNA with the ability to multiply itself in a genome would not be easily purged out in organisms with low N_e. This should also imply that there will be a negative correlation between N_e and genome size, assuming that genome size increase arises as a result of expansion of neutral DNA sequences. We will see in the next chapter that this is not true for bacterial genomes wherein genome size expansion is often a function of addition of beneficial DNA. However, even for higher eukaryotes in which selfish DNA elements dominate genome content, this relationship has been hard to establish with empirical data, largely because of the complexities and uncertainties in N_e estimation.[105]

We note here that N_e for large eukaryotes such as humans ($N_e \sim 10^4$) is many orders of magnitude smaller than that estimated for many free-living bacterial species for which N_e values are often in the 10^9 range. A slightly deleterious mutation or DNA sequence, say, $s \sim -10^{-6}$, will have a much greater chance of fixation in humans (for which $|s| < 1/N_e$) than in free-living bacteria (for which s is $\gg 1/N_e$). Can we calculate s for a neutral piece of DNA, say one copy of a selfish DNA element such as a transposon? To do this, let us invoke some biochemistry and ask how deleterious a random piece of DNA of no beneficial or adverse function would be to a prokaryotic and to a large eukaryotic mammalian cell.

3.6. The cost of a DNA sequence

Nothing comes for free, but some things are very cheap. This can be true of a DNA sequence as well. A DNA sequence, once integrated into the genome of a cell, has to be replicated, consuming deoxyribonucleotides in the process. If this DNA is a gene, it will have to be transcribed, and this uses up ribonucleotides. And if the DNA codes for a protein, then the transcribed RNA will have to be translated into protein, which uses up amino acids. All these are biosynthetic processes that also expend energy while forming new covalent bonds that extend nucleic acid or protein chains. Energy is also spent in producing monomeric units such as deoxyribonucleotides, ribonucleotides, and amino acids. Thus, assuming that a newly introduced DNA sequence offers no selective advantage to its host, its maintenance involves a certain cost. How costly it is can be measured by the fractional cost of its maintenance against the overall energetic

105 T. Lefebure, C. Morvan, F. Malard, C. Francois, L. Konecny-Dupre, et al., 'Less effective selection leads to larger genomes', *Genome Research* 27 (2017), 1016–1028. https://doi.org/10.1101/gr.212589.116 . This recent work shows a correlation between ineffective selection and genome size increase due to selfish DNA multiplication in some crustacean species with low N_e. The paper also highlights difficulties in establishing this relationship more broadly as a result of problems in N_e estimation.

burden in running the cell, which is a complex machine that is much more than a gene or a genome. For example, charging a mobile phone would not be all that expensive in a large factory. However, the same mobile phone could take up a considerable fraction of the total energy utilised by an outfit running a single battery pack! Thus, to ask if a piece of non-functional DNA imposes a cost that is large enough for it to be lost from a population, we must first estimate the fractional cost it adds to the host cell's energy budget. This will be our s.[106] Once we have this value, we can compare it to the host species' estimated N_e to test whether $|s| \gg 1/N_e$. To do this, we will follow the important and insightful recent work published by Michael Lynch and Georgi Marinov in 2015.[107]

The first question is what is the total energy budget of a cell. There are two components to it. First is basal maintenance, which is the potpourri of things that do not contribute to growth. Second is growth-related processes. These can be estimated by growing cells at a defined density and measuring the rate at which they utilise nutrients available to them. The amount of nutrients consumed can then be expressed in equivalents of ATP (adenosine triphosphate), the energy currency of the cell. ATP produces energy when the bonds it has between phosphate groups are broken. Therefore, the energy produced from ATP can be considered equivalent to energy produced when a phosphate group is released. A fast-growing cell would consume more energy than one growing slowly. Therefore, the total energy consumed by a cell should be measured at various growth rates. In a graph of the rate of ATP spent on the y-axis and growth rate on the x-axis the y-intercept will be a measure of the basal energy expenditure per unit time (called C_M) whereas the slope will be equal to the energy spent on growth related processes (called C_G). Lynch and Marinov estimated both for a variety of organisms covering a whole range of cell sizes and came up with a remarkable finding. Both the maintenance expenditure and growth-related energy expense correlated strongly with a single parameter: the cell volume (V in μm^3). The relationship between energy expenditure and cell volume that they derived from their data are: $C_M = 0.39V^{0.88}$ and $C_G = 26.92V^{0.97}$. The units for these are 10^9 ATP molecules or high energy phosphate bonds per cell. Larger the cell, higher the total energy budget (called C_T). Now,

$$C_T = C_G + tC_M \text{, where } t \text{ refers to the time from birth to cell division.}$$

This multiplication factor for C_M is important because C_M estimated above is per unit time (hr).

For the purpose of this discussion, which will consider very approximate estimates that can be quickly done with pen and scrap paper, let us simplify the relationships above and take $C_T = (27 + 0.4t)V \times 10^9$ ATP molecules per cell. For a bacterial cell with $V \sim 1\mu m^3$ and a cell division time of 1hr,

106 If the gene were to be beneficial to the cell then s will be the difference between the benefit it confers and the cost of its maintenance. For a detrimental gene, say coding for a toxic protein, s will have a negative value whose magnitude will be the sum of the effect its toxicity has and its maintenance cost.

107 M. Lynch and G. Marinov, 'The bioenergetic costs of a gene', *Proceedings of the National Academy of Sciences USA* 112 (2015), 15690–15695. https://doi.org/10.1073/pnas.1514974112. This section is built on the findings reported in this paper.

$$C_{T,\text{bacteria}} = (27 + 0.4) \times 10^9 = 27.4 \times 10^9 \text{ ATP molecules per cell}$$

$$C_{T,\text{bacteria}} \sim 27 \times 10^9 \text{ ATP molecules per cell}$$

For a large eukaryotic cell, say mammalian, with $V \sim 2{,}500\mu m^3$ and a cell division time of 20 hrs,

$$C_{T,\text{mammal}} = (27 + 0.4{\times}20) \times 2500 \times 10^9 = 87.5 \times 10^{12} \text{ ATP molecules per cell}$$

$$C_{T,\text{mammal}} \sim 88 \times 10^{12} \text{ ATP molecules per cell}$$

Thus, a large mammalian cell needs over 1,000 times as much energy as a bacterial cell to survive and grow over the course of its lifetime from birth to reproduction. Now that we have an estimate of the total energy expenditure of bacterial and mammalian cells, let us try and estimate what the fractional additional cost a gene would contribute to these cells.

The cost of a protein-coding gene can be divided into three parts: cost of replication and that of transcription and translation. Firstly, replication. The primary energy expenditure during replication arises from the ATP molecules spent in producing deoxyribonucleic acids. These costs differ only slightly across the four types of deoxyribonucleotides and average at ~50 ATP molecules per deoxyribonucleotide. While a DNA polymer is being synthesised by attaching the monomeric deoxyribonucleotides, two phosphate molecules are lost per attachment, but this cost is subsumed in the production costs for these monomers. Additional costs of 1 phosphate per attachment arise from the energy required to unwind the DNA before replication can proceed. Thus, in the simplest scenario, which applies to bacteria, the cost of synthesising a *double-stranded* DNA of length L_g is

$$C_{\text{DNA,bacteria}} \sim ((2 \times 50) + 1)L_g$$

$$C_{\text{DNA,bacteria}} \sim 101L_g$$

In eukaryotes, there are extra costs involved, primarily those involved in the synthesis of histone proteins that are intimately attached to the DNA and help compact the DNA within the confines of a nucleus. And also assuming that the eukaryotic cell is diploid, Lynch and Marinov calculate the total cost of DNA replication in a eukaryotic cell to be

$$C_{\text{DNA,eukaryote}} \sim 526L_g$$

Let us assume an average gene size of 1,000bp. For a bacterial cell, the cost of replicating a gene is ~10^5 ATP molecules. The cost of replicating this gene as a fraction of the cell's total energy expenditure is ~4×10^{-6}. Or in other words, $s_{\text{DNA,1000bp,bacteria}} \sim 4 \times 10^{-6}$. Given that the effective population size $N_e \sim 10^9$ for many free-living bacteria, $|s| \gg 1/N_e$. Thus, the mere cost of replicating a 1,000bp DNA (or for that matter, even a 10–100bp DNA) would overwhelm the drift barrier imposed by population size in bacteria, which argues strongly in favour of highly streamlined genomes. In contrast, for a mammalian cell $s_{DNA,1000bp,mammal} \sim -6 \times 10^{-9}$ for a 1,000bp sequence. This would be much less than $1/N_e$ for both mammals ($N_e \sim 10^4$ for humans) and invertebrates ($N_e \sim 10^6$), suggesting that neutral drift would play a major role in the fate of a piece of non-functional DNA. Whereas the probability that a 1,000bp neutral DNA would be fixed in a bacterial genome is for all practical purposes zero at $< 10^{-180}$, that for humans can be as high as 5×10^{-5}.

One could argue that 85% of a human genome, at ~10^9 bp, if non-functional would add up to a very large cost even for the low N_e for our species. However, much of the extra non-functional DNA has accumulated in small chunks by the multiplication of shorter selfish DNA elements over a period of time at a rate ranging from one every thousand to one every 20 births depending on the transposon[108], each chunk's fate being determined by neutral drift. This rate at which active transposons multiply is in most, if not all, cases higher than the rate at which they are lost (see below for general rate of gene loss in the human lineage). This would mean that purely random processes, or neutral drift, would favour expansion over loss. Losing all accumulated DNA in one or a few steps appears rather impossible, leaving these cells only with the option of finding other ways to manage the cost of carrying all the non-functional DNA. For instance, as mentioned earlier, much of the non-functional DNA in human cells is in the form of *inactivated* transposons. The inactivation of these transposons by mutation appears to have happened in waves following large expansions of selfish DNA sequences. This can be interpreted as an evolutionary mechanism to reduce the costs that can be incurred as a result of the transposon continuing to multiply itself many folds! For example, one type of transposon is found in hundreds of thousands of copies in the human genome. This, in the first instance, speaks to the powerful ability of the transposon to multiply itself. However, 99.9% of these copies have been inactivated in extant genomes, thus greatly limiting any further damage these can cause.[109] That these have not been lost from the genome might be an indication that mutations occur at a much higher rate (~10 per year)[110] than gene loss (~10 per million years ago)[111] in these cells. In contrast, in bacteria the considerable cost that even a single copy of a transposon can add to the cell hugely limits the extent of its spread, thus keeping these genomes streamlined.

The above calculation assumes that this piece of DNA is not expressed via transcription and translation. The costs of these two steps for a protein-coding piece

108 L. Guio and J. Gonzalez, 'New insights on the evolution of genome content: population dynamics of transposable elements in flies and humans', in M. Anisimova (ed.) *Evolutionary Genomics: statistical and computational methods, Methods in Molecular Biology* series (New York: Humana Press, 2019). https://doi.org/10.1007/978-1-4939-9074-0_16

109 G. Bourque, K.H. Burns, M. Gehring, V. Gorbunova, A. Seluanov, et al., 'Ten things you should know about transposable elements', *Genome Biology* 19 (2018), 199. https://doi.org/10.1186/s13059-018-1577-z

110 Ohta and Kimura, 1971.

111 Z.-Y. Wen, Y.-J. Kang, L. Ke, D.-C. Yang, and G. Gao, 'Genome-Wide Identification of Gene Loss Events Suggests Loss Relics as a Potential Source of Functional lncRNAs in Humans', *Molecular Biology and Evolution* 40 (2023), msad103. https://doi.org/10.1093/molbev/msad103

of DNA will quickly add up as we will see now. The cost of transcription or making RNA from a DNA template of a gene has two components. The first is the nucleotide cost for the steady state number of RNA molecules that a cell has. The cost of making ribonucleotides is less than that of deoxyribonucleotides. Lynch and Marinov estimate this to be ~46 ATP per ribonucleotide. If N_R is the number of RNA molecules and L_R the length of the RNA, then the energy spent in making these RNA molecules is $46N_RL_R$. The second cost component arises from the cost of ongoing RNA synthesis. At steady state, the rate of RNA synthesis is equal to the rate of its decay (δ_R) which has been measured in several organisms. The cost of ongoing RNA synthesis per base is two phosphate molecules, which are lost at each step of RNA chain elongation. Then, for RNA of length L_R being produced at a rate of δ_R per hour, the cost of ongoing synthesis is $2N_RL_R\delta_Rt$. We do not include the ribonucleotide cost for these because these molecules are recycled over timescales shorter than the time from birth to cell division. Thus the total cost of RNA synthesis for bacteria becomes

$$C_{\text{RNA,bacteria}} = 2N_RL_R(23 + \delta_Rt)$$

Lynch and Marinov showed that the mean number of RNA molecules per cell increases with cell volume. For a 1μm³ bacterial cell, median $N_R \sim 2$ molecules per cell. For bacteria $\delta_R \sim 10$ per hour. We again take t to be 1 hr. Thus, the cost of a typical bacterial RNA transcript produced from a 1000bp gene becomes $\sim 2 \times 10^5$.

As a fraction of the total cellular energy expenditure $s_{\text{RNA,1000bp,bacteria}} \sim -5 \times 10^{-6}$, making |s| again much greater than $1/N_e$. The fractional cost only increases for genes that get transcribed to high levels.

For higher eukaryotes, transcript production is a bit more involved. It includes the cost of various kinds of post-transcriptional RNA processing including splicing during which large segments of a *pre-mRNA* are cut out to produce a final *mature RNA*. Taking all these into account, Lynch and Marinov estimate the cost of a transcript in eukaryotes to be

$$C_{\text{RNA,mammal}} = N_R(46L_{R,\text{mature}} + 2.17L_{R,\text{pre}}\delta_Rt)$$

For a 2,500μm³ mammalian cell, $N_R \sim 10$, $L_{R,\text{ mature}} = 2{,}000$, $L_{R,\text{ mature}} = 7{,}000$, $\delta_R = 3$ per hour, $t = 20$ hours. Here the transcript cost becomes $\sim 2 \times 10^7$ ATP molecules per cell. As a fractional cost of the total cellular energy requirement, . This, though 100 times higher than that for replication, is still much smaller than $1/N_e$, especially for vertebrates. Thus transcription of a typical gene would be costly enough to overwhelm the drift barrier imposed by N_e for bacteria, but not so for multicellular eukaryotes, vertebrates in particular. It is however possible that in invertebrates, the cost of a highly transcribed gene could be large enough to cross the neutral drift barrier. The situation could be similar even in vertebrates if all their non-functional DNA were to be expressed at average or above average levels. This would be true, if only to a somewhat lesser extent, even if the selfish gene were to be shorter without the large difference between pre- and mature-mRNA. These point to the need for investment in mechanisms that can maintain such non-functional genes at low transcript levels even in eukaryotes.

Finally, similar calculations for protein synthesis, accounting for steady state protein levels and synthesis rates and costs of amino acid synthesis, showed that

$$C_{\text{protein}} = N_p L_p (24 + 5t\delta_p)$$

where N_p is the steady state number of molecules of a protein, which, like RNA, scales with cell volume such that it is ~500 molecules for bacteria and ~10,000 for a eukaryotic cell; L_p is the length of the protein, which is usually ~300 and; δ_p is the rate of protein decay, which we will set to 1 per hour for all organisms. Assuming these numbers, and our usual values for t, $s_{\text{protein, 300 AA,bacteria}} \sim -1.5 \times 10^{-4}$ and $s_{\text{protein, 300 AA,mammal}} \sim -4 \times 10^{-6}$. These show that protein synthesis is expensive, not only in bacteria but also in many situations in eukaryotes. The $|s|$ for an average protein exceeds $1/N_e$ for invertebrates, and is only ~15 times lower than $1/N_e$ for vertebrates. Very likely that a highly expressed protein, even if non-functional, would be expensive enough to amount to a large burden even for vertebrates, indicating that management of such DNA, if first transcribed to high levels, would require additional mechanisms for keeping them untranslated.

In summary, some bacterial genomes approach the genome size expected of a minimal cell. However, these bacteria tend to be highly limited in their capabilities and live in obligate symbiotic relationships with larger hosts and other bacteria. However, bacterial genomes span two orders of magnitude in genome sizes, but remain highly dense in their information content. They differ in this respect from eukaryotic genomes, most remarkably those of multicellular eukaryotes, which carry very large tracts of non-functional DNA. Much of non-functional DNA in eukaryotes comes under the umbrella term *selfish DNA*. This important difference in the information density between bacteria and multicellular eukaryotes very likely stems from the differences between the two in population size. The low population size of eukaryotes renders selection against non-functional and even slightly deleterious mutations or DNA segments too weak for these to be purged out. In contrast, even a single slightly deleterious gene can be easily deleted from the population in bacteria with their large population sizes. Theory suggests that there should be a negative relationship between population size and genome size. Despite great uncertainties in estimation of population size parameters, there is some empirical evidence that this might hold in some eukaryotes, though convincing proof across species remains hard to find. However, this is not the case in bacteria. So how do bacterial genomes evolve? How do they contract? How do they expand? What are the mechanisms that drive these movements? Over to Chapter 4.

4. The ebb and flow of bacterial genomes

4.1. Losing DNA

The outcome of evolution is a function of, first, the rate at which stretches of DNA sequence or mutations are gained or lost and, second, the influence of neutral drift and natural selection on their maintenance. The genome size of eukaryotes is driven by the accumulation of selfish genetic elements, which cannot be purged out by selection rendered ineffective by the low population sizes of these organisms. This predicts that there would be a negative correlation between effective population size (N_e) and genome size, assuming that autonomously-multiplying selfish genetic elements dominate eukaryotic genome evolution. While this has been hard to conclusively establish across a vast range of species, at least in part because of difficulties in accurately estimating N_e, the theory has been largely accepted. However, does this apply to bacteria? Bacterial genomes rarely display low gene density unlike genomes of eukaryotes. How do these genomes evolve? What roles do natural selection and neutral drift play in determining their sizes?

Before we get to any overarching theory of or patterns in bacterial genome evolution, let us ask which of the two opposing forces—gene gain or loss—dominates bacterial genome evolution? We had noted in passing in Chapter 3 that the first two bacterial genomes to be sequenced, those of *Haemophilus influenzae* and *Mycoplasma genitalium*, had "undergone much gene elimination since their divergence event." We had also extensively discussed very small endosymbiont genomes. Is gene loss or gene elimination the norm in bacterial genomes unlike in eukaryotes?

As early as in 1989, several years before the first bacterial genome was sequenced, W.G. Weisberg and co-workers including Carl Woese (see Chapter 2), built a phylogeny of a group of bacteria called *Mollicutes* to which the genus *Mycoplasma* belongs.[1] We had encountered *Mycoplasma genitalium* and *Mycoplasma mycoides* in the previous chapter. *M. genitalium* has what is in all probability the smallest cellular genome for an organism capable of life outside a host, at under 0.6 Mbp. Its faster-growing relative *M. mycoides* has an ~1 Mbp genome. Both are small genomes for bacteria. As a general rule, *Mollicutes*, and not just the *Mycoplasma* among them, are small cells

1 W.G. Weisberg, J.G. Tully, D.L. Rose, J.P. Petzel, H. Oyaizu, et al., 'A phylogenetic analysis of the Mycoplasmas: basis for their classification', *Journal of Bacteriology* 171 (1989), 6455–6467. https://doi.org/10.1128/jb.171.12.6455-6467.1989

 https://doi.org/10.11647/OBP.0446.04

with relatively small genomes. For example, bacteria belonging to the Mollicute genus *Ureaplasma* have genomes averaging at around 1 Mbp, as do the *Phytoplasma*. In contrast, the closest relatives of *Mollicutes* in Weisberg and co-workers' data—namely *Bacillus and Lactobacillus*—have genomes that are 4–5 times larger! Though the genome sizes of these bacteria were not established at the time Weisberg and colleagues built their phylogeny, we can use their data in retrospect to infer how genome size might have evolved in Mollicutes. There are two options. The last common ancestor of Mollicutes and their closest relatives had a small genome, which then gained genes in the lineage towards *Bacillus* and *Lactobacillus*. Alternatively, the ancestor was closer to the larger relatives, but lost large chunks of DNA on the way to evolving the present-day Mollicutes. Which of these two scenarios does the phylogeny support better?

Both Mollicutes and their immediate relatives are extant bacteria. However, the phylogeny shows that *Bacillus/Lactobacillus* lineage sits closer to the common ancestor than do the Mollicutes. The phylogeny does not by itself consider genome sizes and is instead based on the sequence of just the rRNA, a "molecular chronometer"[2] that is representative of the phylogeny of the species and not just of one gene.[3] Nevertheless, the fact that two branches leading out from the common ancestor are of unequal lengths, and that the lineage leading to the small genome of the Mollicutes is further from the common ancestor than that of *Bacillus/Lactobacillus*, would prompt the reasonable hypothesis that the ancestor had a genome size closer to that of *Bacillus/Lactobacillus*. This then leads to the suggestion that the evolution of Mollicutes and the *Mycoplasma* involved considerable DNA loss. Let us ask if we can see similar patterns for multiple bacteria across the Tree of Life.

Fig. 4.1 shows the bacterial phylogenetic tree highlighting the genome sizes of bacteria. Observe the following nodes. (1) The common ancestor between the large genomes of *E. coli/Salmonella enterica* (~5 Mbp) and the small, endosymbiont genomes of *Buchnera* that we had discussed in depth in the previous chapter (~0.5 Mbp); (2) The common ancestor of bacteria of the genera *Rhizobium/Bradyrhizobium*, which have very large genomes at ~8 Mbp, and those of the parasitic pathogens of the genus *Rickettsia* (>1 Mbp); (3) The common ancestor of the pathogenic genus *Chlamydia* with small ~1 Mbp genomes and its closest relatives *Bacterioidetes* (~5 Mbp) and *Porphorymonas* (~2.3 Mbp). In each of these cases, the clade leading to the small ~1 Mbp genomes is more distant from the common ancestor than its relative carrying the several times larger genomes. These suggest that the common ancestors in each of these lineages had relatively large genomes and that the small genomes of *Buchnera, Rickettsia*, and *Chlamydia* have evolved by way of DNA loss from their ancestors. And there is no reason to suppose that this phenomenon of *genome reduction* is not ongoing.

---

2 See discussion of the landmark work of Carl Woese in Chapter 2.

3 Today though, many species phylogenies are built using the sequences of not one, but several conserved genes.

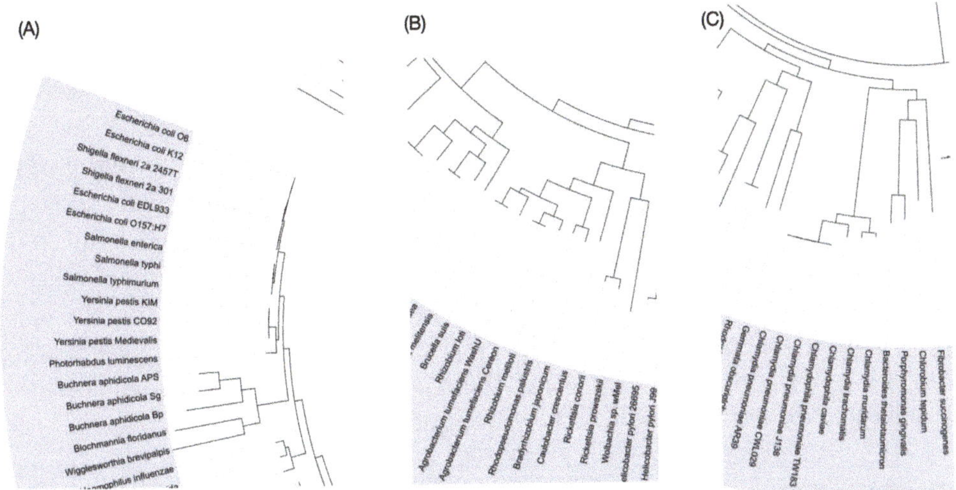

Fig. 4.1. Trees showing bacteria undergoing genome reduction. This figure shows extracts from the iTOL Tree of Life (https://itol.embl.de), showing (A) the *E. coli – Buchnera* branch, (B) the *Rhizobium / /Bradyrhizobium – Rickettsia* branch, and (C) the *Bacteroidetes/Porphorymonas – Chlamydia* branch. Latest iTol paper: I. Letunic and P. Bork, 'Interactive Tree of Life (iTOL) v6: recent updates to the phylogenetic tree display and annotation tool', *Nucleic Acids Research* 52 (2024), W78–W82.

Quoting the work by Weisberg among others, Siv Andersson and Charles Kurland argued, in 1998, that intracellular parasites appear to have evolved from free-living bacteria with genome sizes 4–5 times larger, pointing to genome reduction as a major force in the evolution of bacterial genomes.[4] This encouraged Alex Mira, Howard Ochman, and Nancy Moran to investigate this phenomenon more thoroughly at the turn of the century when ~40 bacterial genomes were fully available.[5] Mira and colleagues sought to estimate the rates at which small deletions and insertions occur in different clades of bacteria. To do so, they considered what are known as *pseudogenes*. Pseudogenes are recognisable as genes, but have undergone mutations—primarily those that introduce premature translation STOP signals—leading to non-functional gene relics. Assuming that the very formation of these pseudogenes in a clade is the result of relaxed selection for maintaining them in these organisms, any further mutations—including insertions and deletions—within these genes occur free from the constraints of selection, and therefore reflect true mutation rates. Mira and co-workers compared the sequences of pseudogenes in ~10 different groups of bacteria with those of their intact versions present in other, closely related bacterial genomes and estimated the rates at which short insertions and deletions occur. Across most if not all the groups of bacteria they studied, the number of deletions as well as the total size of deleted DNA produced by these deletions greatly exceed the number and size of insertions. In some clades, the number of deletions is as

4 S.G.E. Andersson and C.G. Kurland, 'Reductive evolution of resident genomes', *Trends in Microbiology* 6 (1998), 263–268. https://doi.org/10.1016/s0966-842x(98)01312-2

5 A. Mira, H. Ochman, and N.A. Moran, 'Deletional bias and the evolution of bacterial genomes', *Trends in Genetics* 17 (2001), 589–596. https://doi.org/10.1016/s0168-9525(01)02447-7

high as four times that of insertions, and the total length of deleted DNA some 100 times greater. Thus, early analysis of a small set of bacterial genomes indicated that bacterial genome evolution is predominantly driven by DNA loss. However, it has been shown that for pseudogenes in particular, the rate of deletions is several times higher than insertions even in mammals. Nevertheless, unlike in bacteria, the absolute rate at which pseudogenes decay in mammals appears to be too slow to compensate for the rapid multiplication of selfish DNA that results in genome expansion.[6]

In the years around and since the work by Mira and colleagues, examples of genome reduction in several clades were published. Let us look at some of these. We are already familiar with *Buchnera*, which has existed in an endosymbiotic relationship with aphids for over 250 million years. The relationship is tight. Different species of aphids carry different types of *Buchnera* such that the host and endosymbiont have evolved together. *Buchnera* genomes are small and range from over 0.4 Mbp to around 0.65 Mbp. *Buchnera* are relatives of more cosmopolitan bacteria including *E. coli*. So how did the small genome of *Buchnera* evolve from this genus's common ancestor with *E. coli*, and how did the evolution of the range of *Buchnera* genomes itself proceed from the more recent common ancestor of all *Buchnera*?

Shigenobu and colleagues who sequenced the first *Buchnera* genome,[7] as well as Moran and Mira soon after,[8] attempted to computationally reconstruct the genome of the last common ancestor of *Buchnera* and its sister clade comprising of bacteria such as *E. coli* and *Klebsiella*. Here we will follow the results published by Moran and Mira whose paper provides more details on the reconstruction and its evolutionary implications. To reconstruct the genome of the last common ancestor between *Buchnera* and its sister clade comprising of *E. coli*, Moran and Mira took what is known as the parsimony approach. A parsimony approach tries to minimise the number of events or changes required to explain the topology of the phylogenetic tree.[9] For example, let us assume that a gene, *X*, is present in both *E. coli* and in *Buchnera*. The most parsimonious route to this state is from an ancestor that also had this gene, which has been retained in both lineages. It is also entirely possible that the ancestor did not have the gene *X*, which was somehow acquired independently in both lineages; however, this, requiring two changes of state, is less likely. Let us assume that another gene, *Y*, is absent in *Buchnera* but present in *E. coli*. Now, a gain in the lineage to *E. coli* is not at first instance less likely than a loss enroute to *Buchnera*, assuming that the rates of gene loss and gain are not very different. But if *Y* is also present in, say, the bacterium *Vibrio cholerae*, which shares a common ancestor with the lineage leading to both *Buchnera*

6 D. Graur, Y. Shuali, and W.H. Li, 'Deletions in processed pseudogenes accumulate faster in rodents than in humans', *Journal of Molecular Evolution* 28 (1989), 279–285. https://doi.org/10.1007/bf02103423

7 S. Shigenobu, H. Watanabe, M. Hattori, Y. Sakaki, and H. Ishikawa, 'Genome sequence of the endocellular bacterial symbiont of aphids Buchnera sp. APS', *Nature* 407 (2000), 81–86. https://doi.org/10.1038/35024074

8 N. Moran and A. Mira, 'The process of genome shrinkage in the obligate symbiont *Buchnera aphidicola*', *Genome Biology* 2 (2001), research0054.1. https://doi.org/10.1186/gb-2001-2-12-research0054

9 For an overview of the methods used in building phylogenies, including those based on parsimony, see J. Felsenstein, *Inferring Phylogenies* (Sunderland: Sinauer, 2003).

and *E. coli*, then the most parsimonious inference would be the presence of gene *Y* in the common ancestor of *Buchnera* and *E. coli* followed by its loss in the line towards *Buchnera*. The parsimony approach works when the phylogeny is shallow, i.e., the evolutionary distance between the species being compared is so small that multiple events or changes of state actually become less likely. However, parsimony stops being valid when the evolutionary distance between the species being compared increases. For instance, in the example of the gene *X* discussed above, if the compared species were at two extreme ends of the phylogenetic tree separated by, say, a billion years or more of evolution, independent acquisition of *X* in both lineages is much less unlikely, making the idea of parsimony inappropriate.

Moran and Mira, arguing that the phylogeny under consideration here is shallow enough for parsimony to be applicable to most genes,[10] used this principle to reconstruct a common ancestor of *E. coli* and *Buchnera* that comprised ~2,400 genes. If the ancestral genome also had a gene density typical of bacteria, it would have been ~2.5 Mbp long, ~four times larger than the *Buchnera* genome. This implies that nearly three quarters of the gene complement of the ancestral genome was lost along the lineage leading to *Buchnera*. Many deletions were of fragments comprising multiple genes, sometimes as long as 10–11 genes long. Among the genes that were lost are those involved in DNA repair. A study published in 2009 by Moran and colleagues also showed that the rate of substitution mutations in *Buchnera* is ~10 times higher than that documented for any other bacterial species.[11] It is likely that loss of DNA repair functions might have led to accelerated rates of mutation. Moran and Mira further observed that not all losses were complete gene deletions. Some are recognisable as pseudogenes, which have degraded from their intact *E. coli*-like versions to different extents through processes involving mutations and insertions/deletions of a few base pairs. A few retain strong similarities to their respective *E. coli* orthologs[12] whereas others are severely shortened versions. These analyses indicated a preference for deletion over insertion within pseudogenes, consistent with the work by Mira and colleagues quoted earlier,[13] which had shown a >15-fold greater frequency of deletions over insertions in *Buchnera*. Thus, the evolution of *Buchnera* from its last common ancestor with the lineage to *E. coli* appears to be characterised by DNA loss, in the form of both the deletion of large fragments of DNA and of short sequences that eventually cause the removal of pseudogenes from the genome.

10 Also note that in the early 2000s, computational power as well as phylogenetic algorithms were not sufficiently advanced for effective implementation of more sophisticated approaches for phylogeny reconstruction on genome-wide scales.

11 N.A. Moran, H.J. McLaughlin, and R. Sorek, 'The dynamics and time scale of genomic erosion in symbiotic bacteria', *Science* 323 (2009), 379–383. https://doi.org/10.1126/science.1167140

In an earlier study based on a few genomes, Itoh and colleagues showed that the rate of evolution in the *Buchnera* lineage is approximately twice that in the *E. coli* lineage, qualitatively consistent with the above study's findings. See T. Itoh, W. Martin, and M. Nei, 'Acceleration of genomic evolution caused by enhanced mutation rate in endocellular symbionts', *Proceedings of the National Academy of Sciences USA* 99 (2002), 12944–12948. https://doi.org/10.1073/pnas.192449699

12 See Chapter 3 and E.V. Koonin, 'Orthologs, paralogs and evolutionary genomics', *Annual Review of Genetics* 39 (2005), 309–338. https://doi.org/10.1146/annurev.genet.39.073003.114725

13 Mira et al., 2001.

As discussed in Chapter 3, the sizes of *Buchnera* genomes vary from less than 0.42 Mbp to nearly 0.65 Mbp. *Buchnera* enjoying an endosymbiotic relationship with what is called the *Lachninae* subfamily of their aphid insect hosts tend to have genome sizes at the smaller end of the spectrum, consistent with the presence of a second endosymbiont in these insects, whereas those living in the *Aphidinae* subfamily carry larger genomes. In a recent study, Rebecca Chong and colleagues compared the genomes of ~40 different *Buchnera* genomes from a variety of aphid hosts, and spanning the full range of genome sizes, to ask how the species has evolved since its divergence from its own common ancestor.[14] They again used a parsimony-based approach to build an ancestral *Buchnera* genome encoding ~650 genes from which genome reductions of various extents have happened in different kinds of *Buchnera*. Of these ~650 genes, only ~250 are common across the ~40 extant *Buchnera* genomes, showing large variation in gene repertoire across hosts. As expected from our discussion of minimal genomes in Chapter 3, genes involved in core genetic processes—DNA replication, transcription, and translation—are retained across the genus. Consistent with the role of *Buchnera* in the supply of essential amino acids to their hosts, most metabolic pathways for the synthesis of essential amino acids are also retained across the species. In contrast, genes involved in processes such as response to stress, which were already absent to a large extent in the *Buchnera* ancestor, continue to undergo further deletions. Thus, the evolution of *Buchnera*, which began with genome reduction after divergence from its common ancestor with *E. coli*, continues to be defined by gene loss even as the genus has diversified across a variety of aphid hosts.

We have emphasised several times that bacterial genomes are gene dense and that genes comprise ~85–90% of a typical bacterial genome. A striking exception to this rule is the genome of the leprosy-causing bacterium *Mycobacterium leprae*. It belongs to the same genus as *Mycobacterium tuberculosis* which, as the name suggests, causes tuberculosis (TB) in humans. The *M. tuberculosis* genome is ~4.4 Mbp long and codes for ~4,000 genes, displaying a gene density that is typical for bacteria. In contrast, the *M. leprae* genome—which is much smaller at ~3.2 Mbp—codes for only ~1,600 genes.[15] Thus, nearly half of the *M. leprae* genome is devoid of genes (Fig. 4.2). In fact, the *M. leprae* genome carries over 1,100 pseudogenes with some similarity to intact genes in *M. tuberculosis*. This brings up two interesting aspects about the *M. leprae* genome. First, it has undergone genome reduction starting from its common ancestor with *M. tuberculosis*. Second, unlike *Buchnera*, where much of the gene loss is recognisable as DNA loss, much of the *M. leprae* genome that is devoid of genes is still present in some *pseudogenised* form. Unlike in *Buchnera*, DNA loss has not yet caught up with *pseudogenisation* in *M. leprae*. This apparent difference in the gene loss process between *Buchnera* and *M. leprae* makes it

14 R.A. Chong, H. Park, and N.A. Moran, 'Genome evolution of the obligate endosymbiont *Buchnera aphidicola*', *Molecular Biology and Evolution* 36 (2019), 1481–1489. https://doi.org/10.1093/molbev/msz082

15 S.T. Cole, K. Eiglmeier, J. Parkhill, K.D. James, N.R. Thomson, et al., 'Massive gene decay in the leprosy bacillus', *Nature* 409 (2001), 1007–1011. https://doi.org/10.1038/35059006

worthwhile for us to try and understand the evolutionary history of the *M. leprae* genome starting from its common ancestor with *M. tuberculosis*. Did pseudogenisation and gene loss occur in one or more waves? Or did they occur in a staggered manner in many small steps leading to the extant *M. leprae*?

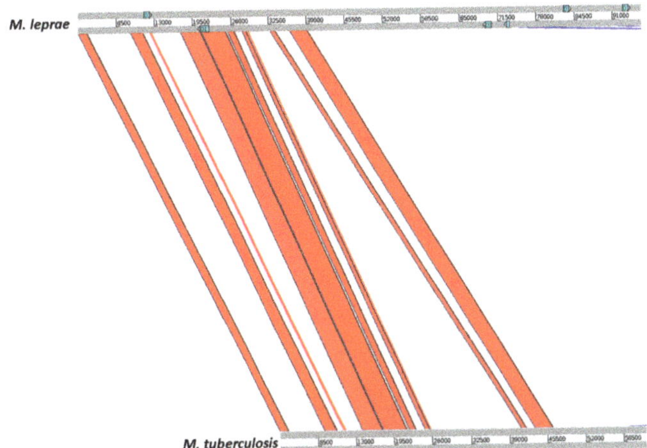

Fig. 4.2. Pseudogenes in *M. leprae*. This figure shows a small fragment of the *M. leprae* genome (top) and the corresponding segment of *M. tuberculosis* genome (bottom). Pseudogenes in *M. leprae* are shown in cyan. There are no pseudogenes for this region in *M. tuberculosis*. Figure produced by Ganesh Muthu using the Artemis Comparison Tool (https://www.sanger.ac.uk/tool/artemis-comparison-tool-act/).

In an early work, Madan Babu proposed that the number of translation STOP codons present within the body of a pseudogene would be correlated to the time at which pseudogenisation happened.[16] If a single event created the first STOP codon within the body of a gene, thus converting it into a pseudogene, then the loss of selection preserving the rest of the gene sequence would result in additional mutations introducing STOP codons at some constant rate. The longer the time elapsed since the first pseudogenisation event, the greater the number of STOP codons accumulated. Madan Babu used this idea to date the first pseudogenisation event for ~1,100 pseudogenes in *M. leprae* and suggested that pseudogenisation occurred in two waves. While this method for dating pseudogenisation events at first glance sounds intuitive and reasonable, it is problematic. STOP codon introduction often happens through insertions or deletions of one or a few base pairs of DNA. What this usually does is introduce not a single STOP codon, but instead create several downstream of the site of insertion or deletion. Thus, several STOP codons can be introduced by a single mutational event. Therefore, it is unlikely that there will be a strong correlation between the age of a pseudogene and the number of STOP codons it carries. Recognising this, Gómez-Valero and colleagues revisited this question a

16 M. M. Babu, 'Did the loss of sigma factors initiate pseudogene accumulation in *M. leprae*?', *Trends in Genetics* 11 (2003), 59–61. https://doi.org/10.1016/s0966-842x(02)00031-8

few years later and produced a more comprehensive timeline of gene decay in *M. leprae*.[17]

Gómez-Valero and co-workers used parsimony to assemble a ~3000-gene ancestor of *M. leprae* and *M. tuberculosis*. Just over 400 of these were fully lost in the *M. leprae* lineage with no detectable homolog in *M. tuberculosis*. These genes were, in all likelihood, lost in chunks with an average of ~3 kbp per segment. Gómez-Valero sought to trace the evolutionary history of the other ~1,100 genes that existed as pseudogenes in *M. leprae*. These genes have a recognisable homolog in *M. tuberculosis*. Gómez-Valero et al.'s approach to dating a pseudogenisation event was as follows. Every pseudogene in the *M. leprae* genome would have had two phases in its history since divergence from the common ancestor with *M. tuberculosis*. The first would be its life as an intact gene and the second as a pseudogene. When did the transition from the first phase to the second happen? Or what fraction of its 'life' since divergence from the common ancestor has the pseudogene spent in its degraded state? The rates at which consequential mutations—as defined by *non-synonymous* mutations, which change the amino acid at the mutated site—occur in a gene would be different between its intact and its pseudogenised state. In its intact state, non-synonymous changes in the *M. leprae* lineage would be disfavoured, and the rate at which they happen would be a function of the corresponding rate for the *M. tuberculosis* lineage. It is more convenient to calculate this rate for the relatively uncomplicated *M. tuberculosis* lineage in which uncertainties over times of pseudogenisation, and therefore temporal changes in rates of non-synonymous mutations, do not exist. This rate can vary across genes depending on the strength of selection for maintaining a given gene's sequence. In the pseudogenised state, all genes can be expected to undergo a rate of '*non-synonymous*'[18] mutation that is close to the average rate of *synonymous* mutations in the *M. leprae* lineage, under the assumption that synonymous sites and pseudogenes are equally free from natural selection. Both these rates can be quantified from the sequence data. The total number of mutations in the *M. leprae* lineage would be a sum of the two rates each multiplied by the time spent in the respective state. If the total time elapsed since divergence from the common ancestor is set to 1 and the fraction of time a gene has spent as a pseudogene is p then obviously the fraction of time spent as an intact gene would be $1 - p$.

With the various rates available from the sequence data, p can be calculated for each pseudogene in *M. leprae*, which is what Gómez-Valero and colleagues did. They found that the distribution of p across pseudogenes followed a single normal distribution. The variation in p across pseudogenes was small. This indicated to Gómez-Valero et al. that pseudogenisation was initiated by a single major event of great evolutionary importance, which resulted in mass inactivation of genes via independent molecular

17 L. Gómez-Valero, E.P.C. Rocha, A. Latorre, and F.J. Silva, 'Reconstructing the ancestor of Mycobacterium leprae: The dynamics of gene loss and genome reduction', *Genome Research* 17 (2007), 1178–1185. https://doi.org/10.1101/gr.6360207

18 The terms 'synonymous' and 'non-synonymous' do not really apply well to pseudogenes! However, the use of this term is for convenience and can be taken to represent the effect a mutation would have on an intact version of the pseudogene.

events over a short period of time. The value of p is small (~0.13), and corresponds to ~20 million years ago. This suggests that pseudogenisation of genes in *M. leprae* is a relatively recent event. This might have been driven by an *M. leprae* ancestor occupying a restricted niche in neuronal cells. Given that this event happened long before the evolution of the *Homo* genus, the question remains whether this event originally occurred in armadillos,[19] which appear to be among the very few known extant natural hosts for *M. leprae*, besides humans and nonhuman primates.[20]

Yet another example of a bacterial clade showing signs of genome reduction is that of the genus *Chlamydia*. These belong to a larger group of bacteria called *Planctomycetes-Verrucomicrobia-Chlamydiae* (PVC) superphylum. Many bacteria in this superphylum are free-living and carry large genomes, often over 4–5 Mbp. However, *Chlamydiales*, the larger group to which the genus *Chlamydia* belongs, are obligate endosymbionts of eukaryotic cells from amoeba to animal cells and typically carry reduced ~1–2 Mbp genomes. These might have diverged from their last common ancestor with the rest of the PVC bacteria as early as 1–2 billion years ago. Among these bacteria, *Chlamydia trachomatis* has evolved as a parasitic human pathogen, diverging from other *Chlamydiae* more recently at only 6 million years ago.[21] Olga Kamneva and co-workers reconstructed the ancestral genome of the PVC superphylum as well as those of various sub-clades within this superphylum including the *Chlamydiales*.[22] For this deep phylogeny, they adopted the *Maximum Likelihood Estimation* to reconstruct ancestral genomes. This approach does not necessarily favour parsimony, but instead finds ancestral states and state changes that best fit the observed relationships between all the genomes being compared. The analysis showed that the last common ancestor of the PVC likely carried ~3,000 genes, and that this genome had decayed to ~800 genes in the common ancestor of the genus *Chlamydia*.

Up till this point, the progress of evolution is similar to that of *Buchnera* in which the ~2,400 gene ancestor of *Buchnera* and *E. coli* lost three-quarters of its genome to produce an ~650-gene ancestor of all *Buchnera*. In the *Buchnera* lineage, further genome degradation occurred and is occurring.[23] However, the two *Chlamydia* genomes included

---

19 These animals first emerged at the Cretaceous-Tertiary boundary some 66 million years ago. See F. Delsuc, F.M. Catzeflis, M.J. Stanhope, and E.J. Douzery, 'The evolution of armadillos, anteaters and sloths depicted by nuclear and mitochondrial phylogenies: implications for the status of the enigmatic fossil Eurotamandua', *Proceedings of the Royal Society B* 268 (2001), 1605–1615. https://doi.org/10.1098/rspb.2001.1702

20 There is some evidence that nonhuman primates might have acquired *M. leprae* infection from humans rather than the other way round. See T. Honap, L.-A. Pfister, G. Housman, S. Mills, R.P. Tarara, et al., 'Mycobacterium leprae genomes from naturally infected nonhuman primates', *PLoS Neglected Tropical Diseases* 12 (2018), e0006190. https://doi.org/10.1371/journal.pntd.0006190

21 A. Nunes and J.P. Gomes, 'Evolution, phylogeny, and molecular epidemiology of Chlamydia', *Infection, Genetics and Evolution* 23 (2014), 49–64. https://doi.org/10.1016/j.meegid.2014.01.029

22 O. Kamneva, S.J. Knight, D.A. Liberles, and N.L. Ward, 'Analysis of genome content evolution in PVC bacterial super-phylum: assessment of candidate genes associated with cellular organization and lifestyle', *Genome Biology and Evolution* 4 (2012), 1375–1390. https://doi.org/10.1093/gbe/evs113

23 Chong et al., 2019. Note the following caveat. The ancestral reconstruction procedure used in this study, by taking the union of the gene sets across all sequenced *Buchnera* genomes, would have by definition forced the ancestor to have a genome larger than any individual *Buchnera*. Though I believe

in the phylogeny constructed by Kamneva et al. are larger than their immediate-ancestral genome. This was further confirmed by a much more recent study by Jennah Dharamshi and co-workers,[24] which included the genomes of several *Chlamydia* and the closely-related *Chlamydophila* species in its analysis. Both studies also noted that a bacterium within the *Chlamydiae*, but belonging to the genus *Parachlamydia*, carried ~2,800 genes. This bacterium is also an endosymbiont of an amoeba. Contrary to the expectation that the evolution of endosymbionts should be characterised by genome reduction, ancestral state reconstruction showed substantial gene gain in *Parachlamydia* compared to the common ancestor of all *Chlamydiales*. The ancestor of all *Chlamydiales* seems to have been a *facultative anaerobe*, which could survive in both oxic and anoxic conditions.[25] However, downstream of this ancestor, various genera of the *Chlamydiales* diverged in their oxygen tolerance capacity. Gene gain in *Parachlamydia* reflects an expansion of genes involved in oxygen tolerance and aerobic metabolism. Thus, while it is true that several endosymbionts and parasitic bacteria evolve predominantly by genome reduction, additions of genes for adaptations particular to their lifestyles cannot be ruled out.

Finally, genome reduction is not unique to bacteria with parasitic or endosymbiotic lifestyles. Some free-living, ocean-dwelling *cyanobacteria* of the genera *Prochlorococcus* and *Pelagibacter* also show evidence of genome reduction. Let us consider *Prochlorococcus* here. Most *Prochlorococcus* genomes code for genomes in the ~1.7 Mbp range, with some lineages but not others showing strong signs of reduction. We will keep in mind that *Prochlorococcus* (and *Pelagibacter*) genomes show a gene density that is even higher than what is typical for most bacteria. This genus is closely related to the genus *Synechococcus*, whose genomes exceed 2 Mbp and often are over 2.5 Mbp. By reconstructing the ~2,000 gene-encoding genome of the common ancestor, Zhiyi Sun and Jeffrey Blanchard showed that rapid gene loss, with a net loss of ~420 genes, occurred in the *Prochlorococcus* lineage soon after its divergence from the common ancestor.[26, 27] However, similar to what we have noted for *Chlamydia*, more recent evolution within the *Prochlorococcus* lineage is characterised by net gene gain (Fig. 4.3).

---

that this is a reasonable assumption, the claim that *Buchnera* lost genes in its recent evolution would be better supported if ancestral reconstruction using approaches such as Maximum Likelihood Estimation also reinforced this idea. To my knowledge, this is yet to be described in the literature.

24 J.E. Dharamshi, S. Köstlbacher, M.E. Schön, A. Collingro, T.J.C. Ettema, and M. Horn, 'Gene gain facilitated endosymbiotic evolution of Chlamydiae', *Nature Microbiology* 8 (2023), 40–54. https://doi.org/10.1038/s41564-022-01284-9

25 Ibid.

26 Z. Sun and J.L. Blanchard, 'Strong genome-wide selection early in the evolution of Prochlorococcus resulted in a reduced genome through the loss of a large number of small effect genes', *PLoS One* 9 (2014), e88837. https://doi.org/10.1371/journal.pone.0088837

27 In providing these numbers I am neglecting two *Prochlorococcus* isolates that have relatively large genomes of >2.7 Mbp.

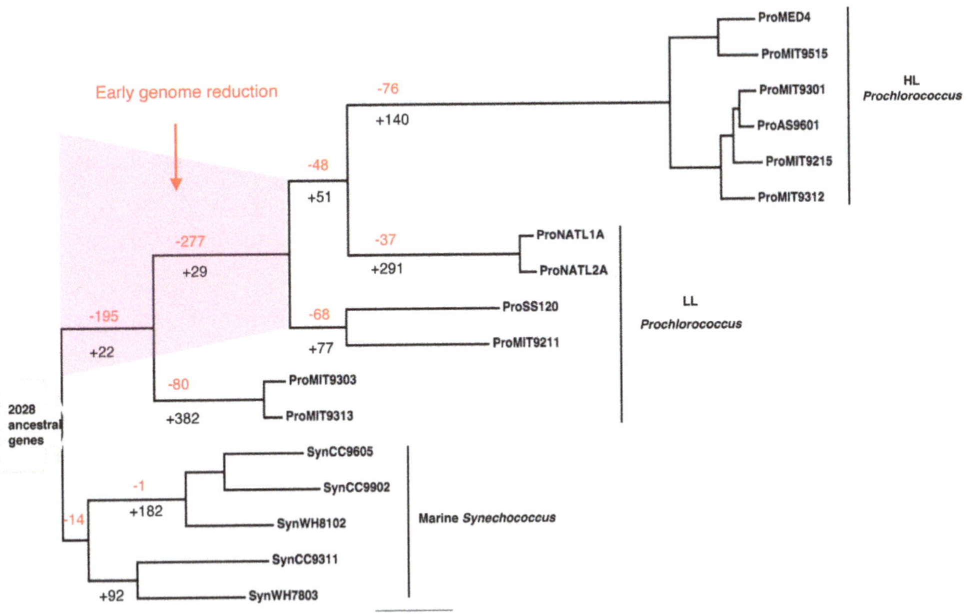

Fig. 4.3. Gene loss in *Prochlorococcus*. This figure shows the number of genes gained (black, positive integers) and lost (red, negative integers) across different branches in the *Prochlorococcus-Synechococcus* tree. Originally published as Figure 2 in Z. Sun and J.L. Blanchard, 'Strong genome-wide selection early in the evolution of Prochlorococcus resulted in a reduced genome through the loss of a large number of small effect genes', *PLoS One* 9 (2014), e88837, under Creative Commons Attribution License.

Genome reductions happen not only in natural populations of bacteria, but have also been demonstrated in bacterial populations evolving in the laboratory. A.I. Nilsson and co-workers evolved populations of *Salmonella* in the laboratory by growing them for a few generations and randomly picking one colony from the resultant population for the next round of growth.[28] Several hundred cycles of growth and random colony picking were performed and the genome was analysed for DNA loss at regular intervals. These researchers estimated that *Salmonella* underwent gene loss at the rate of 0.05 base pairs per generation per chromosome. In contrast, mutant *Salmonella* lacking a repair system, which is also absent in the reduced genomes of endosymbionts like *Buchnera*, underwent deletions 50-fold faster. Deletions spanned a range of lengths, from as small as 1 kbp to over 200 kbp.

A second example of genome reduction in the lab was noted as part of a classic experiment in which Richard Lenski and colleagues evolved *E. coli* for over 60,000 generations in the laboratory.[29] This experiment is called LTEE for Long-Term

28 A.I. Nilsson, S. Koskiniemi, S. Eriksson, E. Kugelberg, J.C.D. Hinton, and D.I. Andersson, 'Bacterial genome size reduction by experimental evolution', *Proceedings of the National Academy of Sciences USA* 102 (2005), 12112–12116. https://doi.org/10.1073/pnas.0503654102

29 R.E. Lenski, 'Experimental evolution and the dynamics of adaptation and genome evolution in microbial populations', *ISME Journal* 11 (2017), 2181–2194. https://doi.org/10.1038/ismej.2017.69 This experiment is still ongoing.

Experimental Evolution. His lab had maintained 12 populations of *E. coli* in the lab for over 50,000 generations when Olivier Tenaillon and colleagues published a report describing the results of their efforts at sequencing the genomes of individual colonies[30] from each population at various time points over the course of evolution.[31] Across their entire dataset, they found 500 new insertions of transposable elements, ~2,000 short insertions and deletions, over 250 large deletions, and ~50 large duplications. The end result of these gains and losses was a net reduction in genome size by an average of ~63 kbp at the end of 50,000 generations, or the loss of ~1.2 bp per generation.[32] Thus, in the absence of selection under standardised laboratory conditions, which typically do not permit gene gains easily (see next section below), bacterial chromosomes can lose DNA at fairly high rates.

4.2. Gaining DNA

Both parasitic/endosymbiotic and free-living bacteria have undergone phases of extensive gene loss in their evolution. However, the evolution of some bacteria with reduced genomes has not always been about losses but also involves some gene gain, mostly in relatively recent times. We saw examples of this in *Chlamydia* and in *Prochlorococcus*. While bacteria such as *Buchnera* have undergone genome reduction, the numbers of genes in reconstructed ancestral genomes discussed earlier will immediately suggest that their relatives *E. coli* and the non-*Chlamydia* PVC clade often gained large numbers of genes compared to their respective common ancestors. The reconstruction of the common ancestor of *M. tuberculosis* and *M.* leprae, while highlighting gene loss in the *M. leprae* lineage, also points to gene gain in *M.* tuberculosis. How and to what extent do bacteria gain new genes?

Some of the reduced genomes that we are now familiar with are not entirely free of gene gains despite their small size. Let us examine these first. Dharamshi and colleagues identified a total of ~1,500 gene gains across the various *Chlamydiae* genomes present in their analysis. Some of these were gained anew or, in the biologist's parlance, *de novo*. These do not have any known homologs outside the *Chlamydiae* clade despite the large collection of protein and gene sequences available in public databases today. They may be gene innovations within *Chlamydiae*. Such innovations can happen through mutations of random sequences or, as shown by Margaret Dayhoff,[33] by duplication, concatenation, and further evolution of short sequence fragments. In bacteria, *de novo* emergence of short protein sequences from random sequences has been experimentally demonstrated.[34] In eukaryotes, the large amounts of non-functional DNA can at

---

30 Each colony having grown from a single bacterium and therefore likely to be genetically homogenous.

31 O. Tenaillon, J.E. Barrick, N. Ribeck, D.E. Deatherage, J.L. Blanchard, et al., 'Tempo and mode of genome evolution in a 50,000-generation experiment', *Nature* 536 (2016), 165–170. https://doi.org/10.1038/nature18959

32 The loss was not linear over time. It was slower over the first 10,000 generations and accelerated later.

33 See Chapter 3.

34 M. Knopp, J.S. Gudmundsdottir, T. Nilsson, F. König, O. Warsi, F. Rajer, P. Ädelroth, D.I. Andersson,

times act as substrates for the evolution of new protein-coding genes.[35] However, in gene-dense bacteria, the extent to which large proteins can evolve *de novo* over short timescales is not clear. Many other sequences gained by *Chlamydiae* showed similarity to genes in distant members of the PVC clade, or even to bacteria that lie *beyond* the PVC clade. It is also not impossible that a subset of genes thought to have arisen *de novo* have distant homologs in organisms we know little about and whose sequences are not available in current databases.

A gene that is exclusively vertically inherited from parental to daughter cells will show strong sequence similarity to its ortholog in closely-related organisms. As the evolutionary distance between two organisms increases, the similarity in the sequence of the two orthologous genes will decrease. Further, the occurrence of this gene will be similar among species from within a lineage than across lineages. In other words, the presence or absence of this gene shows a strong *phylogenetic signal*. On the other hand, genes that are present in a smattering of organisms across the bacterial tree pose a problem. It is possible that such a gene is still vertically inherited but, because of habitat changes experienced by some bacteria, lost in some members of a lineage but not others. This would have had to happen many times across the bacterial kingdom. Assuming that this pattern of this gene's presence can be fully explained by gene loss, one might still expect the sequence similarity between orthologs to be much greater for closely related than for distantly related bacteria if mutation rates for this gene in different lineages are similar. However, a subset of genes gained in some members of *Chlamydiae* showed strong similarity to orthologs in very distantly related bacteria, while being absent in most other *Chlamydiae* or even other PVC bacteria.[36] A phylogenetic tree made using this subset of genes would be highly incongruent with the species tree. The best explanation for such a pattern is that these genes are *not* inherited vertically from parent to progeny, but *across* species by *horizontal transfer*. In other words, a bacterium gains a gene from a source that is not its parent but from a foreign source, a phenomenon that flies in the face of any classical genetic mechanism of heredity. Dharamshi and colleagues estimate that as many as 30% of genes in *Chlamydiae* may have been acquired *horizontally*.[37]

Identifying horizontally-acquired genes using the gold-standard method of testing for phylogenetic incongruence is computationally intensive and is therefore not easily achieved across the board. However, there are more indirect ways to detect recently-acquired genes of 'foreign' origin with some confidence. The base composition of DNA varies widely across bacterial genomes—from as low as under 20% guanine+cytosine

---

'De novo emergence of peptides that confer antibiotic resistance', *mBio* 10 (2019), e00837–19. https://doi.org/10.1128/mbio.00837-19

35 Reviewed in A. McLysaght and D. Guerzoni, 'New genes from non-coding sequence: the role of de novo protein-coding genes in eukaryotic evolutionary innovation', *Philosophical Transactions of the Royal Society B* 370 (2015), 20140332. https://doi.org/10.1098/rstb.2014.0332

36 Dharamshi et al.

37 In the general discussion at the start of this paragraph, the term lineage is loosely defined. At what phylogenetic depth incongruities in gene presence/absence patterns arise would depend on when the gene was horizontally acquired.

(G+C) to as high as 75%. Most genomes have a characteristic average genomic G+C content with some variation across segments. Stretches of DNA that display a G+C content that is very different from the flanking 'average-for-the-host-genome' DNA were probably acquired from a foreign source such as a bacterial genome with a different characteristic base composition.[38] A caveat to this rule is that a foreign piece of DNA once integrated into a host would, over time, slowly ameliorate such that its base composition starts to reflect that of the host. This means that a base composition-driven approach for detecting horizontally acquired DNA works best for recent transfer events. H. Luo and colleagues used this approach to show that in some *Prochlorococcus*, the cyanobacterium with reduced genomes that we briefly discussed earlier, a whopping 40% of the genome might have been horizontally-acquired recently.[39]

Thus, even in bacteria whose evolution is dominated by gene loss and genome decay, gene gain resulting in genome growth is prominent, and in a large number of cases it is a result of horizontal acquisition of foreign DNA. Let us move on to large bacterial genomes and ask how they have grown and how the balance between gene gain and loss has played out here.

Escherichia coli is a complex species. In the history of molecular biology, this bacterium occupies the pride of place, having been the host of the bacteriophages that contributed to the stories of DNA and mRNA.[40] The ease of its genetic manipulation meant that it has remained a major model organism for biological research even in the era of genomes. The type of *E. coli* commonly used in the lab may in fact be lab-adapted now and incapable of doing well in its natural environments! Besides being a model organism, this species encompasses a wide variety of bacterial traits. Some 'commensal' types of *E. coli* are relatively harmless inhabitants of the human gut. Some cause diseases of the intestinal tract of humans and other animals and birds. Some others infect the urinary tract. Yet others may persist in the wider environment. The genomes of *E. coli* varieties are larger than the average bacterial genome. Whereas commensal and lab-adapted *E. coli* strains typically have an ~4.6 Mbp genome, encoding over 4,100 protein-coding genes, many pathogenic types have over ~5.5 Mbp genomes with over 5,300 genes. Thus, this species spans a very large range of genome sizes and gene counts. The sequencing of the genomes of many tens and hundreds of *E. coli* varieties has shown that the gene content varies a lot across *E. coli*. Though the smallest *E. coli* genomes code for ~4,000 genes, only ~2,000 are found across all (or most) *E. coli* types.[41] This implies that the *E. coli* species has a very small *core* genome.

---

38 In reality, the base composition itself does not work best as a discriminator of indigenous and foreign genes. It is more the composition of longer sequence motifs from 3-base to 7-base stretches that appears to work best. See G.S. Vernikos and J. Parkhill, 'Interpolated variable order motifs for identification of horizontally acquired DNA: revisiting the Salmonella pathogenicity islands', *Bioinformatics* 22 (2006), 2196–2203. https://doi.org/10.1093/bioinformatics/btl369

39 H. Luo, R. Friedman, J. Tang, A.L. Hughes, 'Genome Reduction by Deletion of Paralogs in the Marine Cyanobacterium Prochlorococcus', *Molecular Biology and Evolution* 28 (2011), 2751–2760. https://doi.org/10.1093/molbev/msr081

40 See Chapter 2.

41 Several estimates of the core- and pan-genomes of *E. coli* have been published over the years. This

On the other hand, the *pan genome*, which is in effect the union of genes found in at least one member of the species, is large and open: every new *E. coli* genome that gets sequenced is likely to reveal several tens of genes not known in any other *E. coli*. How did this extreme diversity in genome size and gene content evolve in this species?

The evolution of *E. coli* since its divergence from its common ancestor with the genus *Salmonella* is characterised not by persistent genome growth, but by a balance between gene gain and loss, as demonstrated by Howard Ochman and Isaac Jones way back at the turn of the century.[42] This continues to hold true even when the evolution of *E. coli* from its ancestor to another member of the same genus, *Escherichia fergusonii*, is considered. In the latter case, Touchon et al. in 2009 estimated that the common ancestor codes for ~4,000 genes.[43] This implies that the lineage leading to the commensal types of *E. coli* would be defined by a tight balance between gene loss and gain whereas those leading to various pathogenic clades would be often characterised by an excess of gene gain (Fig. 4.4). That different clades of *E. coli* experience different rates of gene gain and loss was further confirmed by a more recent analysis of ~10,000 *E. coli* genomes.[44]

Jeffrey Lawrence and Howard Ochman demonstrated that gene gain in the first-sequenced laboratory strain of *E. coli* was dominated by horizontal gene acquisition.[45] They showed that ~750 out of the ~4,300 protein-coding genes in the genome had a base composition that was distinct from the genomic average, and likely to have originated from foreign sources. The authors estimated that *E. coli*, since its divergence from *Salmonella*, has been acquiring DNA horizontally at the rate of ~16 kbp per million years. Given that this species diverged from *Salmonella* some 100 million years ago, it must have acquired ~1.6 Mbp of DNA from foreign sources over this period. Many horizontally-acquired genes are also lost post-acquisition, whereas a few have survived for longer periods. Some horizontally-acquired genes are transposons. However, unlike in eukaryotic genomes, these transposons are not allowed to multiply themselves to great extents in bacterial genomes due to the strength of selection that the cost of each copy would impose on the host.

is the most recent that I have come across: S.-C. Park, K. Lee, Y.O. Kim, S. Won, and J. Chun, 'Large-Scale Genomics Reveals the Genetic Characteristics of Seven Species and Importance of Phylogenetic Distance for Estimating Pan-Genome Size', *Frontiers in Microbiology* 10 (2019), 834. https://doi.org/10.3389/fmicb.2019.00834

42 H. Ochman and I. Jones, 'Evolutionary dynamics of full genome content in *Escherichia coli*', *EMBO Journal* 19 (2000), 6637–6643. https://doi.org/10.1093/emboj/19.24.6637

43 M. Touchon, C. Hoede, O. Tenaillon, V. Barbe, S. Baeriswyl, et al., 'Organised genome dynamics in the Escherichia coli species results in highly diverse adaptive paths', *PLoS Genetics* 5 (2009), e1000344. https://doi.org/10.1371/journal.pgen.1000344

44 K. Abram, Z. Udaondo, C. Bleker, V. Wanchai, T.M. Wassenaar, M.S. Robeson 2nd, and D.W. Ussery, 'Mash-based analyses of Escherichia coli genomes reveal 14 distinct phylogroups', *Communications Biology* 4 (2021), 117. https://doi.org/10.1038/s42003-020-01626-5

45 J.G. Lawrence and H. Ochman, 'Molecular archaeology of the *Escherichia coli* genome', *Proceedings of the National Academy of Sciences USA* 95 (1998), 9413–9417. https://doi.org/10.1073/pnas.95.16.9413

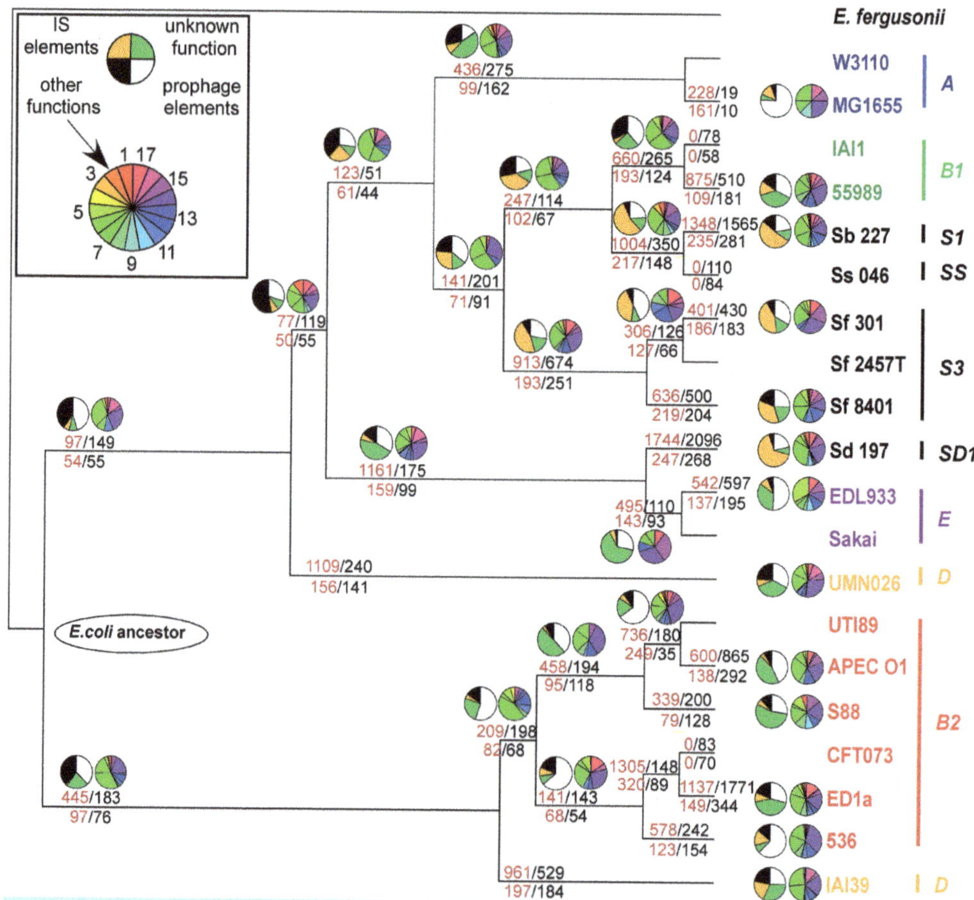

Fig. 4.4. Gene gain and loss in *E. coli*. Figure shows a tree showing the relationships between different *E. coli* and *Shigella* (which is considered as part of *E. coli*). Each branch is labelled with the number of genes gained (red) and the number lost (black) along the branch. The numbers on top of each branch represent the number of genes gained/lost, whereas the numbers at the bottom are the number of inferred gene gain/loss events. A single event can cause the gain or loss of multiple genes. The pie charts show the functions of genes gained in each branch. For a colour key, refer to Supplementary Table 5 in the original source cited below. Originally published as Figure 7 in M. Touchon, C. Hoede, O. Tenaillon, V. Barbe, S. Baeriswyl, et al., 'Organised genome dynamics in the Escherichia coli species results in highly diverse adaptive paths', *PLoS Genetics* 5 (2009), e1000344, under Creative Commons Attribution License.

Many other horizontally-acquired segments in the *E. coli* genome carry signatures of bacteriophage origin. This arises from the fact that bacteriophages do not always kill their host bacteria. Sometimes, some bacteriophages integrate their genome into specific parts of the host bacterial genome in a process called *lysogeny*. These integrated bacteriophage genomic DNA can later be excised out of the host genome, get active, form viruses, and kill their host cell. Alternatively, mutational processes can deprive the bacteriophage DNA of its ability to code for active virus particles or even to be easily detached from the host genome. The ability of such 'cryptic' bacteriophage DNA to become an integral part of the host genome over long periods of time would depend on its *s* and its relationship to N_e. There is evidence that *s* is in fact positive for some integrated bacteriophage genes, which turn out to be beneficial to host survival under some environmental conditions.[46] Bacteriophage DNA not only helps transfer beneficial DNA, but is first and foremost a pathogenic agent that can lead to the killing of the host bacterium. Naturally, bacteria have evolved a whole suite of mechanisms to defend themselves against bacteriophage DNA. Among these is CRISPR, which has gained considerable popularity for its applications in genetic engineering. Curiously, however, many such defence mechanisms against potentially pathogenic foreign DNA might themselves have been acquired horizontally![47] Overall, horizontally-acquired genes—many of which are of bacteriophage origin—play a huge role in defining the genome size and gene content of *E. coli* and, as we will see shortly, many other bacterial species.

The genus *Streptomyces* carries among the largest bacterial genomes known: up to ~12 Mbp, encoding in excess of 9,000 genes, but extending downwards to ~5 Mbp. This genus was the source of the antibiotic streptomycin whose discovery won Selman Waksman his Nobel Prize in 1952. Bradon McDonald and Cameron Currie, using the genomes of over 120 *Streptomyces* genomes, performed a detailed analysis of the evolution of this genus since its divergence, some 380 million years ago, from its common ancestor with the related genus *Kitasatospora*.[48] Of the ~40,000 genes[49] found in at least one *Streptomyces* genome, only ~1,000 are conserved across all the genomes. This implies a high prevalence of gene gain and loss events in this lineage. Across the vast timescale over which this genus has evolved, McDonald and Currie detected over 320,000 events of horizontal gene transfer. Nearly all genes involved in the synthesis of the so-called secondary metabolites, of which streptomycin is an example, have undergone at least one horizontal gene transfer event. Remarkably, even genes involved in core processes such as replication and translation showed signatures of horizontal transfer. This raises questions on the accuracy of species tree reconstructions, which are often based on the assumption of the predominant role of vertical gene transfer from parent to progeny in the evolution of genes involved in core genetic processes. Note however that horizontal transfer events for such core genes, whose disruption can be highly disadvantageous to the host cell, were detected only at nodes involving short evolutionary distances, suggesting that such potentially detrimental events are often reversed under natural selection. Similarly, horizontal transfers of selfish genetic elements such as transposons also lasted only over short evolutionary distances,

46 X. Wang, Y. Kim, Q. Ma, S.H. Hong, K. Pokusaeva, J.M. Sturino, and T.K. Wood, 'Cryptic prophages help bacteria cope with adverse environments', *Nature Communications* 1 (2010), 147. https://doi.org/10.1038/ncomms1146

47 E.V. Koonin, K.S. Makarova, and Y.I. Wolf, 'Evolutionary Genomics of Defense Systems in Archaea and Bacteria', *Annual Review of Microbiology* 71 (2017), 233–261. https://doi.org/10.1146/annurev-micro-090816-093830

48 B.R. McDonald and C. Currie, 'Lateral gene transfer dynamics in the ancient bacterial genus Streptomyces', *mBio* 8 (2017), e00644–17. https://doi.org/10.1128/mbio.00644-17

49 The more appropriate term would be 'gene families'. This would refer to groups of genes that show high sequence similarity and, for practical purposes, considered as a single functional group.

again indicating that these genes are also lost rapidly. Ultimately, ~10 horizontal gene transfer events per million years of evolution resulted in the transferred genes being maintained over the long term. Finally, unlike in the examples described earlier here, most transfer events in *Streptomyces* were within the genus, and often within the same sub-lineage of *Streptomyces*. Overall, there was a negative correlation between the rate of horizontal gene transfer and evolutionary distance, suggesting that transfer rates were higher between closely related genomes (Fig. 4.5). Thus, while many genes in *Streptomyces* involved horizontal transfers at some point in their evolutionary history, the paucity of acquisitions from distant bacterial lineages shows that this genus is genetically coherent.

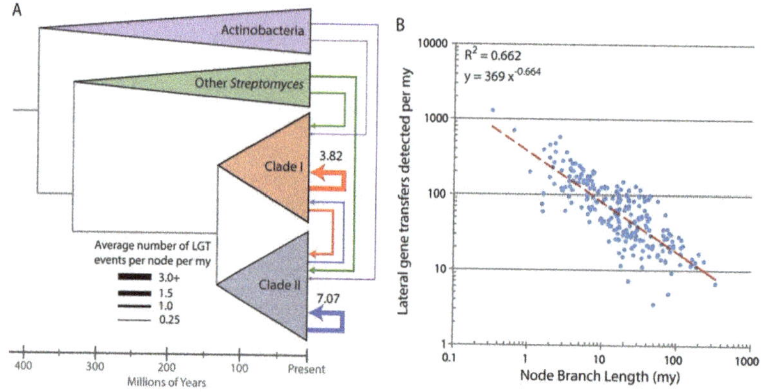

Fig. 4.5. Horizontal gene transfer in *Streptomyces*. This figure shows the rate of horizontal gene transfer in *Streptomyces* as (A) a simplified phylogenetic tree of the species, showing rates of transfers within and to a lesser extent across clades, and (B) as a scatter plot showing the rate of horizontal gene transfer (y-axis) against the phylogenetic relatedness between two *Streptomyces* genomes (x-axis). Originally published as Figure 3 in B.R. McDonald and C. Currie, 'Lateral gene transfer dynamics in the ancient bacterial genus *Streptomyces*', *mBio* 8 (2017), e00644–17, CC BY 4.0.

Over the last 2–3 decades, the role of horizontal gene transfer in determining the evolution of certain interesting gene functions and traits of many bacteria has become increasingly apparent. The most visible among these functions is antimicrobial resistance (AMR).[50] As we have seen in Chapter 1, antibiotic discoveries have always been shortly followed by discoveries of bacteria developing resistance to these drugs. However, this phenomenon and its extent was not always appreciated in its entirety. In the last three decades, antibiotic resistance has raised its hood and is threatening the very success of antimicrobial therapy with antibiotics. Grave predictions of a return to the pre-antibiotic era have been made in many quarters. The phenomenon is in the news regularly and multiple 'challenges' and calls for funding are issued in many

50 S.R. Partridge, S.M. Kwong, N. Firth, and S.O. Jensen, 'Mobile genetic elements associated with antimicrobial resistance', *Clinical Microbiology Reviews* 31 (2018), e00088–17. https://doi.org/10.1128/cmr.00088-17

countries for the development of new antibiotics that can circumvent resistance. The bottom line, and a sobering thought, is that resistance to antibiotics, if "biochemically possible ... will occur."[51]

Whether antibiotic resistance is biochemically possible or not depends on how antibiotics act and whether processes that change the genome can disrupt the mechanisms by which antibiotics act. Antibiotics typically work by deranging essential cellular processes—processes that would usually be encoded by any theoretical minimal cellular genome. However, these work by specifically inhibiting bacterial versions of these processes, thus limiting toxicity to the intended targets of the therapy. Penicillin works by inhibiting the action of an enzyme that catalyses a reaction involved in the formation of the bacterial cell wall. Streptomycin and tetracycline bind to the bacterial ribosome and inhibit translation. Rifampicin, a commonly used first-line antibiotic against *M. tuberculosis*, works by stopping transcription. Other drugs inhibit the DNA unwinding processes that enable both replication and transcription. Some antibiotics target enzymes performing metabolic reactions involved in nucleotide biosynthesis, thus again blocking the supply of raw materials for replication and transcription. Other antibiotics do not target proteins but may bind, for example, to the cell wall itself and prevent its proper formation; a powerful antibiotic that works in this manner is vancomycin. At the end of the day, all antibiotics are chemical molecules, their targets are chemical molecules, and both can be modified in one way or another by chemical processes, thus reducing the interaction between the antibiotic and its target. For example, an enzyme that can degrade an antibiotic into a non-toxic form can confer resistance. So can a protein that does not modify the antibiotic but is able to transport it out of the cell where it can do no harm. Any protein that can modify or protect a target of an antibiotic—in such a manner that the drug no longer interacts with the target—will generate resistance. Finally, mutations in a protein target of antibiotics can render the protein impervious to the antibiotic. These processes are all biochemically possible and offer a strongly positive s in environments containing antibiotics, and so they occur.[52]

Enzymes that take penicillin as a substrate and break it down into inactive products were discovered by Ernst Chain and colleagues even as they were working on the purification and scaled-up production of the antibiotic in the early 1940s! These enzymes belong to a broader class of proteins called beta-lactamases. Penicillin is only one example of a group of antibiotics called beta-lactam antibiotics. Penicillin itself may have been rendered redundant by resistance today, but other beta-lactam

---

51 J. Davies and D. Davies, 'Origins and evolution of antibiotic resistance', *Microbiology and Molecular Biology Reviews* 74 (2010), 417–433, p. 422. https://doi.org/10.1128/mmbr.00016-10

52 Recent scientific literature on antibiotic resistance is unsurprisingly vast. It is not possible to even present a representative picture of developments in this field here. I am taking the liberty of briefly presenting a few fundamental principles by which antibiotic resistance evolves, using far fewer good examples than there are. The emphasis here is on horizontal gene transfer, and we will follow one or two examples of resistance development by mutation later in this chapter. Those interested in the details are urged to refer to the Davies and Davies, 2010 review quoted above and to follow up with primary literature cited therein.

antibiotics such as the omnipresent amoxicillin are still in routine use. There are probably several hundred families of beta-lactamases in circulation today, each able to render some subset of beta-lactam antibiotics ineffective. A search of the Uniprot database, a resource for protein sequence information, for "beta-lactamase" returns over 380,000 unique sequences![53] The ubiquitous presence of these enzymes means that each prescription for amoxicillin is usually accompanied by one for clavulanic acid, a molecule that in turn inhibits some beta-lactamases! How did an enzyme, comprising of a precise sequence of amino acids and forming a structure capable of degrading beta-lactam antibiotics, evolve so soon after the introduction of these antibiotics in medicine? The short answer is that it didn't.

Heather Allen and co-workers sequenced DNA isolated directly from the soil in a remote Alaskan landscape and found examples of diverse classes of beta-lactamases in bacteria found in this pristine environment, probably with very limited antibiotic exposure.[54] These genes, when introduced artificially into *E. coli*, immediately conferred resistance. In an earlier work, Miriam Barlow and Barry Hall had built phylogenetic trees of a class of beta-lactamases and used these trees to trace their ancestry.[55] Barlow and Hall first noted that genes encoding beta-lactamases are found both on chromosomes and on *plasmids*. Plasmids are extrachromosomal circular DNA molecules that can replicate independently of the chromosome and also transfer across bacteria. Thus, plasmids are a major vehicle for horizontal gene transfer.

While a range of computational methods (see above) are needed to infer whether a gene present on the chromosome was acquired horizontally or not, the presence of a gene on a plasmid can be readily taken as evidence of its horizontal acquisition. The movement of a gene from a chromosome onto a plasmid would prime it for horizontal transfer across bacteria, sometimes resulting in the wide dissemination of the gene to distant parts of the phylogenetic tree. Plasmids also evolve in their own way. Genes that enhance the ability of the plasmid to propagate, including antibiotic resistance genes in environments replete with antibiotics, would be welcome additions to many plasmids; accordingly, some bacteria carry superpower plasmids that carry multiple antibiotic resistance-conferring genes all in one place![56] Some plasmids can also get integrated into the host chromosome, like bacteriophage DNA, and replicate stably along with the chromosome while also retaining the ability to be excised from the chromosome and become mobile once again. Barlow and Hall showed that the phylogenetic tree of beta-lactamases was highly discordant with the species tree and

53 As on 14 July 2023. Searched using the Uniref100 filter.

54 H.K. Allen, L.A. Moe, J. Rodbumrer, A. Gaarder, and J. Handelsman, 'Functional metagenomics reveals diverse β-lactamases in a remote Alaskan soil', *ISME Journal* 3 (2010), 243–251. https://doi.org/10.1038/ismej.2008.86

55 M. Barlow and B. Hall, 'Phylogenetic analysis shows that the OXA beta-lactamase genes have been on plasmids for millions of years', *Journal of Molecular Evolution* 55 (2002), 314–321. https://doi.org/10.1007/s00239-002-2328-y

56 See, for example, C. Hennequin, V. Ravet, and F. Robin, 'Plasmids carrying DHA-1 β-lactamases', *European Journal of Clinical Microbiology and Infectious Diseases* 37 (2018), 1197–1208. https://doi.org/10.1007/s10096-018-3231-9

concluded that gram-positive bacteria likely acquired an ancestor of beta-lactamases from a clade of the phylogenetically distant gram-negative bacteria. Further, they showed that mobilisation of the beta-lactamase ancestors by their movement from chromosomes to plasmids had occurred several times in evolutionary history. These mobilisation events could have resulted in this gene spreading widely. However, these were not merely recent events triggered by human use of beta-lactam antibiotics, although recent events of antibiotic abuse have caused the rapid spread of beta-lactam resistance in antibiotic-rich environments. Instead, one of the mobilisation events likely occurred over 100 million years ago and another nearly 50 million years ago, long before the first humans were born on Earth! Finally, plasmids are not the only vehicles for horizontal transfer of beta-lactamases. There is considerable evidence that beta-lactamase genes can be transferred as part of transposons as well,[57] and some bacteriophages also carry beta-lactamase genes in their DNA.[58]

What use did ancient bacteria have for beta-lactamases? What use do bacteria in the soil have for these enzymes? As noted by the early pioneers in antibiotic discovery, many antibiotics are natural products and are produced and present in the soil even in the absence of human action. Even if the concentration of antibiotics in pristine soil is well-below therapeutic levels, these may be toxic enough for bacteria to evolve mechanisms of resistance under the principles of natural selection.[59] There is also the idea that antibiotics may serve as molecules enabling 'communication' among bacteria.[60] Are soil beta-lactamases signal jammers used by other bacteria to disrupt such communication? Whatever may be the answer, ancient beta-lactamases have provided the substrate for the evolution of modern beta-lactamases with their unique abilities to deal with high therapeutic concentrations of beta-lactam antibiotics through a process of mutation and natural selection.[61] These genes in various stages of their evolution have been widely disseminated via horizontal gene transfer.

The principle that antibiotic resistance-determining genes are disseminated by horizontal gene transfer is not limited to beta-lactamases and extends to a large spectrum of resistance genes representing a variety of mechanisms of action. One such example is tetracycline, an antibiotic that acts by inhibiting translation by binding to the ribosome.[62] Resistance to tetracycline is achieved not only by enzymes that inactivate

57 For a very early work, see M. Lafond, F. Couture, G. Vezina, and R.C. Levesque, 1989. 'Evolutionary perspectives on multiresistance beta-lactamase transposons', *Journal of Bacteriology* 171 (1989), 6423–6429. https://doi.org/10.1128/jb.171.12.6423-6429.1989

58 M. Muniesa, A. García, E. Miró, B. Mirelis, G. Prats, J. Jofre, and F. Navarro, 'Bacteriophages and diffusion of β-lactamase genes', *Emerging Infectious Diseases* 10 (2004), 1134–1137. https://doi.org/10.3201/eid1006.030472

59 D.I. Andersson and D. Hughes, 'Evolution of antibiotic resistance at non-lethal drug concentrations', *Drug Resistance Updates* 15 (2012), 162–172. https://doi.org/10.1016/j.drup.2012.03.005

60 G. Yim, H.H. Wang, and J. Davies, 'Antibiotics as signalling molecules', *Philosophical Transactions of the Royal Society B* 362 (2007), 1195–1200. https://doi.org/10.1098/rstb.2007.2044

61 M.L.M. Salverda, J.A.G.M. de Visser, and M. Barlow, 'Natural evolution of TEM-1 β-lactamase: experimental reconstruction and clinical relevance', *FEMS Microbiology Reviews* 34 (2010), 1015–1036. https://doi.org/10.1111/j.1574-6976.2010.00222.x

62 I. Chopra and M. Roberts, 'Tetracycline antibiotics: mode of action, applications, molecular biology,

the antibiotic, but also by transporter proteins, called efflux pumps, that push the antibiotic out of the cell and proteins which bind to and protect the ribosome from the inhibitory action of tetracycline. Tetracycline efflux pumps have been found encoded in a variety of plasmids, many of which also carry genes for resistance to other antibiotics, as well as on transposons. Some of these plasmids have been integrated into the host chromosome as well. Again, genes encoding ribosome protection proteins have also been found encoded on transposons and plasmids. The sole representative of proteins that can modify and inactivate tetracycline and its more modern relative tigecycline has also been described as being mobilised from transposons and plasmids.[63]

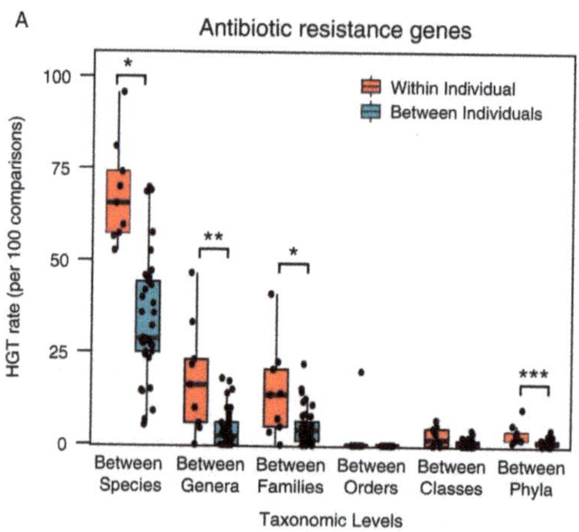

Fig. 4.6. Within-host horizontal gene transfer. This figure shows that the rate of horizontal gene transfer between members of bacterial communities within the same human host is significantly greater than that across hosts. Originally published as Figure 2A in A.G. Kent, A.C. Vill, Q. Shi, M.J. Satlin, and I.L. Brito, 'Widespread transfer of mobile antibiotic resistance genes within individual gut microbiomes revealed through bacterial Hi-C', *Nature Communications* 11 (2020), 4379, CC BY 4.0.

Thus, antibiotic resistance is a popular bacterial characteristic that is often gained by genomes via horizontal gene transfer. A very recent study sequencing bacterial DNA from the human gut by Alyssa Kent and co-workers showed that copies of the same mobilisable DNA element carrying antibiotic resistance genes were present in the genomes of multiple bacteria within the same human host, but less so across hosts[64] (Fig. 4.6). This example provides direct evidence for the power of horizontal gene transfer in spreading genes offering large s rapidly and widely across bacterial species

and epidemiology of bacterial resistance', *Microbiology and Molecular Biology Reviews* 65 (2001), 232–260. https://doi.org/10.1128/mmbr.65.2.232-260.2001

63 J. Sun, C. Chen, C.-Y. Cui, Y. Zhang, X. Liu, et al., 'Plasmid-encoded tet(X) genes that confer high-level tigecycline resistance in *Escherichia coli*', *Nature Microbiology* 4 (2019), 1457–1464. https://doi.org/10.1038/s41564-019-0496-4

64 A.G. Kent, A.C. Vill, Q. Shi, M.J. Satlin, and I.L. Brito, 'Widespread transfer of mobile antibiotic resistance genes within individual gut microbiomes revealed through bacterial Hi-C', *Nature Communications* 11 (2020), 4379. https://doi.org/10.1038/s41467-020-18164-7

residing within a local environment. Because many antibiotics are broad spectrum and can kill bacteria fairly non-specifically, the large *s* offered by antibiotic resistance genes is not species-specific, but applies to many bacterial species in a world where antibiotics are commonplace. Thus, a genetic innovation made in one bacterial species can contribute to the success of a very distant species, making horizontal transfer a favoured vehicle for its dissemination. In the absence of horizontal gene transfer, each bacterial species susceptible to antibiotics in its environment will have to evolve its own resistance mechanisms from scratch or finetune existing ancient mechanisms, so that they can deal with a sudden onslaught of large concentrations of many antibiotics, by mutation and selection—something that cannot be achieved rapidly on a bacteria-wide scale.

A second bacterial trait that is often determined by horizontally-acquired genes is virulence or pathogenicity. A classic example illustrating this is the virulence of *Vibrio cholerae*, the bacterium that causes cholera. Cholera is characterised by extreme diarrhoea and afflicts several million people, leading to thousands of deaths each year. In the 19th century, before the germ theory of disease was established, John Snow traced the origin of a cholera epidemic in London to a water pump; shutting down the pump helped to contain the epidemic. The bacterium itself was discovered independently in the second half of the 19th century by Filipo Pacini and Robert Koch.[65] Historically, there have been seven cholera pandemics worldwide, affecting every continent except Antarctica. The seventh pandemic, which began in 1961, is ongoing. Two *'biotypes'* of the cholera bacterium have been described. The sixth pandemic (1899–1923) and presumably all others prior to it were caused by what is called the *Classical* biotype, whereas the driver of the seventh pandemic belongs to the *ElTor* biotype.[66] *V. cholerae* causes disease primarily by producing and secreting a protein toxin called the cholera toxin, which was discovered in Calcutta by Shambhunath De in 1951.[67] The toxin binds to and eventually disrupts the electrolyte balance in the host intestinal cells leading to diarrhoea.

A study published by Mathew Waldor and George Mekalanos showed that the segment of DNA coding for the cholera toxin was self-transmissible across *V. cholerae* hosts.[68] The genes coding for the toxin (called ctxA and ctxB) have unusually low G+C content for the *V. cholerae* genome (Fig. 4.7). It could be excised out of the host chromosome, replicate into many copies like a plasmid, and produce bacteriophage particles capable of escaping from the host cell and transmitting the genes to other non-toxigenic *V. cholerae* bacteria. Most interestingly, the bacteriophage particles

65 D. Lippi, E. Gotuzzo, and S. Caini, 'Cholera', *Microbiology Spectrum* 4 (2016), PoH-0012–2015. https://doi.org/10.1128/microbiolspec.poh-0012-2015

66 M. Claeson and R. Waldman, 'Cholera through history', *Brittanica* (2024). https://www.britannica.com/science/cholera/Cholera-through-history

67 A. Sen and J.K. Sarkar, 'Life and work of Shambhu Nath De', *Resonance* (2012), 943–954. https://doi.org/10.1007/s12045-012-0108-6

68 M.K. Waldor and J.J. Mekalanos, 'Lysogenic conversion by a filamentous phage encoding cholera toxin', *Science* 272 (1996), 1910–1914. https://doi.org/10.1126/science.272.5270.1910

formed are of a *filamentous* type which, uniquely among bacteriophages, do not lyse the host cell and are instead excreted out! Thus, the major virulence factor for cholera is transmitted horizontally among *V. cholerae* by a filamentous bacteriophage that imposes little burden on the host cell by not killing it. Bacteriophages, as noted by d'Herelle[69] himself, are specific to a narrow range of hosts. Thus, the scope for bacteriophage-mediated horizontal transfer of genes would mostly be across members of the same species or genus of bacteria.

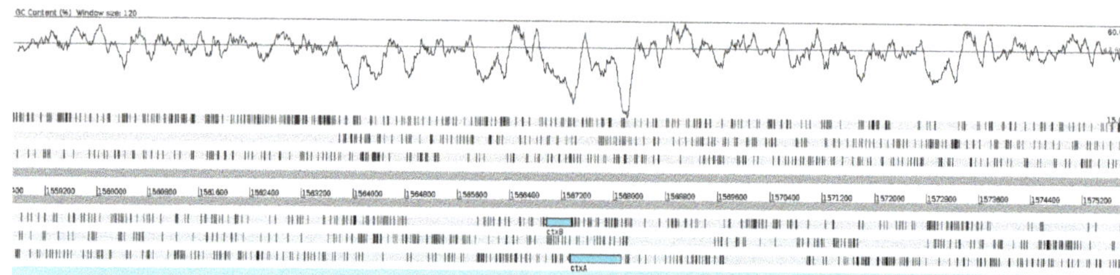

Fig. 4.7. Cholera toxin in *V. cholerae*. This figure represents a short fragment of the genome of *V. cholerae* ElTor. The fragment includes the cholera toxin genes (ctxA and ctxB, highlighted in cyan). The track at the top gives shows the local G+C content, with the horizontal line marking the genomic average (~47%). Note the extraordinarily low G+C content of ctxA and ctxB compared with the genomic average.

Specificity of a bacteriophage to a host arises from the host expressing certain *receptors* for the bacteriophage on its host surface. A protein called '*Tcp*', forming a surface structure called a pilus, is the receptor for the filamentous bacteriophage encoding the cholera toxin. It turns out that this receptor is also encoded on a segment of the chromosome that might be of bacteriophage origin.[70] Thus, the evolution of the virulent *V. cholerae* might have been initiated by the acquisition of the bacteriophage DNA encoding Tcp receptors, which would have then made the acquisition of the genes for the cholera toxin possible. Both these steps are products of horizontal gene transfer. *V. cholerae* strains with the cholera toxin but lacking the receptor have been described. One explanation for their presence would be that the DNA encoding Tcp was lost after the acquisition of the cholera toxin genes. Another possibility is that the cholera toxin can be transmitted from the *V. cholerae* strain to another by other bacteriophages that do not require the pilus for infection,[71] by a mechanism called *generalised transduction* in which a bacteriophage transmits not (only) bacteriophage DNA but any suitably-sized fragment of the bacterial chromosome horizontally to recipient cells.

A very recent study analysed the genomes of multiple members of the *V. cholerae* species to trace the evolution of the causative agent of the seventh pandemic, which, as

69 See Chapter 2.

70 D.K.R. Karaolis, J.A. Johnson, C.C. Bailey, E.C. Boedecker, J.B. Kaper, and P.R. Reeves, 'A Vibrio cholerae pathogenicity island associated with epidemic and pandemic strains', *Proceedings of the National Academy of Sciences USA* 95 (1998), 3134–3139. https://doi.org/10.1073/pnas.95.6.3134

71 E.F. Boyd and M.K. Waldor, 'Alternative mechanism of cholera toxin acquisition by Vibrio cholerae: generalized transduction of CTXΦ by bacteriophage CP-T1', *Infection and Immunity* 67 (1999), 5898–5905. https://doi.org/10.1128/iai.67.11.5898-5905.1999

noted above, belongs to the ElTor biotype of the species.[72] This study was motivated by the sequencing of the genomes of several pre-pandemic ElTor isolates, which—in the first half of the 20th century—were discovered in the Middle East. Isolates dating back to the turn of the century did not appear to have caused cholera in their hosts, whereas those isolated in the mid-20th century caused severe cholera but did not transmit from one host to another as well as the pandemic strains. The analysis of these genomes suggested several stages in the evolution of the seventh-pandemic ElTor strain, mostly driven by horizontal gene transfer. Around 1902–1903, these isolates acquired an ElTor form of the Tcp pilus receptor of the bacteriophage carrying the cholera toxin. Shortly after, a version of the cholera toxin that was similar to those circulating in the sixth-pandemic Classical biotype was acquired. These events would have made the non-pathogenic ElTor *V. cholerae* ancestor virulent. Much later, between 1939 and 1954, the ElTor version of the cholera toxin was acquired. This version of the cholera toxin differs from its Classical relative in the protein subunit responsible for the binding of the toxin to the host cell; one can surmise that the ElTor version is better at this than the Classical variant. At around the same time, two more genetic elements, termed VSP-I and VSP-II, were acquired. These elements together help to protect *V. cholerae* from predatory bacteriophages circulating in the human gut[73] and thus enhance its ability to survive, cause disease, and transmit. Thus, the emergence of the seventh-pandemic ElTor *V. cholera* seems to have been decided ultimately by two horizontal gene transfer events: (i) the conversion of the Classical-type cholera toxin to the ElTor type which might bind more efficiently to intestinal cells, and (ii) the acquisition of additional genetic elements that at the very least enhance the bacterium's defence against predatory bacteriophages.

V. cholerae is not all about cholera. It is also about cannibalism. *V. cholerae* lives a major chunk of its life not causing disease in humans, but in an aquatic environment in association with zooplanktons. The surface on which *V. cholerae* lives is rich in the molecule chitin.[74] While growing on chitin-rich surfaces, a population of *V. cholerae* expresses a secretion system called the Type 6 Secretion System on the cell surface. The secretion system is almost certainly horizontally acquired. This secretion system attaches the host cell to other neighbouring bacterial (including other susceptible *V. cholerae* and eukaryotic) cells using machinery that resembles the part of a bacteriophage that attaches itself to prey bacteria. Once attached, the predatory bacterium injects various proteins into the prey cell through the secretion system. These proteins eventually kill the prey. Cells that produce the secretion system themselves are protected from such a fate.

72 D. Hu, B. Liu, L. Feng, P. Ding, X. Guo, M. Wang, B. Cao, P.R. Reeves, and L. Wang, 'Origins of the current seventh cholera pandemic', *Proceedings of the National Academy of Sciences USA* 113 (2016), E7730–E7739. https://doi.org/10.1073/pnas.1608732113

73 B.J. O'Hara, M. Alam, and W.-L. Ng, 'The Vibrio cholerae Seventh Pandemic Islands act in tandem to defend against a circulating phage', *PLoS Genetics* 18 (2022), e1010250. https://doi.org/10.1371/journal.pgen.1010250

74 E.K. Lipp, A. Huq, and R.R. Colwell, 'Effects of global climate on infectious disease: the cholera model', *Clinical Microbiology Reviews* 15 (2002), 757–770. https://doi.org/10.1128/cmr.15.4.757-770.2002

The cellular contents of the prey, including its DNA, are released into the environment. Now, the segment of DNA that encodes the secretion system also codes for a set of genes that enables *V. cholerae* to take up DNA from the environment. This means that when the secretion system is expressed and operates to kill prey bacteria, the killer cells are also primed to take up any DNA that might be released into the environment.[75] Some of this acquired DNA might have played a role in evolving virulence. Even if the DNA that is taken up does not become part of the host genome, it can be broken down and used for nutrition, a phenomenon demonstrated in various bacteria.[76] Thus, a horizontally-acquired DNA segment enables horizontal acquisition of more DNA while *V. cholerae* are growing in their natural environments. In summary, the evolution of *V. cholerae*—its virulence in particular—is heavily driven by horizontal gene transfer. This can be mediated by multiple mechanisms, from bacteriophages that do not kill their host cells to fratricide that enables consumption of DNA released into the environment. This also illustrates the general phenomenon in which the acquisition of one segment of DNA by horizontal gene transfer opens the floodgates and enables the acquisition and establishment of others, which together define the traits of the host bacterium!

Another example of horizontal gene transfer determining virulence is the case of the genus *Salmonella*. This genus diverged from its common ancestor with *E. coli* over a 100 million years ago. Since then, it has diverged into two species, *S. enterica* and *S. bongori*. *S. enterica* now comprises of many *'serotypes'* representing a range of host specificities and intestinal and systemic pathologies including typhoid in humans.[77] Shortly after divergence from *E. coli*, *Salmonella* acquired a genetic element called the SPI-1 pathogenicity island through a plasmid or a bacteriophage. All varieties of *Salmonella* encode this genetic element, which is responsible for the bacterium successfully colonising the host intestine. After divergence from *S. bongori*, *S. enterica* acquired another genetic element called the SPI-2 pathogenicity island. This genetic element is integrated into the host chromosome at a site usually associated with the integration of bacteriophage DNA. Thus, it is very likely that SPI-2 is of bacteriophage origin. SPI-2 contributes hugely to post-intestinal stages of *S. enterica* pathology, to such an extent that the loss of this element reduces virulence as much as 10,000-fold! Further genetic diversification of the *S. enterica* lineage into various host specificities and disease outcomes appears to have involved several additional mutations and acquisitions. One example is the acquisition of what is called the *'spv operon'* via plasmid-mediated horizontal gene transfer in some sub-lineages of *S. enterica*. This operon appears to play a role in non-typhoidal systemic

75 S. Borgeaud, L.C. Metzger, T. Scrignari, and M. Blokesch, 'The type VI secretion system of Vibrio cholerae fosters horizontal gene transfer', *Science* 347 (2015), 63–67. https://doi.org/10.1126/science.1260064

76 S.E. Finkel and R. Kolter, 'DNA as a nutrient: novel role for bacterial competence gene homologs', *Journal of Bacteriology* 183 (2001), 6288–6293. https://doi.org/10.1128/jb.183.21.6288-6293.2001; E.S. Antonova and B.J. Hammer, 'Genetics of natural competence in Vibrio cholerae and other Vibrios', *Microbiology Spectrum* 3 (2015), VE-0010–2014. https://doi.org/10.1128/microbiolspec.ve-0010-2014

77 A.J. Baumler, R.M. Tsolis, T.A. Ficht, and L.G. Adams, 'Evolution of Host Adaptation in Salmonella enterica', *Infection and Immunity* 66 (1998), 4579–4587. https://doi.org/10.1128/iai.66.10.4579-4587.1998

disease caused by *S. enterica*. Thus, the emergence of virulence at various stages in the evolution of *Salmonella* appears to have been driven primarily by horizontal gene transfer.

Horizontal gene transfer plays important roles not only in the evolution of processes that impact pathogen success, but also metabolism, including core metabolic processes that sustain all life! A large fraction of life forms on Earth today need oxygen. The oxygen that is depleted as a result of life processes is replenished by the biochemical process of photosynthesis, most famously performed by plants. During photosynthesis, plants use light energy to eventually oxidise water to release oxygen. For the photosynthesising cell, this process generates energy in the form of ATP molecules. Early life evolved in a world devoid of oxygen. The first organisms to perform photosynthesis were not plants, which evolved much later. Bacteria belonging to the cyanobacterial clade were and are photosynthetic organisms.[78] In fact, the plant and algal cell organelle chloroplast, in which photosynthesis occurs, is a highly reduced cyanobacterial endosymbiont of plant cells,[79] at least from the point of view of its DNA. Cyanobacteria are an ancient bacterial lineage but are not the only lineage capable of performing photosynthesis. The ability to perform photosynthesis is distributed very sparsely across the bacterial tree with at least five groups of bacteria capable of performing this process. A very early analysis of the complete genomes of bacteria from these five photosynthetic lineages by Jason Raymond and colleagues showed that gene trees built from the sequences of genes involved in phototrophy/photosynthesis were inconsistent with the species tree assembled from ribosomal RNA sequences.[80] This suggested that the ability to perform photosynthesis might have been disseminated among these bacteria by horizontal gene transfer. Studies published a few years later by Debbie Lindell et al., reporting the sequencing of bacteriophages that infect the cyanobacteria *Prochlorococcus* and *Synechococcus*, showed that genes that are part of the *light harvesting machinery* in these bacteria are encoded in the bacteriophage DNA (Fig. 4.8).[81] While these findings do not necessarily show that components of the

78 Cyanobacterial photosynthesis is a complex process involving a profound array of molecules. It is indeed very likely that the first photosynthetic organisms were not cyanobacteria as they are now, but organisms that could perform a more basic phototrophic metabolism.

79 Recent evidence suggests that the DNA of extant chloroplasts might be a chimaera derived from multiple cyanobacterial lineages. N. Sato, 'Are Cyanobacteria an ancestor of chloroplasts or just one of the gene donors for plants and algae?', *Genes* (*Basel*) 12 (2021), 823. https://doi.org/10.3390/genes12060823

80 J. Raymond, O. Zhaxybayeva, J.P. Gogarten, S.Y. Gerdes, and R.E. Blankenship, 'Whole-genome analysis of photosynthetic prokaryotes', *Science* 298 (2002), 1616–1620. https://doi.org/10.1126/science.1075558

81 D. Lindell, M.B. Sullivan, Z.I. Johnson, A.C. Tolonen, F. Rohwer, and S.W. Chisholm, 'Transfer of photosynthesis genes to and from Prochlorococcus viruses', *Proceedings of the National Academy of Sciences USA* 101 (2004), 11013–11018. https://doi.org/10.1073/pnas.0401526101; M.B. Sullivan, D. Lindell, J.A. Lee, L.R. Thompson, J.P. Bielawski, and S.W. Chisholm, 'Prevalence and evolution of core Photosystem II genes in marine cyanobacterial viruses and their hosts', *PLoS Biology* 4 (2006), e234. https://doi.org/10.1371/journal.pbio.0040234; G. Zeidner, J.P. Bielawski, M. Shmoish, D.J. Scanlan, G. Sabehi, and O. Béjà, 'Potential photosynthesis gene recombination between Prochlorococcus and Synechococcus via viral intermediates', *Environmental Microbiology* 7 (2005), 1505–1513. https://doi.org/10.1111/j.1462-2920.2005.00833.x

photosynthetic apparatus originally evolved in bacteriophages, these indicate that the bacteriophages might have, at some point in evolutionary history, captured a piece of bacterial DNA encoding these genes; these sequences might have undergone further changes in the bacteriophage DNA while being transferred to other cyanobacteria. For example, a gene encoding a protein called *Hli* that protects the photosynthetic machinery from excess excitation under bright light is present in multiple but divergent copies in some *Prochlorococcus* but not others. The Hli sequences present in the bacteriophages appear to be closest not to the original, ancestral copy of Hli that is conserved across *Prochlorococcus* but to the extra copies present in a subset of *Prochlorococcus*. The divergence in sequence between the ancestral and the newer copies of Hli seems to have occurred in the bacteriophage.[82] Thus, expansion of the Hli family of proteins in some *Prochlorococcus* is driven by bacteriophage-mediated horizontal gene transfer. Encoding multiple copies of the Hli family proteins might be a particular adaptation seen in *Prochlorococcus* exposed to high light intensities in the open ocean.[83] Allowing such an advantageous expansion of this family in their *Prochlorococcus* hosts in turn allows bacteriophages to successfully replicate their DNA as part of the host chromosome.

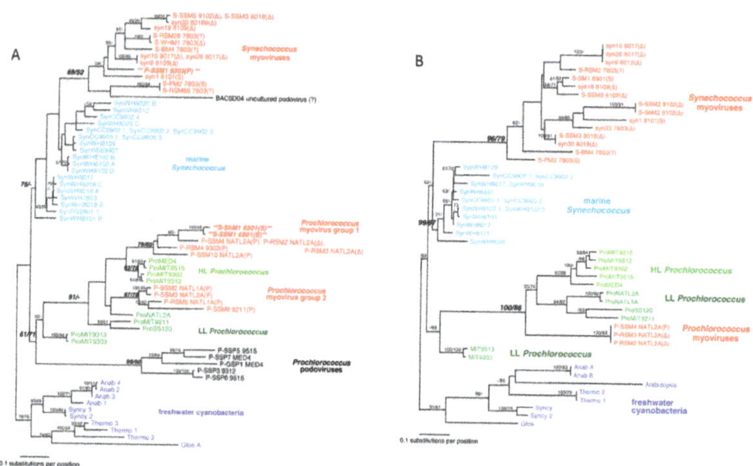

Fig. 4.8. Horizontal gene transfer of photosynthesis genes. This figure shows gene trees for two genes involved in photosynthesis, (A) psbA and (B) psbD, in *Prochlorococcus* and *Synechococcus*. These trees show the close relationship between these genes found in bacteriophages and those in the bacteria. These suggest transfer of these genes from the bacteria to the bacteriophage. This can potentially provoke further transfers to other bacteria through the bacteriophage. Originally published as Figures 1 and 2 in M.B. Sullivan, D. Lindell, J.A. Lee, L.R. Thompson, J.P. Bielawski, and S.W. Chisholm, 'Prevalence and evolution of core Photosystem II genes in marine cyanobacterial viruses and their hosts', *PLoS Biology* 4 (2006), e234, under Creative Commons Attribution License.

82 Ibid.

83 W.R. Hess, G. Rocap, C.S. Ting, F. Larimer, S. Stilwagen, J. Lamerdin, and S.W. Chisholm, 'The photosynthetic apparatus of Prochlorococcus: Insights through comparative genomics', *Photosynthesis Research* 70 (2001), 53–71. https://doi.org/10.1023/a:1013835924610

There are now several more examples of horizontally-acquired genes involved in metabolism, though that of something as fundamental as photosynthesis might be the most dramatic. In fact, metabolic genes may in general be among those most susceptible to horizontal transfers. These include members of the bacterial nitrogen fixation pathway, the process by which atmospheric nitrogen is converted into nitrates and nitrites, which can then be absorbed as nutrients by plants. Yet another example features the bacterium *Myxococcus xanthus*. A group of ~100,000 *M. xanthus* bacteria aggregate in a structured manner to form what are called fruiting bodies during starvation. These structures protect them from extreme stresses. When faced with nutrient-replete conditions, these fruiting bodies germinate into cells that grow and divide normally. An analysis of genes involved in the development of fruiting bodies showed that the sequences of several genes encoding metabolic enzymes are similar to relatives in very distant bacteria but not to those from the clade to which *M. xanthus* belongs.[84] Thus, several metabolic enzymes contributing to fruiting body development in *M. xanthus* show signatures of horizontal acquisition. We have also mentioned earlier in this chapter that many metabolic systems in *Streptomyces*, including those responsible for the synthesis of antibiotics, were horizontally acquired. It has also been suggested that nearly all metabolic innovations in the *E. coli* lineage since its divergence from *Salmonella* some 100 million years ago were driven by horizontal gene transfer.[85]

Finally, genes that enable adaptation of bacteria to their environments—regulators of gene expression and proteins involved in DNA repair, which determine adaptive trajectories by regulating the rate and types of mutations—are also often horizontally acquired. We will discuss the evolution of regulators of transcription in detail in Chapter 5. Here, we briefly describe the results of a study by Mohak Sharda and colleagues from my lab, which investigated how a certain DNA repair system called non-homologous end joining repair evolves in bacteria.[86] The major mechanism of DNA repair in bacteria involves what is called homologous recombination, in which damaged DNA is repaired using the corresponding DNA segment from another copy of the chromosome. This works well in bacteria in which DNA replication begins soon after the birth of a new cell such that a second copy of the chromosome is available for repair mediated by homologous recombination to work. On the other hand, non-homologous end-joining, which is a commonly used mechanism for DNA repair in eukaryotes, is far less used by bacteria. In fact, less than a quarter of bacterial genomes code for the non-homologous end-joining repair pathway. Bacteria that code for this pathway tend to be slow-growing and carry large genomes with high G+C content. A

84 B. Goldman, S. Bhat, and L.J. Shimkets, 'Genome evolution and the emergence of fruiting body development in *Myxococcus xanthus*', *PLoS One* 2 (2007), e1329. https://doi.org/10.1371/journal.pone.0001329

85 C. Pál, B. Papp, and M.J. Lercher, 'Adaptive evolution of bacterial metabolic networks by horizontal gene transfer', *Nature Genetics* 37 (2005), 1372–1375. https://doi.org/10.1038/ng1686

86 M. Sharda, A. Badrinarayanan, and A.S.N. Seshasayee, 'Evolutionary and comparative analysis of bacterial non-homologous end-joining repair', *Genome Biology and Evolution* 12 (2020) 2450–2466. https://doi.org/10.1093/gbe/evaa223

second copy of the chromosome may not be always available in slow-growing bacteria, rendering homologous recombination less effective. Large genomes with high G+C content could make DNA replication take longer to complete, which can also limit the availability of a full second copy of the chromosome. Finally, the phylogeny of genes involved in this repair pathway was highly incongruent with the species phylogeny. Many of these horizontal transfer events occurred not only between bacteria, but also between bacteria and archaea.[87]

Beyond examples and specific functions, how prominent across the board is horizontal gene transfer in bacterial evolution? Tal Dagan and colleagues estimated that ~80% of all genes (up to 95% and not less than ~65%) in the bacterial kingdom might have experienced horizontal gene transfer at some point in time![88] They also showed that the number of gene transfer events or the number of genes transferred per event is several times higher between pairs of organisms that are closely related than between distantly-related organisms. Examples include high levels of transfer within *Prochlorococcus* and *Synechococcus* lineages, wherein bacteriophage-mediated exchanges of photosynthesis-related genes described above are particularly notable. Horizontal gene transfers between distant bacteria are possible and may play important roles in determining the gene repertoire of large bacterial genomes as argued by Otto Cordero and Paul Hogeweg.[89] Cordero and Hogeweg showed that the number of protein families showing signatures of distant horizontal transfer is positively correlated with a measure of genome size.[90] This correlation was not linear, but was super-linear. In other words, larger genomes code for a higher *proportion* of genes acquired horizontally from distantly-related sources than smaller genomes.[91] One possible explanation for this is that complex environments like the soil, in which bacteria with large genomes thrive, support a high phylogenetic diversity of bacteria. Close encounters between phylogenetically distant bacteria in the same environment would favour gene transfer between them. Thus, a process that contributes to genome growth occurs frequently in environments that also call for their inhabitants to be endowed with large numbers of genes.

---

87 Ibid.

88 T. Dagan, Y. Artzy-Randrup, and W. Martin, 'Modular networks and cumulative impact of lateral transfer in prokaryotic genome evolution', *Proceedings of the National Academy of Sciences USA* 105 (2008), 10039–10044. https://doi.org/10.1073/pnas.0800679105

89 O.X. Cordero and P. Hogeweg, 'The impact of long-distance horizontal gene transfer on prokaryotic genome size', *Proceedings of the National Academy of Sciences USA* 106 (2009), 21748–21753. https://doi.org/10.1073/pnas.0907584106

90 Cordero and Hogeweg correlated the number of horizontally transferred gene families to the total number of families. Since the latter would be linearly correlated to genome size, we simplify the correlation to that between horizontally transferred families and genome size.

91 This appears to be inconsistent with the findings of Mcdonald and Currie which showed that most gene transfers within *Streptomyces* are across very close relatives. Unfortunately, the difference in the methods, the phylogenetic resolution, and reporting standards used in the two studies are very different. This makes a direct comparison of the two results fraught. It is indeed a concern and an opportunity that each study detecting horizontal transfers on a genome-wide scale uses a different method. Thus, we look at each study in isolation and discuss its findings as an idea and a distinct possibility.

It has been argued in the past that rampant horizontal gene transfer can render species trees—which are built from sequences of molecular chronometer genes that should reflect vertical descent and not horizontal transfer—unreliable.[92] This appears to be supported, at least superficially, by Dagan and colleagues' estimates that ~80% of all genes show some signs of horizontal transfer.[93] However, as shown by R.G. Beiko and coworkers, a fraction of genes involved in core processes do seem to be largely refractory to horizontal gene transfer.[94] For example, core components of the translation machinery show strong signatures of vertical descent. And as discussed earlier in the case of *Streptomyces*, horizontally acquired genes in core functions are often lost rapidly, possibly under selection. Thus, despite the major role of horizontal gene transfer in bacterial evolution, bacterial species trees built out of core, molecular chronometer genes remain largely coherent. Even if these genes have undergone some horizontal transfers, the signature of vertical descent is still strong enough to produce reliable species trees. What horizontal gene transfer adds though is an extra layer of connections or "highways"[95] between distantly related bacterial clades, thus creating a complex network of gene transfers on top of a species tree. The relative contribution of tree-like and highway-like modes of evolution differs across clades of bacteria, with tree-like vertical inheritance overall being the single most dominant mode.[96]

A bacterium can gain new genes by de novo innovation, horizontal gene transfer, and, thirdly, by gene duplication. While transposons have a powerful ability to duplicate themselves many times, causing massive genome expansions especially in eukaryotes, any gene can potentially get duplicated, creating *paralogs*,[97] with some probability. Once duplicated, selection on maintaining the gene sequence gets relaxed for one of the copies. Sometimes mutations in one of the copies can result in the creation of a new function, a phenomenon known as *neofunctionalization*. One can intuitively propose that gene families, which comprise of multiple genes with similar but not identical sequences, arise by gene duplication followed by neofunctionalization. However, this needs more careful evaluation. For example, we have seen that the *Hli* multigene family in *Prochlorococcus* arose in fact from multiple bacteriophage-mediated horizontal gene transfer events, and that sequence divergence of a progenitor *Hli* gene likely happened in the bacteriophage DNA and not in the bacterial genome. Sometimes a second copy of a gene may not generate a new function: instead, additional copies of a gene can just help the cell produce more of the corresponding protein. For instance, rapid growth requires very high transcription of ribosomal RNA genes, levels that cannot

92 W.F. Doolittle, 'Phylogenetic classification and the universal tree', *Science* 284 (1999), 2124–2129. https://doi.org/10.1126/science.284.5423.2124

93 Dagan et al.

94 R.G. Beiko, T.J. Harlow, and M.A. Ragan, 2005. 'Highways of gene sharing in prokaryotes', *Proceedings of the National Academy of Sciences USA* 102 (2005), 14332–14337. https://doi.org/10.1073/pnas.0504068102

95 Ibid. Term introduced in the title.

96 P. Puigbo, Y.I. Wolf, and E.V. Koonin, 'The Tree and Net Components of Prokaryote Evolution', *Genome Biology and Evolution* 2 (2010), 745–756. https://doi.org/10.1093/gbe/evq062

97 See Chapter 3.

be reached from a single copy of these genes. Thus, faster growing bacteria encode multiple copies of ribosomal RNA genes. These probably arose by duplication, but the role of horizontal gene transfer across members of the same species cannot be ruled out either.

Duplications can also allow an organism to deploy a capability in novel situations without requiring one copy of the gene to develop a new function. An example was discovered as part of Richard Lenski and group's LTEE experiment evolving *E. coli* in the lab for over 60,000 generations. The medium in which the bacteria were growing contained citrate, a molecule *E. coli* cannot utilise for nutrition under the aerobic conditions used in the experiment. After about 31,000 generations, a small fraction of the *E. coli* populations maintained in Lenski's laboratory evolved the ability to utilise the citrate.[98] It turned out that this novel ability was conferred by a duplication of a gene called *citT* involved in citrate metabolism.[99] The second copy of *citT* was positioned downstream of the original *citT*, but in a sequence context that allowed this gene to be transcribed under aerobic conditions. No point mutation could confer this ability to utilise citrate under aerobic conditions to *E. coli*. The new copy could not have been acquired by horizontal transfer, which was not possible under the laboratory conditions in which the bacteria were grown. Thus, duplication appears to be the only option. The fact that this trait evolved after several tens of thousands of generations and in only one of several evolving populations also indicates that this duplication was a rare event.

Anecdotal examples of these kinds now invoke the question of the relative prevalence and importance of horizontal gene transfer and duplication in bacterial genome evolution. Todd Treangen and Eduardo Rocha set out to answer this question by examining the origin of members of multigene families in closely-related bacterial genomes.[100] Such families could have arisen by duplication (paralogs) or by horizontal transfer from closely-related bacteria (*xenologs*). To distinguish between these two possibilities, Treangen and Rocha used two parameters. First, paralogs—especially those arising from *recent* duplications, something that is forced by restricting the scope of their study to closely-related genomes—will show high sequence similarity. Second, duplications are often tandem, i.e., the two copies are likely to be found next to each other. Though it is possible that rearrangements that push these copies apart occur, these are less likely over the short evolutionary distances used by Treangen and Rocha. The study revealed that more than 90% of gene family expansions in bacteria arise by horizontal gene transfer and not duplication (Fig. 4.9). Furthermore, xenologs

98 Z.D. Blount, C.Z. Borland, R.E. Lenski, 'Historical contingency and the evolution of a key innovation in an experimental population of *Escherichia coli*', *Proceedings of the National Academy of Sciences USA* 105 (2008), 7899–7906. https://doi.org/10.1073/pnas.0803151105

99 Z.D. Blount, J.E. Barrick, C.J. Davidson, R.E. Lenski, 'Genomic analysis of a key innovation in an experimental *Escherichia coli* population', *Nature* 489 (2012), 513–518. https://doi.org/10.1038/nature11514

100 T.J. Treangen and E.P.C. Rocha, 'Horizontal transfer, not duplication, drives the expansion of protein families in prokaryotes', *PLoS Genetics* 7 (2011), e1001284. https://doi.org/10.1371/journal.pgen.1001284

are longer-lived than paralogs. Paralogs, however, are expressed at higher levels than xenologs. Finally, the percentage of mutations that are non-synonymous, and therefore particularly likely to be of consequence, is much higher in xenologs than in paralogs. These together indicate that horizontal gene transfer is the dominant force for gene gain and genome expansion in bacteria. Duplications are often discussed from the perspective of neofunctionalization. However, as the examples of ribosomal RNA and *citT* and the results of Treangen and Rocha's work suggest, the role of duplications in bacteria may not be about neofunctionalization, but about providing an extra copy to enable higher transcription, or transcription in newer contexts.

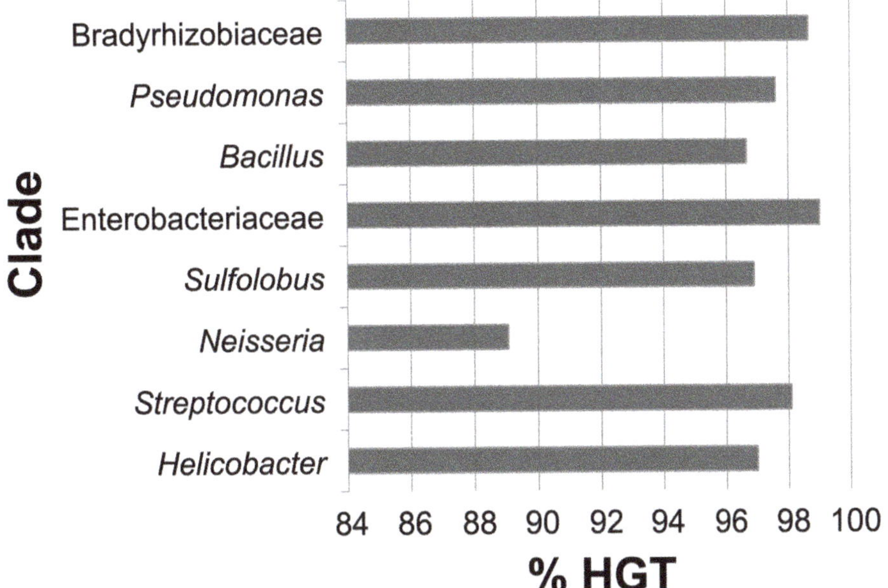

Fig. 4.9. Dominance of horizontal gene transfer in bacterial gene family expansions. This figure shows that a majority of genes representing protein family expansions in bacterial genomes were gained by horizontal gene transfer. Originally published as Figure 2 in T.J. Treangen and E.P.C. Rocha, 'Horizontal transfer, not duplication, drives the expansion of protein families in prokaryotes', *PLoS Genetics* 7 (2011), e1001284, under Creative Commons Attribution License.

Whereas the genomics era has underlined the importance of horizontal gene transfer in bacterial evolution, our knowledge of this phenomenon is not recent. In retrospect, the pneumococcal type transformation discovered by Frederick Griffith[101] in the late 1920s, which led to the demonstration of DNA being the genetic material, is a result of horizontal gene transfer of virulence-related genes from the dead, virulent pneumococci to its benign relatives. The early conflict on the nature of the bacteriophage phenomenon between Felix d'Herelle and the Belgian group in the 1920s led by Jules Bordet might have been a result of bacteriophage lysogeny, the phenomenon by which bacteriophage DNA gets integrated into the host DNA. It would take another three

101 See Chapter 2.

decades and André Lwoff's Nobel Prize-winning work[102] to explain this phenomenon. In 1946, Joshua Lederberg and Edward Tatum showed that *E. coli* that could not grow in the absence of specific nutrients would start doing so in the presence of other *E. coli* that could.[103] The ability to do so, once metabolite transfer between the two through the growth medium could be ruled out, would necessarily require transfer of genetic material from the latter to the former, presumably also requiring physical contact between the two. This phenomenon came to be known as conjugation. Shortly after, in 1952, Norton Zinder and Lederberg showed that a gene exchange phenomenon they described in *Salmonella* occurred in the absence of direct physical contact between bacteria, and that it was mediated by a filtrable agent with characteristics consistent with those of a bacteriophage.[104] Thus, the phenomenon of transduction by which bacteriophages act as vehicles of horizontal gene transfer was recognised. Many later discoveries in bacterial genetics came to be through the use of transformation, conjugation, and transduction as genetic tools by researchers. The fact that antibiotic resistance could be transferred across bacteria was discovered as early as in 1959![105] Thus, horizontal gene transfer not only plays a central role in the evolution of bacteria but has also enabled key discoveries that drove molecular biology in the 20th century. However, the extent of its prevalence would not be fully appreciated until the genomic era.

4.3. Changing DNA

While gene loss and gain by horizontal gene transfer are important mechanisms underlying genome growth and contraction, genomes also change in sequence through the humbler mutation, by which we refer to base substitutions, short insertions, and deletions. We have discussed mutations and their consequences briefly in Chapter 3. Here we will consider *advantageous* mutations that allow bacteria to adapt to new circumstances and, in a bit more detail, non-randomness in the process of mutagenesis, which again we briefly mentioned in Chapter 3.

It is often said that most mutations are deleterious to some extent. The fact that genes involved in essential processes show fewer variations than other genes supports this idea. This idea however assumes that most genes, or most base positions in these genes, in extant genomes have reached their optimal states over the course of evolution, which is not necessarily true. Even if they are optimal for a certain circumstance, environmental conditions can change to a different state and remain in this new, unfamiliar state long enough for the bacterium to require changes in the genome to

102 Nobel Prize in Physiology or Medicine, 1965.

103 J. Lederberg and E.L. Tatum, 'Gene recombination in *Escherichia coli*', *Nature* 158 (1946), 558. https://doi.org/10.1038/158558a0

104 N.D. Zinder and J. Lederberg, 'Genetic exchange in Salmonella', *Journal of Bacteriology* 64 (1952), 679–699. https://doi.org/10.1128/jb.64.5.679-699.1952

105 Via P.M. Hawkey, 'The origins and molecular basis of antibiotic resistance', *The BMJ* 5 (1998), 657–660. https://doi.org/10.1136/bmj.317.7159.657

adapt. These alterations to the genome can of course occur through gene loss and gain mechanisms, but there is nothing really to rule out *advantageous* mutations in well-established genes which might emerge to enable the bacterium to adapt under selection.

A major source of information on such advantageous or adaptive mutations is laboratory evolution experiments,[106] akin to the long-term evolution experiment pursued by Richard Lenski's laboratory. These experiments—many of which have been reported in the scientific literature—are typically performed in conditions that do not allow horizontal gene transfer. However, gene loss and duplications, especially of transposable elements already present in the genome, are possible, as shown for deletions in *Salmonella* maintained in the lab by Nilsson and colleagues and in *E. coli* by Lenski's group.[107] However, the predominant mode of evolution in such experimental setups is mutations, often substitutions but also small insertions and deletions.

While sequencing clones of *E. coli* evolved over 50,000 generations in Richard Lenski's laboratory, Tenaillon and colleagues identified well over 14,500 substitutions.[108] Over 96% of these were found in what are called *mutator* lines. These are populations in which mutations in DNA repair pathways were present, resulting in a large increase in genome-wide mutation rates; in these populations, a large majority of mutations would not be adaptive. The question then is what is the prevalence of adaptive or advantageous mutations in non-mutator and pre-mutator (prior to the emergence of mutations in DNA repair and the consequent acceleration in mutation rate) lines? To address this, Tenaillon and colleagues separately measured the rates of synonymous and non-synonymous mutations in these lines. Synonymous sites, despite their roles in gene expression, are under relatively less selection pressure than non-synonymous sites. Rates of synonymous mutations therefore would be a reflection, though not necessarily an accurate measure, of basal mutation rates. Given the knowledge of the genetic code and the rate of synonymous mutation, one can then derive an estimate of the expected rates of non-synonymous mutations in the absence of selection. Any deviation in the observed rate of non-synonymous substitutions from this expectation would indicate selection. A high rate of observed non-synonymous mutations suggests high prevalence of adaptive or advantageous mutations.

Tenaillon and colleagues found that in the first 500 generations, the rate of non-synonymous substitution was over 17 times higher than expected from the rate of synonymous substitutions. This declined to ~3.5 times over 50,000 generations. These findings suggested that early adaptation to the experimental environment was driven by a large excess of adaptive mutations. Though their prevalence decreased over time, a fraction of these still conferred an advantage to the bearer through 50,000 generations and more. Additionally, *intergenic* mutations—mutations in DNA regions

106 As pointed out in Chapter 2, the first laboratory evolution experiment to my knowledge was reported by Frederick Twort way back in 1907.

107 Nilsson et al., 2005.

108 Tenaillon et al., 2016.

between genes—also occurred at a higher rate than synonymous mutations, indicating an adaptive role for these as well. Further, several mutations had arisen independently in multiple populations. This is not something that one would expect under neutral drift, so it corroborates the adaptive nature of these substitutions.

While comparisons of rates of synonymous and non-synonymous mutations work for substitutions, they do not apply to insertions and deletions. To test whether these kinds of mutations were also adaptive, Tenaillon and colleagues compared their rates with those observed in mutation-accumulation experiments in which bacteria are evolved under little or no selection pressures. These are similar to those performed by Nilsson and colleagues in measuring rates of DNA loss in *Salmonella* growing in the laboratory. These analyses showed that the rates of insertions and deletions (as well as those of non-synonymous and intergenic mutations) were several times higher in the long-term experimental evolution populations than in the mutation-accumulation lines. Thus, evolution under selection often results in advantageous mutations, more so early on in evolution when the gap between the maximum achievable fitness and that already achieved is the greatest.

Antibiotic resistance is a characteristic that bacteria can acquire not only by horizontal transfer but also by mutations in the genome. Many laboratory evolution experiments have focused on the discovery of mutations that confer antibiotic resistance. Aalap Mogre in my lab performed one such study, which was published in 2014.[109] He grew *E. coli* in low concentrations of the antibiotic kanamycin that inhibits translation and asked where mutations conferring resistance occurred and whether these mutations would be detrimental to the bacteria if the antibiotic was removed from the medium. He found that mutations in a protein called EF-G, which is part of the translation machinery, confer to the host bacterium a degree of tolerance to the antibiotic. EF-G, as expected from its role in protein synthesis, is a highly conserved, essential protein. The amino acid sites which, when mutated, confer kanamycin tolerance, are highly conserved across bacteria as well. So, does this mutation come with a cost? In other words, does the mutation enable the bacteria to grow in the presence of an antibiotic that inhibits translation, but in the absence of the antibiotic would be detrimental? It turned out that this was not the case, at least in the conditions used in the laboratory. In the absence of the antibiotic, the mutants competed effectively with their parents, which lacked the mutations, for growth. Further, there was little evidence that the mutation compromised protein synthesis in the absence of the antibiotic. Thus, this work showed that adaptive mutations can arise even in highly conserved residues of essential proteins, and that these may not necessarily impose a great cost when the agent of selection in which the mutation arose is removed.

Antibiotic resistance via mutations happens not only in laboratory settings but in the wild as well. A particular example for this is *Mycobacterium tuberculosis*. *M. tuberculosis*, with its unique cell surface, is intrinsically resistant to many antibiotics. Via mutations in the chromosome, the bacterium—which experiences little horizontal transfer—gains resistance

109 A. Mogre, T. Sengupta, R.T. Veetil, P. Ravi, and A.S.N. Seshasayee, 'Genomic analysis reveals distinct concentration-dependent evolutionary trajectories for antibiotic resistance in *Escherichia coli*', *DNA Research* 21 (2014), 711–726. https://doi.org/10.1093/dnares/dsu032

to antibiotics which are effective against it despite intrinsic resistance.[110] Rifampicin is a commonly used anti-*M. tuberculosis* antibiotic. It acts by inhibiting transcription. *M. tuberculosis* gains resistance to rifampicin through mutations in the catalytically active subunit of RNA polymerase, the large multi-subunit enzymatic machinery that performs transcription. In fact, even *E. coli*, in the laboratory, acquires resistance to rifampicin very quickly by finding these RNA polymerase mutations. These mutations reduce the affinity of the RNA polymerase to rifampicin, thus reducing the effectiveness of the drug. With RNA polymerase being an essential enzyme, these mutations may come at a cost. This is in fact the case for several rifampicin resistance-conferring RNA polymerase mutations. However, further evolution allows the bacterium to find *compensatory* mutations, or adaptive mutations that reduce the cost of the resistance-conferring mutation.[111]

Other antibiotics against *M. tuberculosis* are what are called *prodrugs*. They are administered in an inactive form, but enzymes encoded by the bacterium itself convert them into the active substance. Examples of such drugs include pyrazinamide and isoniazid. In these drugs, mutations that abrogate the ability of bacterial enzymes to activate the prodrug can confer resistance. Because the space of inactivating mutations is generally larger than those that specifically modify activity, mutations that destroy the activity of these prodrug-activating enzymes should be easy to find. In the case of pyrazinamide, the enzyme PncA that activates the drug is not essential to the bacterium's survival. Thus, a large variety of inactivating mutations in this gene can lead to effective resistance, making the evolution of resistance to pyrazinamide relatively easy. In contrast, the enzyme KatG that processes isoniazid into its active form is required for the multiplication of the bacterium inside host cells. Therefore, any mutation in this gene that can confer effective resistance without interfering with the bacterium's survival in the host environment should not reduce its primary activity. This limits the space of mutations that are available for isoniazid resistance, but does not shrink it to zero. In fact, a mutation in the gene encoding KatG that confers isoniazid resistance appears to come with little cost. Isoniazid resistance also arises from mutations in the antibiotic's target protein; as well as from mutations that increase the expression of the target gene, which allows a proportion of the protein molecules to remain functional even in the presence of the antibiotic.[112]

Thus, antibiotic resistance in *M. tuberculosis* proceeds via mutations that reduce the effectiveness of the drug in one of multiple ways. The ease of finding such mutations depends on the space available for them, the cost they impose on the bacterium, and the ability of the bacterium to find compensatory adaptive mutations that reverse the cost of a resistance-conferring mutation.

110 S.M. Gygli, S. Borrell, A. Trauner, and S. Gagneaux, 'Antimicrobial resistance in Mycobacterium tuberculosis: mechanistic and evolutionary perspectives', *FEMS Microbiology Reviews* 41 (2017), 354–373. https://doi.org/10.1093/femsre/fux011; E.S. Kavvas, E. Catoiu, N. Mih, J.T. Yurkovich, Y. Seif, N. Dillon, D. Heckmann, A. Anand, L. Yang, V. Nizet, J.M. Monk, and B.O. Palsson, 'Machine learning and structural analysis of Mycobacterium tuberculosis pan-genome identifies genetic signatures of antibiotic resistance', *Nat Commun.* 9 (2018), 4306. https://doi.org/10.1038/s41467-018-06634-y

111 D. Hughes and G. Brandis, 'Rifampicin resistance: fitness costs and the significance of compensatory evolution', *Antibiotics* 2 (2013), 206–216. https://doi.org/10.3390/antibiotics2020206

112 Gygli et al.

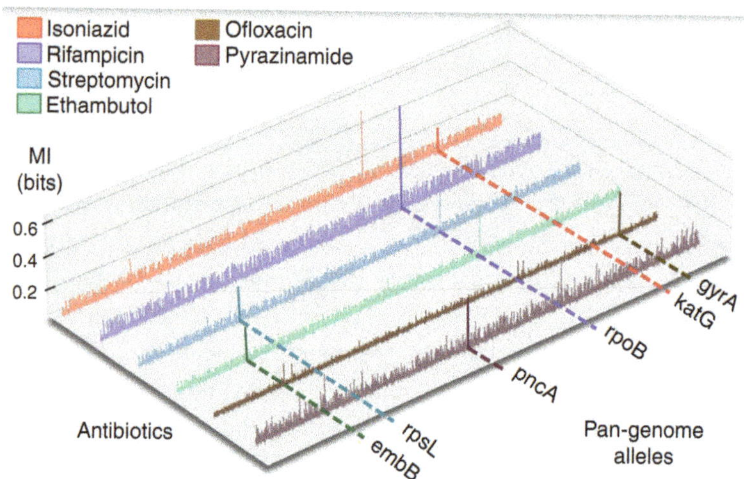

Fig. 4.10 Antibiotic resistance mutations in *Mycobacterium tuberculosis*. This figure shows correlations, as measured by mutual information, between gene sequence variants and resistance to antibiotics in *M. tuberculosis*. High vertical bars indicate high association between the two. One gene, which shows the highest correlation, has been highlighted for each antibiotic. Originally published as Figure 1 in E.S. Kavvas, E. Catoiu, N. Mih, J.T. Yurkovich, Y. Seif, N. Dillon, D. Heckmann, A. Anand, L. Yang, V. Nizet, J.M. Monk, and B.O. Palsson, 'Machine learning and structural analysis of Mycobacterium tuberculosis pan-genome identifies genetic signatures of antibiotic resistance', *Nat Commun* 9 (2018), 4306, CC BY 4.0.

Our discussion on advantageous mutations has so far focussed on those that have clearly emerged under natural selection. These are circumstances in which bacterial populations start off at a huge disadvantage relative to what they can potentially achieve with mutations in the genome. These select for adaptive mutations ($|s| > N_e$). However, a majority of mutations, especially those that occur in the absence of selection ($|s| < N_e$), are unlikely to be advantageous. What proportion of mutations are advantageous is anybody's guess at this point and is likely to vary on a case-by-case basis. However, even in the absence of selection or suboptimal changes in the environment, beneficial mutations can sometimes become accessible. A recent study by Mrudula Sane and colleagues has shown that when the types of mutations *available* to a bacterial population switch from what it has evolved under for a long time, an entirely new space for mutations opens up.[113] This space, despite the absence of any strong selective pressure, includes a high proportion of advantageous mutations. Thus, advantageous mutations often arise in the presence of strong selection pressures but can also emerge in the absence of selection when new mutational spaces open up.

Now that it is clear that mutations can be advantageous, especially under selection, can bacteria tune mutation rates so that they increase the chance of discovering advantageous mutations when circumstances so demand? Back in the 1940s, Luria

113 M. Sane, G.D. Diwan, B.A. Bhat, and D. Agashe, 'Shifts in mutation spectra enhance access to beneficial mutations', *Proceedings of the National Academy of Sciences USA* 120 (2023), e2207355120. https://doi.org/10.1073/pnas.2207355120

and Delbrueck famously used arguments from probabilities and the phenomenon of bacteria developing resistance to death by bacteriophages to show that resistance mutations exist in the bacterial genetic material at low frequencies prior to exposure to bacteriophages, and that the introduction of the killing stress imposes strong selection that results in the resistant mutants taking over the population.[114] This however does not mean that stressful environments *cannot* increase mutation rates and cause an increase in the ability of the bacterial population to find adaptive mutations.

Oliver Tenaillon and colleagues noted the emergence of mutations in genes involved in DNA repair that conferred a mutator characteristic to bacterial populations in the long-term evolution experiment.[115] Mutator lines would not be advantageous in the long run however, as these would, in most cases, eventually accumulate an excess of deleterious mutations. In the long-term evolution experiment for example, the rate at which the mutator lines accumulated mutations declined over time, possibly due to the emergence of *anti-mutator* mutations.[116] However, bacteria can achieve transient mutator states that do not call for mutations in DNA repair genes.[117] For example, a recent work showed that *E. coli* populations exposed to high ethanol stress showed higher mutation rates under high stress conditions. Bacteria that had evolved to show high ethanol tolerance also showed high transient mutation rates (Fig. 4.11).[118] One way by which this can happen is through errors during transcription and/or translation of genes involved in DNA quality control or repair. Another mechanism by which transient mutator states emerge is via what is called the SOS pathway. Some stressful environments, particularly those that result in DNA damage, activate the SOS response in *E. coli*. Among the multitude of genes that are activated in turn by the SOS response are certain error-prone variants of DNA polymerase, the enzyme involved in DNA replication. Increased activity of such error-prone DNA polymerases, which happens in a small proportion of starving *E. coli* cells, can increase mutation rates. Some repair proteins are expressed at very low levels in cells. Such low levels mean that after cell division, some daughter cells may be left without the repair protein and thus experience elevated mutation rates, at least till their own copy of the protein is synthesised.[119]

---

114 S.E. Luria and M. Delbrück, 'Mutations of bacteria from virus sensitivity to virus resistance', *Genetics* 28 (1943), 491–511. https://doi.org/10.1093/genetics/28.6.491

115 Tenaillon et al., 2016.

116 In natural populations, some DNA repair mechanisms are conserved and a limited set is part of the minimal cellular genome. Bacteria such as Buchnera that lack these mechanisms display elevated mutation rates but as we will note later in this chapter, the strength of selection in this endosymbiont appears to be weak. A small proportion of natural isolates of *E. coli* appear to be heritable mutators despite the long-term disadvantage of this trait. See P.L. Foster, 'Adaptive mutation: implications for evolution', *Bioessays* 22 (2000), 1067–1074. https://doi.org/10.1002/1521-1878(200012)22:12%3C1067::aid-bies4%3E3.0.co;2-q

117 Foster, 2000.

118 T. Swings, B. Van den Bergh, S. Wuyts, E. Oeyen, K. Voordeckers, K.J. Verstrepen, M. Fauvart, N. Verstraeten, and J. Michiels, 'Adaptive tuning of mutation rates allows fast response to lethal stress in Escherichia coli', *Elife* 6 (2017), e22939. https://doi.org/10.7554/elife.22939

119 S. Uphoff, N.D. Lord, B. Okumus, L. Potvin-Trottier, D.J. Sherratt, and J. Paulsson, 'Stochastic activation of a DNA damage response causes cell-to-cell mutation rate variation', *Science* 351 (2016), 1094–1097. https://doi.org/10.1126/science.aac9786

Thus, mutation rates need not be constant all the time within a cell or across a population of genetically identical cells. Hypermutable states can arise transiently via physiological mechanisms. Alternatively, and possibly less desirably, such states can become heritable when they are caused by mutations in genes responsible for DNA quality control. The hypermutable state, either transient or heritable, is not adaptive itself, but it increases the chance of the bacterial population finding an adaptive mutation.

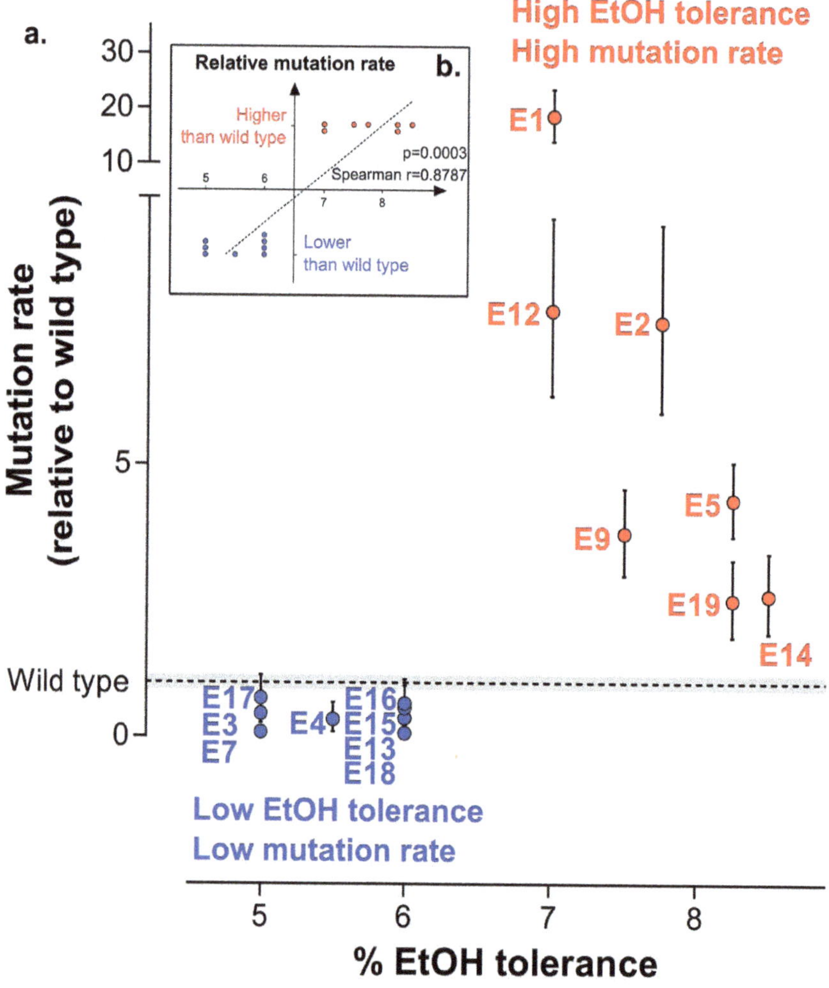

Fig. 4.11 Adaptive mutations and stress. This figure shows that bacteria showing high tolerance to ethanol, obtained after laboratory evolution, also show high mutation rates. Originally published as Figure 3 in T. Swings, B. Van den Bergh, S. Wuyts, E. Oeyen, K. Voordeckers, K.J. Verstrepen, M. Fauvart, N. Verstraeten, and J. Michiels, 'Adaptive tuning of mutation rates allows fast response to lethal stress in Escherichia coli', *Elife* 6 (2017), e22939, CC BY 4.0.

We now accept that mutations can be advantageous or adaptive, and environmental conditions as well as physiological and genetic factors can tune mutation rates to the advantage of the bacterial population. Now, are mutations equally possible across the length of the chromosome, or are different regions of the genome differently susceptible to mutations? Inigo Martincorena and colleagues analysed ~120,000 single site mutations across 34 strains of *E. coli* and measured the nucleotide diversity at synonymous sites for ~3,000 genes conserved in these genomes.[120] The idea behind measuring synonymous diversity again is that this would be a reflection of basal mutation rate, largely independent of selection. The authors found that there is a 50-fold variation in synonymous site diversity across the *E. coli* genome. Genes that are adjacent to each other on the genome, in particular those that are involved in the same function, showed similar synonymous diversity. Genes that are essential for *E. coli* growth and survival and genes that are expressed at high levels showed lower synonymous diversity.

Synonymous sites can also be under selection pressure, though not as much as non-synonymous sites. Given this incredible relationship between synonymous site diversity and gene function, a natural question that arises is whether the observed relationship is explained by selection operating on synonymous sites. While this cannot be fully ruled out, various lines of evidence suggested that the contribution of selection to the observed synonymous site diversity is minimal. One way by which selection acts on synonymous sites is through codon usage bias. Though the genetic code is redundant, not all codons coding for an amino acid are equally abundant in any genome. Changing the codon to another for the same amino acid can reduce the efficiency of translation. Martincorena and colleagues noted that these biases in codon usage did not correlate with synonymous site diversity.[121] A second route by which selection acts on synonymous sites is from the role of the sequence—including that of synonymous sites—in determining how the transcribed mRNA folds into compact structures, which can influence the loading of the ribosome onto the mRNA. Again, these researchers found little correlation between mRNA folding properties and synonymous diversity. Overall, they estimate that selection contributes to only ~5% of the variation in synonymous diversity. Thus, the conclusion from this study is that mutation rates can vary across the chromosome in a manner that exposes the more important genes to lower rates of mutation.

What might be the cause of such variation in mutation rates across the genome? This remains unclear. One possibility, for which there is some evidence, is that proteins which bind to the DNA—for example to compact it such that the DNA molecule can fit inside the confines of the cell and/or regulate transcription—have a say.[122] Bacterial genomes encode several such proteins called *nucleoid associated proteins*. Regions of the DNA bound by these proteins can either be protected from agents that cause mutations, or at the other end of the spectrum they can be inaccessible to DNA repair proteins. In

120 I. Martincorena, A.S.N. Seshasayee, and N.M. Luscombe, 'Evidence of non-random mutation rates suggests an evolutionary risk management strategy', *Nature* 485 (2012) 95–98. https://doi.org/10.1038/nature10995

121 Ibid.

122 There is strong evidence accumulating in favour of this argument in cancer cells, but not so much in bacteria.

the former case, regions bound by these proteins would show reduced mutation rates and in the latter, higher mutation rates. Tobias Warnecke and colleagues measured the rate at which different types of base changes occur and asked how these differed between regions bound and not bound by a set of nucleoid associated proteins.[123] Overall, they found that regions bound by these proteins showed lower mutability than those not bound. However, the difference was relatively small, though apparently larger than one would expect by random chance, for synonymous sites.

Interestingly, the relationship between the binding of these proteins to the DNA and mutability varied with the phase of growth. *E. coli*, when introduced to a fresh growth medium, shows a short lag phase, during which it physiologically adapts to the new environment, followed by a period of exponential growth. As nutrients deplete, growth slows down until there is zero net growth in what is termed as the stationary phase. The effect of these proteins in reducing the mutation rate appeared to be the strongest during the stationary phase. This applied particularly to cytosine-to-thymine mutations. For several other types of mutations, sites bound by nucleoid associated proteins seemed to show higher mutation rates during periods of rapid growth, which is when major mechanisms of DNA repair are fully active. Under these conditions, the binding of the nucleoid associated proteins might be blocking DNA access to the repair machinery. These findings show that nucleoid associated proteins, though not strong predictors of mutation rates, might have small yet important roles in directing mutations to specific regions of the bacterial genome.

There are other means by which certain regions of the genome can experience higher mutation rates than the rest. One way involves what are called *contingency loci*. These contingency loci comprise *tandem repeats* of one or more nucleotides—for example, a stretch of 10 cytosines (a cytosine is repeated in tandem 10 times). During replication, as the DNA unwinds and reanneals with the newly synthesised strand, misalignments can occur between the two strands at such sites so that the newly synthesised strand contains insertions or deletions of one or more 'copies' of the repetitive element. When this happens within a body of a gene, it usually results in the introduction of a premature STOP codon, thus often inactivating the protein. These mutations are highly reversible: any change in the length of the repeat can be reversed during any future round of replication.[124]

Julian Parkhill and colleagues, while sequencing the genome of the intestinal pathogen *Campylobacter jejuni*, found several such contingency loci.[125] Remarkably, they found that expansions or contractions of these sites can occur at a high frequency of one every twenty genomic molecules sequenced. These loci were often found in genes encoding cell

123 T. Warnecke, F. Supek, and B. Lehner, 'Nucleoid-associated proteins affect mutation dynamics in *E. coli* in a growth phase-specific manner', *PLoS Computational Biology* 8 (2012), e1002846. https://doi.org/10.1371/journal.pcbi.1002846

124 Hood, D., and R. Moxon, 'Gene variation and gene regulation in bacterial pathogenesis', in D.A. Hodgson and C.M. Thomas (ed.) *Signals, Switches, Regulons and Cascades* (Cambridge: Cambridge University Press, 2002).

125 J. Parkhill, B.W. Wren, K. Mungall, J.M. Ketley, C. Churcher, et al., 'The genome sequence of the food-borne pathogen *Campylobacter jejuni* reveals hypervariable sequences', *Nature* 400 (2000), 665–668. https://doi.org/10.1038/35001088

surface structures. These structures are *antigenic*, and are usually recognised by the host immune system as foreign. Their recognition by the immune system eventually results in the pathogenic bacterium being targeted for killing. However, the sub-population of cells with expansions or contractions of the contingency loci within these genes would not produce the antigen, or would produce an altered version of it that is no longer recognised by the immune system. This therefore emerges as a major strategy which this bacterium (among others, including *Haemophilus influenzae* and the gastric pathogen *Helicobacter pylori*) uses for immune evasion and thus effective virulence.

A second mechanism by which mutation rates can be increased at certain sites is via certain DNA modifications. In bacteria, enzymes called *DNA methyltransferases* add a methyl group to adenine and cytosine bases. This happens not at all adenines and cytosines—that is, not at random sites. Instead, each methyltransferase recognises short sequence motifs at which it modifies a cytosine or an adenine. In *E. coli*, the adenine methyltransferase Dam methylates the adenine at the GATC (guanine-adenine-thymine-cytosine) motif, whereas the cytosine methyltransferase Dcm methylates the internal cytosine at CCAGG and CCTGG motifs. Of these two kinds of methylation, cytosine methylation is especially relevant to mutation rates. Methylated cytosine can spontaneously convert to thymine at a higher rate than unmethylated cytosine can, thus making this modified base a mutation hotspot.[126]

A recent study by Joshua Cherry showed that the rate of cytosine to thymine mutations is much higher at methylated sites than at unmethylated sites in several bacterial genomes including those of *E. coli* and *Salmonella*, indicating that the high mutability of methylated cytosines might play a role in genome evolution.[127] In 2012, Christina Kahramanoglou and colleagues showed—in a work which my lab contributed to—that most CC[A/T]GG sites in *E. coli* are methylated during the stationary phase, but that a subset are not methylated in the entire population of growing bacteria.[128] A slightly elevated proportion of sites partially methylated during growth were at non-synonymous sites compared to those fully methylated. Partial methylation was associated with an extended CCC[A/T]GG motif. During the stationary phase, mutations at methylated cytosines are corrected by a repair pathway called Vsr, but this is not expressed during growth. These findings suggest that methylation-dependent mutations might be consequential and that *E. coli* minimises the risk posed by such mutations during growth by positioning CC[A/T]GG sites in a manner that decreases the probability of their methylation.

Overall, mutations can be adaptive and mutation rates need not be constant over time or across the length of the genome.

---

126 M.G. Marinus, 'DNA methylation in Escherishia coli', *Annual Review of Genetics* 21 (1987), 113–131.

127 J.L. Cherry, 'Methylation-Induced Hypermutation in Natural Populations of Bacteria', *Journal of Bacteriology* 200 (2018), e00371–18. https://doi.org/10.1128/jb.00371-18

128 C. Kahramanoglou, A.I. Prieto, S. Khedkar, B. Haase, A. Gupta, V. Benes, G.M. Fraser, N.M. Luscombe, and A.S. Seshasayee, 'Genomics of DNA cytosine methylation in Escherichia coli reveals its role in stationary phase transcription', *Nature Communications* 3 (2012), 886. https://doi.org/10.1038/ncomms1878

4.4. The balance among events: gains, losses and mutations

Both gene loss and gene gain predominantly mediated by horizontal transfer influence genome sizes. Early analysis of pseudogenes by Alex Mira and colleagues indicated that DNA loss by deletions outweighs insertions.[129] However, their study was limited to the small subset of genomes that were available some 20 years ago. Further these were restricted to small additions and deletions within pseudogenes. Around the same time, Berend Snel and co-workers showed that within the clade of *Proteobacteria,* to which *E. coli, Salmonella, Buchnera* and *Vibrio* belong, gene loss is prevalent. Starting from the last common Proteobacterial ancestor, there might have been as many as 1,000 genes lost en route to the relatively large genome of *E. coli.*[130] In the years since the publication of these papers, the number of genomes sequenced has increased exponentially, alongside our ability to analyse patterns of gene presence and absence within complex phylogenetic trees.

Pere Puigbó and colleagues, in 2014, performed a comprehensive analysis of the rates of gene loss and gene gain in 35 groups of bacteria.[131] Each group of bacteria comprised of genomes that were closely related. While it is possible, in principle, to perform such an analysis across the bacterial tree, evolutionary reconstructions become less reliable as the phylogeny grows deeper. Thus, Puigbó and colleagues' decision to use closely related bacterial genomes for phylogenetic analysis was driven by their desire to obtain the most reliable estimates of gene loss and gain rates possible. By performing these analyses for different groups of bacteria spread across the tree, the researchers could ask how these estimates vary for different clades.

Puigbó and coworkers defined presence and absence of gene families (which may contain one or more genes, the latter being multigene families), as well as the number of members in each gene family, for all bacterial genomes in their dataset. They defined four kinds of events: (i) *gain*—in which a gene family is gained from scratch; (ii) *loss*—in which a gene family is entirely lost; (iii) *expansion*—in which new members are added to an existing family; and (iv) *reduction*—in which a family loses a subset of its members. Whether expansions were a result of horizontal acquisition or duplication was not of concern to this work. The numbers of each of these four types of events increased with increasing branch length, indicating that these more or less followed a clock-like trajectory. Most events were gains or losses: these overwhelmed family expansions and reductions by an order of magnitude. The rate of gene loss was about three times that of gene gain. Given the large difference between gain/loss and

129 Mira et al., 2001.

130 This is not a net loss. The ancestral genome was predicted to contain only 2,000 genes, compared to the 4,000–5,500 genes in *E. coli.* See B. Snel, P. Bork, and M.A. Huynen, 'Genomes in flux: The evolution of archaeal and Proteobacterial gene content', *Genome Research* 12 (2001), 17–25. https://doi.org/10.1101/gr.176501

131 P. Puigbó, A.E. Lobkovsky, D.M. Kristensen, Y.I. Wolf, and E.V. Koonin, 'Genomes in turmoil: quantification of genome dynamics in prokaryote supergenomes', *BMC Biology* 12 (2014), 66. https://doi.org/10.1186/s12915-014-0066-4

expansion/reduction, the 1.5 times higher rate of expansion over reduction would not be enough to counter the large excess of loss over gain events. Thus, the predominant mode of genome evolution in bacteria is gene loss (Fig. 4.12). Note that the overall preference for gene loss in bacterial evolution appears to be far smaller than one would expect from DNA loss measured in pseudogenes by Mira and colleagues (see previous section).[132] This suggests that the tilt towards DNA loss in non-functional DNA is compensated for to an extent by the gain of functional DNA, often via horizontal gene transfer.

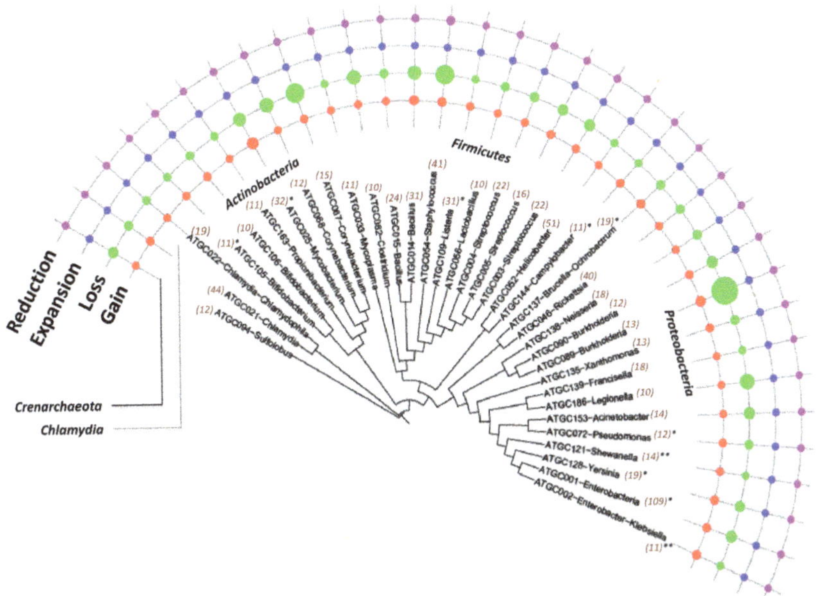

Fig. 4.12 Bacterial genome dynamics. This figure shows excess of gene loss over gene gain events in bacteria, especially in some clades such as *Burkholderia* and *Staphylococcus*. Originally published as Figure 3 in P. Puigbó, A.E. Lobkovsky, D.M. Kristensen, Y.I. Wolf, and E.V. Koonin, 'Genomes in turmoil: quantification of genome dynamics in prokaryote supergenomes', *BMC Biology* 12 (2014), 66, CC BY 4.0.

However, this did not apply across the board, with gene gain being more prominent in some clades. Further, clades showing high rates of gains on average also showed high rates of losses. This suggested that the overall genome 'flux', which is the rate of all gain, loss, expansion, and reduction events added together, is a meaningful parameter. Finally, genome size was not correlated to a balance between growth and loss processes, but was in fact positively correlated with flux.[133] Thus, the work of Puigbó and colleagues suggests that the sum of various gene gain and loss processes

132 Mira et al., 2001.

133 This may appear counterintuitive given that losses are more frequent than gains. However, this may be somewhat explained by the fact that free-living bacteria do show smaller rates of loss than parasitic bacteria with small genomes. Further, correlations are not perfect and there are always large variations, which can very well explain these non-intuitive observations.

together explains the genome size distribution of bacteria, and that genomes that undergo higher genome flux tend to be larger.

Finally, how do rates of mutation compare with those of gene loss and gain? Let us first consider data from the study by Tenaillon and colleagues.[134] Over 50,000 generations, the *E. coli* populations lost a net 63 kbp of DNA, or an average rate of ~1.2 bp per generation. Gene loss did not increase linearly with time and over the first 10,000 generations, they had lost only an average of ~4 kbp of DNA or ~0.4 bp per generation. The number of mutations per non-mutator line was ~80 over 50,000 generations, or ~0.0016 per generation. Over the first 10,000 generations, the rate was higher at ~0.002–0.003 per generation. Though the number and the rate of mutations is higher than that of gains and losses in terms of the events that cause these variations,[135] the number of bases affected by gain and loss events per generation is orders of magnitude higher than those affected by mutations.

The relationship between genome flux and mutation rate need not be consistent across bacterial clades. Some bacterial species undergo extensive horizontal gene transfer, but others evolve primarily through vertical descent. An example of the latter would be antibiotic resistance in *M. tuberculosis*, which is often determined by mutations and their vertical descent. Clonal species can show very little variation in the sequence of conserved genes, but at the same time display a great diversity of traits. Evgeni Bolotin and Ruth Hershberg analysed *genome fluidity*, as determined by differences in gene presence and absence, and sequence divergence by mutation for several pathogenic bacterial species.[136] They found that the degree of change in gene content per non-synonymous mutation varies considerably across species. Clonal species show low levels of sequence divergence but high genome fluidity. A 1% sequence divergence corresponds to a ~26% change in genome fluidity in *M. tuberculosis*, whereas the same level of sequence divergence corresponds to 5% or less fluidity in non-clonal species such as *E. coli* and *Bacillus cereus*. In the absence or relative paucity of horizontal gene transfer, genome fluidity in clonal species is driven almost entirely by gene loss. Puigbó and colleagues, while comparing the rates of gene gain and loss, also noticed that the number of gene gain or loss events are always higher than the rate of substitutions per site or per gene, and that this can vary over nearly an order of magnitude across bacterial clades.[137] Finally, in a recent study, Jaime Iranzo and colleagues showed that across nearly 35 clades of bacteria, gene sequence evolution is delayed compared to genome evolution by gene gain and loss[138] to different extents in different clades.

134 Tenaillon et al., 2016. The discussion here is based on a rough reading of the data presented by the authors in the figures published as part of their paper.

135 Each gain and loss event will affect multiple base pairs whereas mutations target a single base pair.

136 E. Bolotin and R. Hershberg, 'Gene loss dominates as a source of genetic variation within clonal pathogenic bacterial species', *Genome Biology and Evolution* 7 (2015), 2173–2187. https://doi.org/10.1093/gbe/evv135

137 Puigbó et al., 2014.

138 I. Iranzo, Y.I. Wolf, E.V. Koonin, and I. Sela, 'Gene gain and loss push prokaryotes beyond the homologous recombination barrier and accelerate genome sequence divergence', *Nature Communications* 10 (2019), 5376. https://doi.org/10.1038/s41467-019-13429-2

Taken together, gene gain and loss is a major mode by which bacterial genomes evolve. Gene loss events occur at a faster rate than gene gain. In the absence of gene gain by horizontal gene transfer, genomes would lose DNA. However, in many lineages, net gene gain is often driven by horizontal gene transfer. Bacteria that experience high genome flux, as determined by rates of gains and losses, tend to carry larger genomes. Mutations can be adaptive and non-uniform over time and across the length. All three sources of variation play roles in determining genome content to various extents in different bacterial clades.

The predominance of DNA loss in bacterial genomes and the relative under-representation of gene duplication in bacteria presents the intriguing idea that in the absence of horizontal gene transfer, which requires microbial cooperation, bacterial genomes would tend to shrink. That would mean that any given bacterium is hardly self-sufficient when it comes to discovering new functions. If, in an alternate reality (or in extended[139] laboratory evolution of pure cultures), barriers to horizontal gene transfer were strong enough to abolish it, how would selection and drift work together to enable adaptation? Would it then be predominantly driven by mutations? In any case, would genomes *have* to somehow compensate for DNA loss under drift? We don't have an answer, but this is something worth exercising those "little grey cells"[140] for.

4.5. The balance among forces: selection and drift

Eukaryotic genome evolution is dominated by gene gain, whereas bacterial genome evolution is dominated by losses. The events that drive gene gain in eukaryotes are different from those in bacteria—whereas the former gains genes by duplication, especially of selfish genetic elements, the latter often does so by horizontal gene transfer. Eukaryotes with large genomes tend to have smaller population sizes and weak overall selection[141] on their DNA. They tend to carry large amounts of the so-called 'junk DNA'. It is largely accepted that there should be an overall negative correlation between population size or strength of selection and genome size for eukaryotes. Is this true for bacterial genomes as well? Our discussion of genome content so far, in particular the fact that bacterial genomes almost always have high gene density, would indicate that this need not be so. One might intuitively expect that there should be little or no relation between genome size and the average strength of selection acting on the genome sequence. Chih-Horng Kuo and colleagues tested this idea.[142] They measured the strength of selection by calculating the ratio of non-synonymous to synonymous site substitutions (K_a/K_s), a concept we have seen several times till now.

139 Extended far beyond what Richard Lenski's LTEE has achieved so far, and possibly in dynamic environments mimicking natural situations.

140 Agatha Christie, Hercule Poirot.

141 Recall from Chapter 3 that small population sizes would allow genomes to retain DNA that are neutral or even weakly deleterious.

142 C.-H. Kuo, N.A. Moran, and H. Ochman, 'The consequences of genetic drift for bacterial genome complexity', *Genome Research* 19 (2009), 1450–1454. https://doi.org/10.1101/gr.091785.109

A very low rate of non-synonymous substitution relative to synonymous substitution (low K_a/K_s) would indicate high selection for keeping the amino acid sequence of proteins intact. Because the underlying mutation rate for synonymous and non-synonymous sites would be the same, a low K_a/K_s ratio implies that a majority of mutations in non-synonymous sites have been purged out under selection. A K_a/K_s ratio of around one would indicate low selection: non-synonymous sites get mutated as frequently as synonymous sites. In some cases, non-synonymous sites are more mutable than synonymous sites, pointing to selection for the amino acid sequence of proteins to change, indicating a predominance of adaptive mutations as described by Tenaillon and colleagues in analysing genomes from Lenski's evolving populations of *E. coli.*[143] Because adaptive mutations are relatively rare over long evolutionary timescales, K_a/K_s would be inversely proportional to the strength of selection. Most bacterial genomes have an average K_a/K_s ratio $\ll 1$, implying that non-synonymous mutations are lost under selection.

Kuo et al. measured the strength of selection for ~40 pairs of closely-related bacteria, and argued that low K_a/K_s (or high strength of selection) would point to high population sizes. They found a negative correlation between K_a/K_s and genome size, or a positive correlation between the strength of selection or population size and genome size—opposite to what has been noted for eukaryotes (Fig. 4.13). The very small range over which gene density of (free-living) bacteria varies was also correlated with the strength of selection, as would be expected.

9 out of 10 genomes with high K_a/K_s ratios in Kuo et al.'s work were "insect endosymbionts ... extremophiles ... vector-borne pathogens ... and human pathogens with limited transmission routes".[144] Thus, several small genomes, particularly those reflecting restricted habitats such as endosymbionts, show a relatively low strength of selection, suggesting that their evolution involved more neutral drift than for bacteria with larger genomes and more cosmopolitan lifestyles. This contrasts with eukaryotes, in which neutral drift causes genomes to grow in size. This may be explained by the tendency of bacterial genomes to lose DNA and that of eukaryotic genomes to expand through extensive gene duplication. Horizontal gene transfer, the major mechanism of genome expansion in bacteria, may be unavailable to endosymbionts in their very tight niches inside eukaryotic cells. They also often lack the proteins required to integrate any incoming DNA with the chromosome.

Note from Chapter 3 that Kimura's model posits that neutral or even weakly advantageous DNA can be lost from populations with very low N_e with a similar probability to disadvantageous DNA. It has been estimated that, within a host, the N_e of *Buchnera* could only be around 10–20, which is well within the region where mere drift can cause the loss of slightly beneficial genes.[145] This does not even take into

143 Tenaillon, 2016.

144 Kuo et al. 2009, p. 1451.

145 J. Perreau, B. Zhang, G.P. Maeda, and N.A. Moran, 'Strong within-host selection in a maternally inherited obligate symbiont: Buchnera and aphids', *Proceedings of the National Academy of Sciences USA*

account the predominance of DNA loss in bacterial genomes. Thus, the predominance of DNA loss, coupled with reduced selection arising from extremely low population sizes that can even cause very slightly advantageous DNA to be lost by chance, would result in small genome sizes.

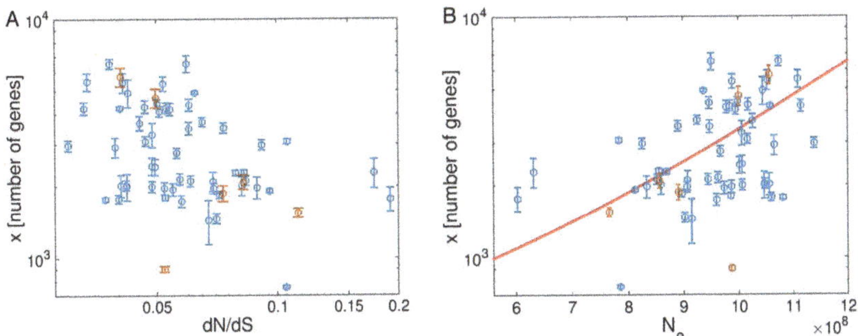

Fig. 4.13 Relationship between population size and number of genes in a genome. This figure shows the relationship between the number of genes encoded in a bacterial genome and K_a/K_s ratio (also referred to as dN/dS ratio), and estimated N_e. Originally published as Figure 1 in I. Sela, Y.I. Wolf, and E.V. Kooni, 'Theory of prokaryotic genome evolution', *Proceedings of the National Academy of Sciences USA* 113 (2016), 11399–11407, under the PNAS Open Access option.

Genome reduction is not all about loss of selection and low population sizes. Among the ~40 genome pairs investigated by Kuo et al., ~30 showed very low K_a/K_s ratios. Most such genomes were also large, typical of most bacterial lineages. However, three pairs carried small genomes at less than 2 Mbp. These included *C. jejuni*, which we have seen in the context of hypermutable contingency loci; the cyanobacteria *Prochlorococcus*, one of our examples for both gene loss and gene gain by horizontal gene transfer; and *Mycoplasma*, which carry the smallest genomes for bacteria that can be maintained outside a biotic host. It appears that genome reduction in these groups of bacteria is not driven by neutral drift and selection may have had a larger role to play than in say, endosymbionts.

Let us look at the *Prochlorococcus* case. *Prochlorococcus* are free-living marine cyanobacteria, as different from *Buchnera* as chalk is from cheese. As discussed earlier in this chapter, many *Prochlorococcus* genomes are reduced compared to their relatives, *Synechococcus*. There may be as many as 10^{27} Prochlorococcus cells on Earth. The effective population size of this genus may be around 10^9, two orders of magnitude greater than that of even *E. coli*.[146] The bioenergetic costs of maintaining even a 10-base piece of non-functional DNA,[147] even without transcribing it, would be expensive

118 (2021), e2102467118. https://doi.org/10.1073/pnas.2102467118

146 B. Batut, C. Knibbe, G. Marais, and V. Daubin, 'Reductive genome evolution at both ends of the bacterial population size spectrum', *Nature Reviews Microbiology* 12 (2014), 841–850. https://doi.org/10.1038/nrmicro3331. The section on genome streamlining in *Prochlorococcus* is in essence a summary of the key findings and opinions discussed in this review.

147 See Chapter 3, M. Lynch and G. Marinov, 'The bioenergetic costs of a gene', *Proceedings of the National Academy of Sciences USA* 112 (2015), 15690–15695. https://doi.org/10.1073/pnas.1514974112

for a population with such a high N_e. Unlike *Buchnera* and other endosymbionts, *Prochlorococcus* have retained the machinery to integrate foreign DNA. Thus, horizontal gene transfer is possible, and we have already noted that there have been several gene gains in various *Prochlorococcus* lineages, a notable example being the bacteriophage-mediated expansion of gene families involved in photosynthesis. Thus, DNA loss can be compensated for by horizontal gene acquisition. Genomes of free-living bacteria with reduced genomes also tend to have very short intergenic sequences, making them extremely gene-rich even for bacteria. The prevailing idea therefore is that *Prochlorococcus* have not lost DNA under neutral drift, as in *Buchnera*, but might have undergone genome streamlining under selection.

The ocean is a nutrient poor environment in which the relative cost of maintaining any piece of non-functional DNA gets magnified. Furthermore, the *Prochlorococcus* cells are small and, because this results in a high cell surface to volume ratio, it is believed that this is an adaptation that allows more efficient nutrient uptake. Genes that have been lost in *Prochlorococcus* often have a higher K_a/K_s ratio than those that are retained.[148] In fact, there has been a stepwise decrease in the average K_a/K_s ratio of *Prochlorococcus* genes since the divergence of this genus from *Synechococcus*.[149] *Prochlorococcus* lineages have also lost genes for the synthesis of expensive 'common goods', molecules whose function in a subset of cells benefits the whole bacterial community including cells that do not produce these molecules. As long as bacteria live in a community and there are community members that produce the common goods, non-producers can benefit by saving the energy expenditure involved in contributing their bit to the pool of the common good. An example is the gene *KatG*, encoding the protein that detoxifies external hydrogen peroxide, a reactive oxygen species. *Synechocoocus* produce KatG, whereas the genome-reduced *Procholorococcus* do not and presumably benefit by living in a community with KatG-producing *Synechococcus*. Taken together, genome reduction in *Prochlorococcus* appears to show signatures of selection.

Not everything about the evolution of *Prochlorococcus* is immediately indicative of selection or streamlining. For example, like *Buchnera, Prochlorococcus* have also lost several DNA repair genes. The consequent elevated mutation rate might also explain DNA loss and genome reduction. Let us also recall that genome reduction occurred in *Prochlorococcus* fairly early since their divergence from *Synechococcus*, and that more recent genome fluidity events have involved gene gain. Hao Zhang and colleagues, in two papers,[150] have suggested that genome reduction in *Prochlorococcus* happened during a catastrophic glaciation event some 650 million years ago and that during this period, the N_e of *Prochlorococcus* might have been low, of the order of 10^4 or 10^5. This

148 Note that there is a typographical error in the relevant section in Batut et al., 2014.

149 Sun and Blanchard, 2014.

150 H. Zhang, Y. Sun, Q. Zeng, S.A. Crowe, and H. Luo, 'Snowball Earth, population bottleneck and Prochlorococcus evolution', *Proceedings of the Royal Society B* 288 (2021), 20211956. https://doi.org/10.1098/rspb.2021.1956; H. Zhang, F.L. Hellweger, and H. Luo, 'Genome reduction occurred in early Prochlorococcus with an unusually low effective population size', *BioRxiv* (2023). https://doi.org/10.1101/2023.06.25.546417. Published more recently in *ISME Journal* 18 (2024), wrad035.

would therefore suggest that even in *Prochlorococcus*, genome reduction occurred under drift, and that arguments of genome streamlining based on present day population sizes of *Prochlorococcus* may be misleading.

In sum, however, the evolution of bacterial genomes, especially genome growth, occurs under selection. Itamar Sela and co-workers used Kimura's formula for the probability that a gene would be fixed in a population,[151] and rates of gene gains and losses, to calculate the probability that a gene is gained or lost over a given time interval.[152] At every timestep, a genome gains or loses DNA, which is determined by the difference between the two probabilities. These values would depend on the value of *s*, which will vary from gene to gene. Sela and colleagues then compared the distributions of genome sizes predicted from such models to those of extant bacterial genomes and showed that a positive value of *s* for most genes is necessary for the models to recapitulate real genome data. This suggests that, on average, genes acquired by bacterial genomes must offer a benefit to the organism to be maintained, and that selection does drive the evolution of most bacterial genomes. This appears to be true even for bacteria such as *H. pylori*, which was classified by Kuo and colleagues as an organism with high K_a/K_s and hence relatively lower selection. These findings are consistent with Marinov and Lynch's ideas of bioenergetic costs of DNA and the large population sizes of many extant bacteria, which would call for a predominant role for selection in bacteria—at least in those species that are not undergoing genome reduction under population bottlenecks. Even in the endosymbiont *Buchnera*, there is selection. In a recent paper, Julie Perreau and co-workers showed that some genetic variants of *Buchera* can outcompete others during transmission down aphid generations, indicating selection favouring some variants over others.[153] More broadly, across endosymbionts, the fact that there is metabolic complementarity between endosymbionts and hosts points to the role of selection in genome reduction here.[154]

Most selection is conventional and needs no further explanation. But the selection operating on some genes is perverse and arises from *addiction*. Such genes are often acquired by horizontal transfer. Some systems of such genes are addictive to the host bacterium, which cannot lose them once acquired even if they do not offer any conventional selective advantage to the host. In other words, once the system is acquired by a bacterium, its subsequent loss would be toxic. This is exemplified by what are known as toxin-antitoxin systems. As the name indicates, these systems contain a toxin for a gene and an antidote for it. These genes are almost always encoded next to each other. The antitoxin protects the cell against the action of the toxin. Typically, the toxin molecule is more stable under normal cellular conditions than the antitoxin. Therefore,

151 See Chapter 3.

152 I. Sela, Y.I. Wolf, and E.V. Kooni, 'Theory of prokaryotic genome evolution', *Proceedings of the National Academy of Sciences USA* 113 (2016), 11399–11407. https://doi.org/10.1073/pnas.1614083113

153 Perreau et al., 2021.

154 G.M. Bennett, Y. Kwak, and R. Maynard, 'Endosymbioses have shaped the evolution of biological diversity and complexity time and time again', *Genome Biology and Evolution* 6 (2024), evae112. https://doi.org/10.1093/gbe/evae112

when a cell loses both genes post-acquisition, the concentration of the anti-toxin molecule declines more rapidly than that of the toxin over generations. Eventually, in the absence of fresh synthesis of the molecules from the DNA, the concentration of the toxin overwhelms that of the antitoxin, leading to cell death. This phenomenon has been referred to as *post-segregational killing*.[155] In some cases, conventional selective advantages have also been ascribed to these systems.

A particularly interesting example of toxin-antitoxin-like systems is offered by what are known as *restriction-modification* systems. Though these are not conventionally defined as toxin-antitoxin systems, they operate by a similar principle. They are composed of a *restriction enzyme* and a DNA-modifying *DNA methyltransferase*. The restriction enzyme cleaves the DNA at specific sequence motifs, whereas the DNA methyltransferase methylates the DNA at the site that is cleaved by the restriction enzyme. The methylation activity of the DNA methyltransferase protects the DNA from cleavage by the restriction enzyme which would otherwise chop the DNA into small, non-functional pieces, thus killing the host cell. The phenomenon of post-segregational killing ascribed to classical toxin-antitoxin systems applies to restriction-modification systems as well.[156] There exists a vast diversity of restriction-modification systems, each with its own unique target DNA motif.

Years ago, work from our lab showed that restriction-modification systems are usually weakly conserved within bacterial clades.[157] We also found evidence that many of these genes had been horizontally acquired, based on both their atypical base composition and their sporadic occurrence across the bacterial phylogeny. Finally, many of these systems have specifically lost only their restriction component, leaving an *orphan* or *solitary* methyltransferase intact. In some cases, sequences suggestive of highly degraded versions of genes encoding restriction enzymes could be detected close to methyltransferase genes. These orphan methyltransferases showed a level of conservation that would be expected of any average chromosomal gene. The two DNA methyltransferases in *E. coli* that we briefly mentioned earlier in this chapter are orphan methyltransferases, which have evolved functions that offer a conventional selective advantage to the bacterium. This indicates that horizontal acquisition of restriction-modification systems is common, and that bacterial evolution deals with the problem of post-segregational killing presented by restriction-modification systems by selectively losing the restriction component while, over time, the methyltransferase could potentially evolve core functions that are beneficial to the cell (Fig. 4.14).

---

155 J. Guglielmini and L. van Melderen, 'Bacterial toxin-antitoxin systems', *Mobile Genetic Elements* 1 (2011), 283–290. https://doi.org/10.4161/mge.18477

156 N. Handa and I. Kobayashi, 'Post-segregational killing by restriction modification gene complexes: observations of individual cell deaths', *Biochimie* 81 (1999), 931–938. https://doi.org/10.1016/s0300-9084(99)00201-1

157 A.S.N. Seshasayee, P. Singh, and S. Krishna, 'Context-dependent conservation of DNA methyltransferases in bacteria', *Nucleic Acids Research* 40 (2012), 7066–7073. https://doi.org/10.1093/nar/gks390

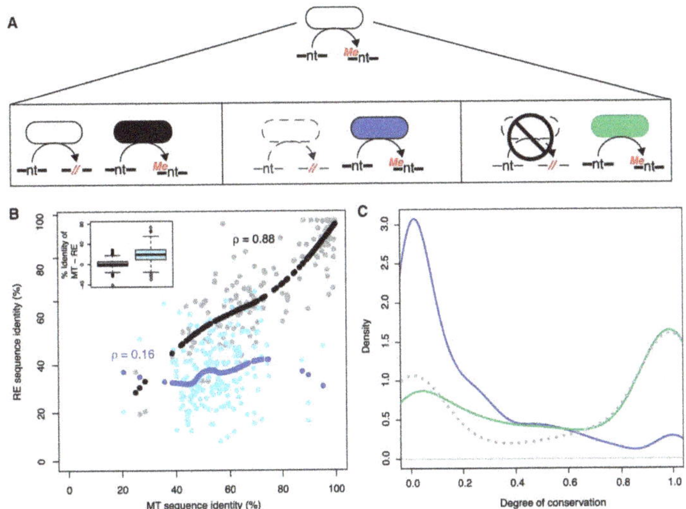

Fig. 4.14 Evolution of restriction-modification systems in bacteria. (A) Restriction-modification systems can be intact, retaining genes for both the restriction enzyme and the methyltransferase; they can be degenerate, carrying a highly divergent (and presumably inactive) restriction enzyme; or they can be an orphan methyltransferase, having lost the restriction enzyme gene. (B) This figure shows decoupled sequence divergence of the components of degenerate (blue) restriction-modification systems compared to intact (grey) systems. (C) This figure shows the degree of conservation of degraded (blue) restriction-modification systems and orphan (green) methyltransferases, showing that the latter are more conserved. Grey dots represent an average of the two. Originally published as part of Figure 3 in A.S.N. Seshasayee, P. Singh, and S. Krishna, 'Context-dependent conservation of DNA methyltransferases in bacteria', *Nucleic Acids Research* 40 (2012), 7066–7073, CC BY-NC 3.0; copyright owned by the author of this book.

Restriction-modification systems are not merely addictive. They also confer a conventional selective advantage to their bearer. They are a form of defence against bacteriophages. Bacteriophage DNA is usually not methylated, making it a suitable target for restriction enzymes. However, the defence mechanism is not failproof. The efficiency with which the restriction enzyme destroys bacteriophage DNA is dependent on the occurrence of the enzyme's target sequence motif in the bacteriophage DNA. It is entirely possible that 4–6 bp motifs are entirely absent from the relatively small genomes of many bacteriophages, in which case a restriction enzyme targeting such a motif would be ineffective. Any bacteriophage with a target motif that somehow escapes this defence mechanism would now carry methylated DNA and would be protected from destruction by other bacteria carrying the same restriction enzyme. There are several strategies by which bacteria might have evolved to counter these limitations.

One is that several restriction enzymes have promiscuous specificity, i.e. they do not specifically recognise and cleave at a specific motif, but can do so more non-specifically. As demonstrated by Kommireddy Vasu and colleagues, these restriction enzymes with promiscuous specificity are particularly advantageous to bacteria challenged by bacteriophages.[158] Curiously, however, the methyltransferase associated with such

158 K. Vasu, E. Nagamalleswari, and V. Nagaraja, 'Promiscuous restriction is a cellular defense strategy

promiscuous restriction enzymes is highly specific to its target motif, necessitating additional cellular mechanisms for protection of the bacterial genome. Remarkably, Matheshwaran Saravanan and co-workers demonstrated that it takes only a single base substitution to make a promiscuous restriction enzyme specific (and vice-versa).[159] Whether this is as easily achieved for the methyltransferase as well is unclear. Thus, the ability to counter the limitations of site specificity to defence against bacteriophages is highly evolvable, as long as mechanisms for the protection of self-DNA are available.

Other strategies are also possible without risking imbalance between the activities of the restriction enzyme and the methyltransferase. Theoretical calculations by Rasmus Eriksen and colleagues showed that under natural conditions that are replete with bacteriophages, bacteria forming multi-species communities should code for multiple restriction-modification systems, and that systems with similar specificity should be shared across multiple members of the community.[160] On the other hand, when bacteriophages are rare, restriction-modification systems often impose a cost that is greater than the benefit they confer. In such situations, bacterial genomes code for few restriction-modification systems, with little sharing of specificity among community members. Consistent with these predictions, genome sequences across thousands of bacteria showed that some bacterial genera encoding large numbers of restriction-modification systems shared similar specificities, whereas those with only a few such genes rarely shared specificities.

In summary, the evolution of bacterial genomes—particularly the expansion of these genomes—is driven by selection. Selection is often conventional but sometimes perverse, arising from addiction. This is in contrast to genome growth in higher eukaryotes, much of which occurs under neutral drift. Genome reduction in bacteria often occurs when population sizes are low and the strength of selection is weak. Free-living bacteria with reduced genomes are often believed to have streamlined their genomes under selection pressure from, for instance, environments that are low in nutrient content, but the role of neutral drift in these so-called streamlining events has also been put forth. The end result of these evolutionary processes is a genetically diverse domain of life that has come to be the most dominant form of cellular life on Earth.

---

that confers fitness advantage to bacteria', *Proceedings of the National Academy of Sciences* 109 (2012), E1287–E1293. https://doi.org/10.1073/pnas.1119226109

159 M. Saravanan, K. Vasu, and V. Nagaraja, 'Evolution of sequence specificity in a restriction endonuclease by a point mutation', *Proceedings of the National Academy of Sciences* 105 (2008), 10344–10347. https://doi.org/10.1073/pnas.0804974105

160 R. Eriksen, N. Malhotra, A.S.N. Seshasayee, K. Sneppen, and S. Krishna, 'Emergence of networks of shared restriction-modification systems in phage-bacteria ecosystems', *Journal of Biosciences* 47 (2022), 38. https://doi.org/10.1007/s12038-022-00274-7

5. Reading and organising the genome

5.1. Expressing the genome and decision making

The genome is a blueprint,[1] and does not by itself get its host cell up and running. The genome must be read and interpreted before it can set in motion many series of connected events that somehow create life. The first step in this process is transcription, during which a gene sequence, a small part of the genome, is read and an RNA transcript with a sequence corresponding to that of the transcribed DNA is produced. Many of these RNA molecules serve as messengers (mRNA), and are further read to create proteins during translation; other RNA molecules, such as the rRNA and tRNA which help the ribosome perform translation, play direct roles in cell function without being translated. The discussion in this chapter will focus on transcription, how it is regulated, how regulators of transcription evolve and the role played by genomics in our understanding of these processes. We will conclude by asking how transcription and the manner in which genes are strung together to form a genome are linked.

Transcription is essentially an enzymatic process that is constrained by the sequence of the DNA being transcribed. The process minimally requires a DNA template, free ribonucleotides that can be linked together to form the RNA chain and an enzyme that can polymerise ribonucleotides to create an RNA sequence that is complementary to the sequence of the DNA template. In addition, the mechanics of transcription requires additional enzymes that help unwind the DNA in front of the machinery that performs transcription, and a host of other proteins that ensure that the process doesn't stall in the middle of a gene and terminates at the right place; these will not be described much in this book. The discovery of the enzyme and that of the fact that transcription is tightly regulated in bacterial cells played important roles in the series of epiphanies that led to the explosion of molecular biology in the 1960s.[2]

In order to transcribe a gene, RNA Polymerase (RNAP), the enzyme that performs transcription, should specifically bind somewhere near the start of the gene. Once this happens, the double-stranded DNA must unwind and the unwound DNA must move

1 A flexible one at that, such that the same sequence can be interpreted in different ways to produce different trait outcomes.

2 J. Hurwitz, 'The Discovery of RNAP', *Journal of Biological Chemistry* 280 (2005), 42477–42485. https://doi.org/10.1074/jbc.x500006200; R.R. Burgess, 'What is in the black box? The discovery of the sigma factor and the subunit structure of *E. coli* RNAP', *Journal of Biological Chemistry* 297 (2021), 101310. https://doi.org/10.1016/j.jbc.2021.101310; M. Lewis, 'A tale of two repressors – a historical perspective', *Journal of Molecular Biology* 409 (2011), 14–27. https://doi.org/10.1016/j.jmb.2011.02.023 See Chapter 2.

 https://doi.org/10.11647/OBP.0446.05

base-by-base relative to the RNAP. As the enzyme reads the DNA bases, ribonucleotides complementary to the base being read should be assembled and attached to the growing, nascent RNA chain. The DNA in front of the RNAP must be kept unwound throughout the process. Finally, the RNAP should drop off the gene and terminate transcription at the end of the gene. The focus of this chapter will be on transcription initiation.

The RNAP is a multi-subunit protein,[3] i.e., it comprises several proteins that assemble together to form a functional enzyme. These subunit proteins include those that perform the enzymatic reaction of linking ribonucleotides together, proteins that ensure that the enzyme stays on the DNA through the length of the gene and assembly factors.[4] The core RNAP, which in *E. coli* has five subunits, is perfectly capable of performing transcription but cannot specifically recognise and initiate transcription at the start of genes. Specific recognition of these transcription start sites requires an exchangeable subunit called the σ-factor (sigma factor; Fig. 5.1). The σ-factor binds to the core RNAP, forming what is called the RNAP holoenzyme. The RNAP holoenzyme then specifically recognises DNA sequences upstream of the start of genes. The σ-factor also helps the RNAP unwind the DNA, thus initiating transcription. The σ-factor usually dissociates from the RNAP complex after initiation.

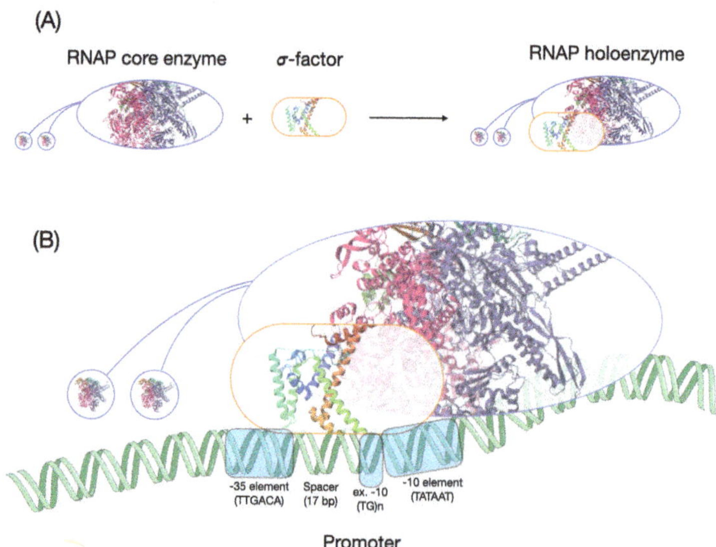

Fig. 5.1. Transcription **initiation**. (A) This figure shows the formation of an RNAP holoenzyme by the binding of the RNAP core enzyme with a σ-factor. The structure of the RNAP inside the oval is from PDB: 7MKP, and that of a fragment of a σ-factor is from PDB: 1SIG. (B) This figure shows the interaction of an RNAP holoenzyme with the promoter. The image of the DNA is from SMART-Servier Medical Art, part of Laboratoires Servier, via Wikimedia Commons, available freely under CC BY-SA 3.0.

3 Note that RNA polymerase from bacteriophage T7 comprises a single subunit. It was discovered about a decade or so after the discovery of the bacterial multi-subunit RNA polymerase.

4 Eukaryotes have multiple types of RNAPs. The eukaryotic RNAP transcribing messenger RNA is larger than the prokaryotic RNAP and has many more subunits.

The DNA sequence that the RNAP holoenzyme recognises is called the promoter. The promoter region is usually A+T-rich. Each gene has its own promoter sequence, but taken together many promoters show some common properties. For example, the bacterial promoter—based on the paradigm established in *E. coli* but shown to be applicable to many other bacterial genomes[5]—is bipartite. There is a six-base −10 element (minus 10) and a six-base −35 element (minus 35). The −10 element is centred 10 bases upstream of the transcription start site (the site at which mRNA synthesis begins) of a gene, and the −35 element is centred 35 bases upstream. The −10 element, when analysed across many genes, has a consensus sequence TATAAT, whereas the −35 element shows a consensus of TTGACA. The specificity-determining σ-factor, when bound to the RNAP, recognises these elements on the DNA. The sequence of the stretch of DNA between the two elements is immaterial. However the length of this spacer is critical to ensure that the −10 and the −35 elements are oriented correctly for the RNAP to bind to the promoter. The precise consensus sequence is not necessary to produce a functional promoter. It is merely a construct that represents the most common base found at each site.

Natural promoters usually differ from the consensus at one or more sites, and the more divergent it is from the consensus element the weaker is its affinity to the RNAP. Therefore, each gene, on the basis of its promoter sequence alone, has its own unique ability to attract RNAP and initiate its own transcription. This creates cross-gene variation in the extent to which a gene can be transcribed. Some promoters do not contain a −35 element, and these sequences carry what is an extended −10 element, which is a slightly longer version of the −10 sequence motif.

Though the sequence of the promoter itself can determine to some extent the expression level of a gene, this does not vary within the lifetime of a cell. Changes in the promoter sequence can happen over generations and, similar to mutations within a gene sequence, its fate can be determined by selection or drift. However, a cell often needs to make decisions in a matter of minutes about which gene to express, and when, within its lifetime. Many bacterial cells experience conditions that change from time to time. Even if their genetic repertoire is sufficient to handle all these environmental conditions, only a subset of their genes would be required under any given condition. Expressing the rest can be costly. As we noted in Chapter 3, expressing a gene under conditions in which the gene offers no selective advantage to the cell can be very costly, especially in bacteria with large population sizes. In addition, there is a constraint that arises from resource availability. The number of free RNAP molecules available to initiate transcription is often limited, because ~80% of all RNAP molecules are involved in transcribing a very small number of genes coding for rRNAs.[6] Therefore,

5 A.M. Huerta, M.P. Francino, E. Morette, and J. Collado-Vides, 'Selection for unequal densities of σ70 promoter-like signals in different regions of large bacterial genomes', *PLoS Genetics* 2 (2006), 185. https://doi.org/10.1371/journal.pgen.0020185

6 D.F. Browning and S.J.W. Busby, 'The regulation of bacterial transcription initiation', *Nature Reviews Microbiology* 2 (2004), 57–65. https://doi.org/10.1038/nrmicro787; I. Bervoets and D. Charlier, 'Diversity, versatility and complexity of bacterial gene regulation mechanisms: opportunities and

the number of RNAP molecules available for transcription is often far less than the number of genes in the genome. Finally, in complex bacterial genomes, the expression of one gene may counteract that of another and such conflicts should necessarily be contained. Thus, regulation of gene expression, or, in other words, taking decisions on which gene to express at any point in time, is important.

Various regulatory systems, or networks, help the cell achieve gene regulation. First, though the sequence of DNA is relatively static, its structure is not. The DNA double helix is usually in what is called a B-form, in which each turn of the DNA has ~10 base pairs. The double helix can unwind or overwind such that the number of base pairs per turn is less than or greater than 10, and, in *E. coli*, this is to a large extent determined by the energy levels available to the cell.[7] DNA that is unwound is said to be negatively supercoiled. As the degree of negative supercoiling decreases and approaches the standard B-form twist, the DNA is said to be more relaxed. *E. coli* DNA is rarely, if ever, positively supercoiled, though this is known to happen in other bacteria. Enzymes under the umbrella name topoisomerase help modulate supercoiling states of DNA. In *E. coli* a topoisomerase called DNA gyrase negatively supercoils DNA, whereas DNA topoisomerase 1 helps relax DNA. When the cellular energy levels are high, the DNA is negatively supercoiled due to high DNA gyrase activity and this permits rapid transcription; during starvation, the DNA becomes relaxed, which can globally suppress transcription.[8] However, this overarching link between DNA supercoiling and transcription does not apply equally to all genes (Fig. 5.2). It has been observed that genes whose expression is preferentially reduced during starvation (or whose expression is high specifically during rapid growth) have a G+C-rich region in their promoters. This might make the promoter harder to unwind because G-C base pairs are more stable than A-T base pairs.[9] Unwinding of such promoters might be facilitated by negative supercoiling, which is favoured during high growth states. This mechanism appears to affect the expression of many genes involved in translation, including that of rRNA, whose transcription at high levels under nutrient stress can be hugely wasteful and damaging. Under nutrient-replete conditions, however, high transcription of such genes is necessary to support growth. On the other hand, promoters that are extraordinarily A+T-rich may be preferentially transcribed during starvation, when the genome in general is less negatively supercoiled.[10] Therefore, the

---

drawbacks for applications in synthetic biology', *FEMS Microbiol Rev.* 43 (2019), 304–339. https://doi.org/10.1093/femsre/fuz001

7 The number of bases per turn is called the twist. It represents how one strand of DNA winds around the other. There is a second component called writhe. This represents the coiling of the entire double helix around itself. We will not discuss this in any detail here.

8 C.J. Dorman, 'DNA supercoiling and transcription in bacteria: a two-way street', *BMC Molecular and Cell Biology* 20 (2019), 26. https://doi.org/10.1186/s12860-019-0211-6

9 R. Forquet, M. Pineau, W. Nasser, S. Reverchon, and S. Meyer, 'Role of the discriminator sequence in the supercoiling sensitivity of bacterial promoters', *mSystems* 6 (2021), e00978–21. https://doi.org/10.1128/msystems.00978-21

10 B.J. Peter, J. Arsuaga, A.M. Brier, A. Khodursky, P.O. Brown, and N. Cozzarelli, 'Genomic transcriptional response to loss of chromosomal supercoiling in *Escherichia coli*', *Genome Biology* 5 (2004), R87. https://doi.org/10.1186/gb-2004-5-11-r87

fact that the structure of the DNA can respond to some cellular conditions and in turn affect the extent to which different genes are transcribed makes the DNA itself an important regulator of gene expression.

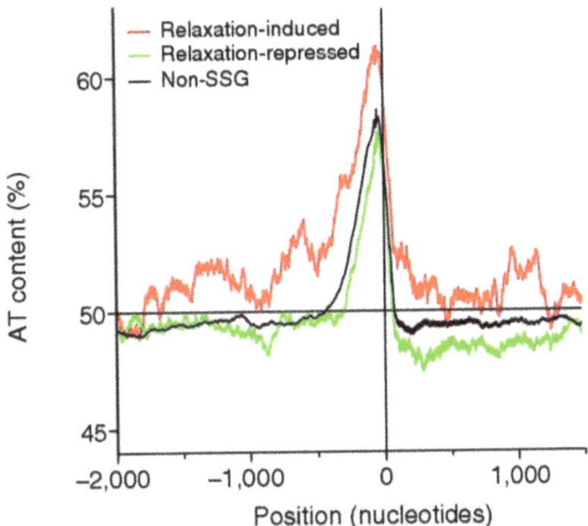

Fig. 5.2. Base composition upstream of genes regulated by DNA supercoiling in *E. coli*. This figure shows that genes that are induced by DNA relaxation are more A+T-rich than the average gene, whereas the reverse holds for genes that are induced by negative supercoiling. Along the x-axis, values to the left of '0' indicate positions upstream of genes, and positions to the right indicate the gene body and further beyond. Originally published as Figure 5B in B.J. Peter, J. Arsuaga, A.M. Brier, A. Khodursky, P.O. Brown, and N. Cozzarelli, 'Genomic transcriptional response to loss of chromosomal supercoiling in *Escherichia coli*', *Genome Biology* 5 (2004), R87, CC BY 2.0.

Yet another component of the core transcriptional machinery that plays a regulatory role is the σ-factor, which helps the RNAP recognise promoters. Many bacterial genomes code for multiple σ-factors. Depending on the relative abundance of the core RNAP and σ-factors, the multitude of σ-factors can all be bound to abundant RNAP molecules, or they compete for the limited real estate presented by an insufficient number of core RNAP molecules. Though protein quantification by different labs support different scenarios, the most comprehensive and recent analysis (to my knowledge) supports the latter.[11] Thus, we now accept that different σ-factors compete with each other for binding to the core RNAP. The outcome of this competition will be determined by the relative abundance or availability and the affinity of each σ-factor to the core RNAP.

Different σ-factors recognise different promoter types. The standard bipartite promoter structure that we described earlier is best recognised by what is called the σD σ-factor, following the nomenclature used for *E. coli*. σD is a 'housekeeping' σ-factor that, by recognising the standard promoter, helps initiate transcription of a majority of genes involved in growth and metabolism that operate in nutrient-rich conditions. A second σ-factor, σ^S in *E. coli*, becomes available in sufficient concentrations as nutrients

11 S.E. Piper, J.E. Mitchell, D.J. Lee, and S.J.W. Busby, 'A global view of *Escherichia coli* Rsd protein and its interactions', *Molecular Biosystems* 5 (2006), 1943–1947. https://doi.org/10.1039/b904955j

deplete and cells enter a period of starvation and stress. This σ-factor helps the RNAP bind to promoters of genes underlying the bacterial response to a variety of stresses, which together form the general stress response. There is evidence that several promoters bound by σ^S-bound RNAP are recognised by this σ-factor when the DNA is relaxed, as it is during starvation, pointing to how DNA structure can contribute to differential promoter recognition by various σ-factors.[12] Because different σ-factors, when bound to the RNAP, can recognise their own set of promoters, the outcome of the competition between these σ-factors for binding to the core RNAP is a major determinant of which genes are expressed. We will return to this aspect of regulation later in this chapter.

Whereas the structure of the DNA and σ-factors (as regulatory molecules) are still part of the machinery that performs transcription, several other 'outside' players fulfil important roles in gene regulation.[13] The most prominent among these are transcription factors (TFs). TFs are DNA-binding proteins. They bind to specific DNA sequence motifs often present around the promoter region. The DNA sites to which TFs bind are sometimes called operators, or simply TF-binding sites. When bound to these sites, TFs can either activate or repress transcription (Fig. 5.3).[14] TFs repress transcription usually by binding close enough to the promoter that they block access to the RNAP; in other words, they sterically hinder RNAP-promoter interactions. By binding to one site near the promoter and another further upstream, they can also loop the intervening DNA and form a strongly repressive structure that prevents RNAP activity. Sometimes they do not block the initial interaction between the enzyme and the DNA, but instead prevent further progress.

The discovery by Arthur Pardee, Francois Jacob and Jacques Monod of a repressor of the set of genes that help *E. coli* metabolise sugar lactose, published in 1959, played a central role in the discovery of messenger RNA.[15] This repressor was isolated a few years later by Benno Muller-Hill[16] and shown to bind specifically to its operator site on DNA. While Monod was working on the induction of lactose metabolism genes, Andre Lwoff was demonstrating the phenomenon of bacteriophage lysogeny. Mark Ptashne's work revealing the central role of repressors of transcription in the maintenance of lysogeny was yet another landmark in the history of gene regulation.[17]

---

12 S. Kusano, Q. Ding, N. Fujita, and A. Ishihama, 'Promoter selectivity of Escherichia coli RNAP E sigma 70 and E sigma 38 holoenzymes. Effect of DNA supercoiling', *Journal of Biological Chemistry* 271 (1996), 1998–2004. https://doi.org/10.1074/jbc.271.4.1998

13 Small molecules such as guanosine tetraphosphate are produced in response to starvation and can bind to the RNAP and repress transcription of growth-related genes. We do not discuss this regulatory arm beyond brief mentions in this book.

14 Browning and Busby, 2004.

15 A.B. Pardee, F. Jacob, and J. Monod, 'The genetic control and cytoplasmic expression of inducibility in the synthesis of b-galactosidase in *E coli*', *Journal of Molecular Biology* 1 (1959), 165–178. https://doi.org/10.1016/b978-0-12-131200-8.50004-6

16 W. Gilbert and B. Mueller-Hill, 'Isolation of the lac repressor', *Proceedings of the National Academy of Sciences USA* 56 (1966), 1891–1898. https://doi.org/10.1073/pnas.56.6.1891

17 Reviewed and described in retrospect in M. Ptashne, *A Genetic Switch: Phage Lambda Revisited* (Plainview, NY: Cold Spring Harbor Laboratory Press, 2004).

The discovery of the repressor-based regulation of lactose metabolism genes also showed that in bacteria several genes encoded in tandem on the genome can be expressed from a single promoter. Such a series of co-transcribed genes is referred to as an operon, a fundamental feature of bacterial genomes: the ~4,000 genes in the *E. coli* genome may be organised into ~2,000 operons. Not all genes are organised into operons; many are singletons. Some operons are short, comprising not more than two or three genes whereas other uber-operons can encompass tens of genes.

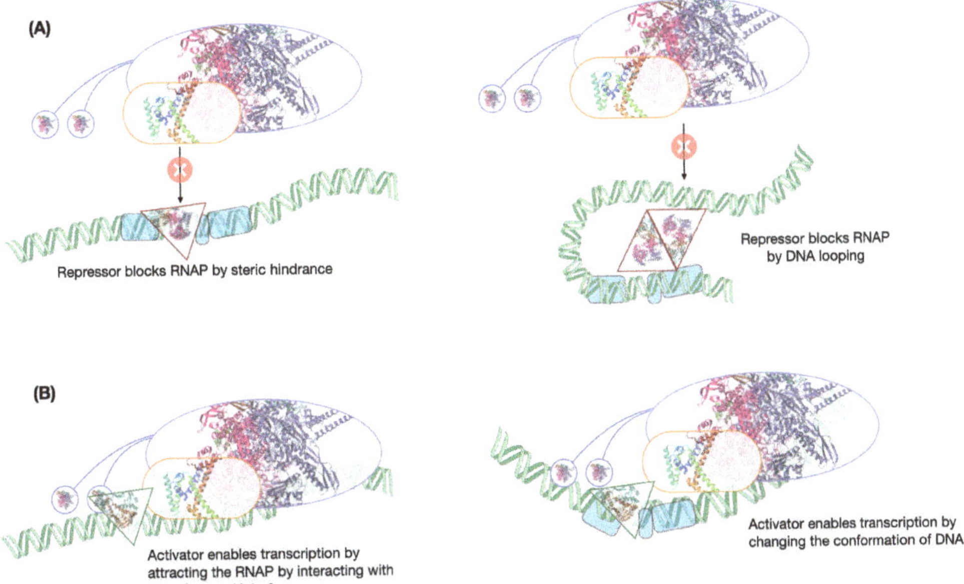

Fig. 5.3. Activation and repression by transcription factors. This figure shows a sample of simple ways by which (A) repressors (red-bordered triangles) and (B) activators (green-bordered triangles) act to regulate transcription initiation. The structure of the lac repressor filling the red triangle is from PDB: 1LB1, and that of CRP filling the green triangle is from PDB: 4N9H. The images of the DNA are from SMART-Servier Medical Art, part of Laboratoires Servier, via Wikimedia Commons, available freely under CC BY-SA 3.0.

Not all regulators of transcription are repressors. Activators normally bind upstream of the promoter and have elements that can attract the RNAP, interacting either with the core RNAP components or with the σ-factor. Ellis Engelsberg, along with his colleagues Joseph Irr, Joseph Power and Nancy Lee, discovered in 1965 that the genes for utilisation of the sugar arabinose in *E. coli* came under the control of an activator.[18] This discovery was initially met with much scepticism because of the deeply entrenched repressor-based model of gene regulation espoused by work of Pardee, Jacob and Monod. In the words of Steven Hahn, "(though) the evidence in 1965 for positive control by AraC was as good or better than the data used to formulate

18 Reviewed from a scientific and historical perspective in S. Hahn, 'Ellis Englesberg and the discovery of positive control in gene regulation', *Genetics* 198 (2014), 455–460. https://doi.org/10.1534/genetics.114.167361

the negative control model, Englesberg needed to accumulate much additional data to answer his critics."[19] Nevertheless, it was only a matter of time before other activator systems were described and Engelsberg stood vindicated. Many TFs can activate transcription of one gene but repress that of another, and whether they activate or repress transcription depends on where they bind relative to the promoter.[20] Some TFs, including Engelsberg's activator, can even perform dual actions on the same target gene by changing its binding site upstream of the gene. One can expect any activator to be able to repress transcription as long as it binds the DNA in such a manner that the RNAP cannot bind to the promoter. The reverse need not be true—a pure repressor cannot activate transcription just because it binds upstream of the promoter—it may not possess an interface to attract the RNAP.

The activity of TFs themselves is often determined by the presence or absence of a signal. For example, a TF that activates transcription of genes responsible for metabolising a sugar as a nutrient will be activated by the presence of the sugar. Such a TF, in addition to being able to bind DNA, will also be able to bind to the sugar to which it responds. The binding of the sugar to the protein will then activate (if the TF is an activator of the sugar metabolism genes) or hinder the TF's ability to bind to the DNA (if the activator is a repressor). Many TFs in bacteria possess such a property. The repressor of lactose metabolism binds to allolactose (similar to lactose). When not bound to allolactose, the repressor binds to the DNA and blocks RNAP activity. The binding of allolactose causes the TF to release the DNA, thus allowing transcription. The activator of arabinose metabolism binds to the DNA both in the presence and absence of the sugar. In the former situation, it acts as an activator but switches to being a repressor in the latter. Other TFs may not directly bind to a signal, but may be activated following a series of reactions that are initiated by a separate signal-sensing protein that, for example, may be located on the cell membrane. Each TF regulates its own set of target genes and the set of TF-target gene interactions constitutes a transcriptional regulatory network. Thus, TFs are proteins whose activities are usually determined by the presence of certain environmental or cellular conditions, in response to which they regulate the transcription of other genes.

5.2. The transcriptional regulatory network of *E. coli*

The *E. coli* genome encodes ~300 TFs for its total complement of ~4,400 genes. Even in this well-studied organism, we do not know all the regulatory connections these TFs make. Over half of these TFs have at least one known target gene,[21] as discovered through biochemical or genetic experiments; the others are predicted to be TFs based on their sequences. In

19 Hahn, 2014.

20 M. M. Babu and S.A. Teichmann, 'Functional determinants of transcription factors in Escherichia coli: protein families and binding sites', *Trends in Genetics* 19 (2003), 75–79. https://doi.org/10.1016/s0168-9525(02)00039-2

21 A target gene of a TF is a gene whose expression is regulated by the TF. Usually, it refers to genes that are directly regulated by the TF which binds to a site upstream of the gene's promoter.

addition to these TFs, *E. coli* has seven σ-factors competing to bind to the core RNAP. These TFs and their target genes or operons together constitute the transcriptional regulatory network. For *E. coli* there are publicly available databases such as RegulonDB[22] and Ecocyc,[23] from which the currently known transcriptional regulatory network can be downloaded. These networks comprise of data from a variety of experiments—from small-scale, detailed studies on how a particular TF binds to an operator to regulate a target gene, to large-scale, bird's-eye view studies that catalogue the list of all genes or operons that are regulated by one or more TFs under a set of growth conditions.

The targets of a TF can be defined in several ways. A gene can be called a target of a TF if the regulator binds upstream of the gene and, when bound, alters the expression state of the gene. This would define direct targets of a TF. Sometimes, the mere binding of a TF to an operator is used to define a target irrespective of whether there is evidence that the binding affects the expression of the gene. This may be appropriate when there is reason to believe that absence of evidence (of an effect on gene expression) is not evidence of absence. On the flip side, some TF-DNA interactions may also be non-functional. Alternatively, genes that change in expression when a TF is deleted can be called targets of the regulator. However, the targets defined may, therefore, not always be bound by the TF, and may change in expression as a result of a cascade of effects initiated far upstream by the direct regulation of a different gene(s) by the TF.

Given such complications in the ways in which a regulatory network can be defined, are such networks even useful to define on a genome-wide scale? In other words, does a regulatory network—built by aggregating data from hundreds to thousands of experiments together encompassing a cocktail of approaches—predict gene expression: the defining, measurable output of the regulatory network? Xin Fang and colleagues recently showed that a transcriptional regulatory network built from data on where TFs bind on the genome agrees well with genes that change in expression when a TF is deleted, and that the regulatory network is good enough to predict the gene expression states of over 85% of operons.[24] Earlier work by Gabor Balazsi and colleagues had shown that groups of genes that are expressed together under a given condition often belonged to coherent, closely-linked parts of the then known regulatory network.[25] Thus, the transcriptional regulatory network—despite being incomplete even for a well-studied model organism such as *E. coli*—serves as a good predictor of gene expression. However, though groups of genes regulated in the same manner may be expressed together, the expression level of a TF may not correlate well with that of its targets, in part because the activity of a TF is not defined entirely by its expression level.[26]

---

22 https://regulondb.ccg.unam.mx/

23 https://www.ecocyc.org/

24 X. Fang, A. Sastry, N. Mih, D. Kim, J. Tan, et al., 'Global transcriptional regulatory network for *Escherichia coli* robustly connects gene expression to TF activities', *Proceedings of the National Academy of Sciences USA* 114 (2017), 10286–10291. https://doi.org/10.1073/pnas.1702581114

25 G. Balazsi, A.-L. Barabasi, and Z.N. Oltvai, 'Topological units of environmental signal processing in the transcriptional regulatory network of *Escherichia coli*', *Proceedings of the National Academy of Sciences USA* 102 (2005), 7841–7846. https://doi.org/10.1073/pnas.0500365102

26 S.J. Larsen, R. Rottger, H.H.H.W. Schmidt, and J. Baumbach, '*E. coli* gene regulatory networks

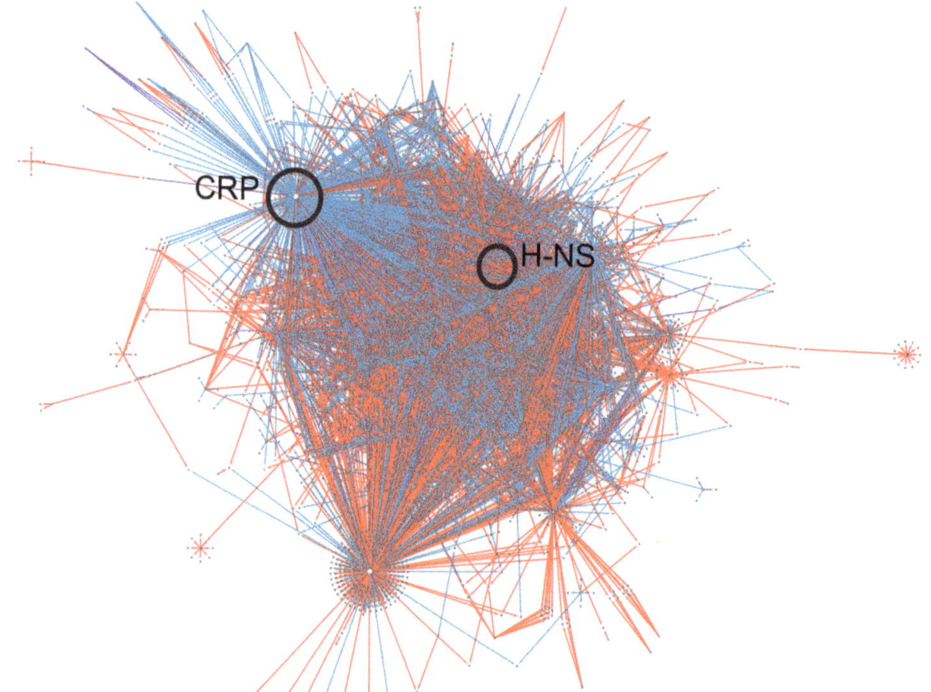

Fig. 5.4. The *E. coli* regulatory network. This figure shows a representation of the *E. coli* transcriptional regulatory network. Each line indicates a regulatory interaction between a regulator (mostly, but not necessarily, TFs) and a target gene. Red lines mark repressive interactions, whereas blue lines indicate activating interactions. Global regulators CRP and FNR are marked. Figure produced by Ganesh Muthu using the regulatory network available in the RegulonDB database (https://regulondb.ccg.unam.mx/) and Cytoscape (y-Force layout; https://cytoscape.org/ and https://www.yworks.com/products/yfiles-layout-algorithms-for-cytoscape).

The first decade of this century saw the publication of several papers describing graph theoretical studies of biological networks. Among these networks are transcriptional regulatory networks. To some extent, these studies were spurred by genome-scale studies of the eukaryote *Saccharomyces cerevisiae*, a yeast. One major work identified binding sites for over 150 TFs encoded by this organism,[27] triggering a large number of studies curating and analysing the vast amounts of data produced by this work. Any network is a graph that draws *edges* connecting points called *nodes*. In what is called a protein-protein interaction network, nodes are proteins and an edge is drawn between two proteins that physically interact with each other. The edges in such a network are not directional: if protein *A* interacts with protein *B*, then *B* also interacts with *A* and there is no direction to how the two proteins interact with each other. A transcriptional regulatory network, which connects TFs to their target genes or their binding sites, is directional: each edge is directed from the TF to a target gene because the TF *regulates*

are inconsistent with gene expression data', *Nucleic Acids Research* 47 (2019), 85–92. https://doi.org/10.1093/nar/gky1176

27 C.T. Harbison, D.B. Gordon, T.I. Lee, N.J. Rinaldi, K.D. Macissac, et al., 'Transcriptional regulatory code of a eukaryotic genome', *Nature* 431 (2004), 99–104. https://doi.org/10.1038/nature02800

the expression of the target gene (Fig. 5.4). Some representations of the transcriptional regulatory network also include σ-factors as regulatory proteins similar to TFs, others do not.

As one would immediately guess, such networks are not exclusive to biology. One can envisage a whole host of networks—such as the internet, electricity grids, and postal networks—and all of these can be directional. In, say, a postal network, which connects two post offices with an edge if letter bundles are sent from one to the other, not all nodes are connected equally. For example, the general or hub post office of a city will receive all letters sent to the city and then forward it to various local offices. The local post offices, despite being in the same city, may not be connected to each other directly. Thus, many post offices will have low connectivity, often being connected both ways with only the city's general post office. The general post office on the other hand is highly connected, but the number of such offices is very small compared to the number of local offices.

Similar trends have been described for biological networks as well. For example, in the *E. coli* transcriptional regulatory network, most TFs regulate only a few genes;[28] the repressor of lactose metabolism regulates only what is called the *lac operon*. On the other hand, a few TFs regulate hundreds of genes! The former, with their limited sphere of influence, are often called *local* TFs, and the latter, in contrast, are referred to as *global* TFs. That said, however, the distribution of the number of targets a TF has is continuous, and therefore it is not entirely obvious where a line demarcating local from global TFs should be drawn. As a result, there have been several definitions of what constitutes a global TF.

Often, an arbitrary threshold number of targets is used to separate global and local TFs. But is the number of targets the only parameter that defines global TFs? Some studies, primarily by Julio Collado-Vides and colleagues, argue otherwise. To follow this line of thinking, we must first understand and visualise the network itself a bit better. The regulatory network is not a disconnected set of TFs regulating their target genes in a one-on-one or a one-on-many manner. Just as a TF can regulate multiple genes, many genes are regulated by multiple TFs. For example, the operon for lactose metabolism is regulated not only by the lactose-responsive repressor but also by a TF CRP that responds indirectly to glucose availability in such a way that CRP becomes active when glucose is limiting. The same CRP acts as a second regulator of arabinose metabolism as well. The lactose operon is expressed only when the repressor is not bound and CRP *is* bound; the arabinose metabolism genes are expressed when the arabinose-responsive TF is bound to its operator sites in an activating configuration *and* CRP is also bound. Genes that determine the decision of the *E. coli* cell to move or to stay put are regulated by several TFs, and biochemical experiments with purified

28 A recent study has suggested that very few TFs regulate only a single target gene: T. Shimada, H. Ogasawara, I. Kobayashi, N. Kobayashi, and A. Ishihama, 'Single-target regulators constitute the minority group of TFs in *Escherichia coli* K-12', *Frontiers in Microbiology* 12 (2021), 697803. https://doi.org/10.3389/fmicb.2021.697803

protein and DNA sequences suggest that several tens of regulators can bind to regions upstream of these genes.[29] It is well-nigh impossible for all these regulators to bind simultaneously to the small stretch of DNA upstream of these genes determining motility/adhesion. However, different small sets of regulators may be active and involved in the regulation of these genes under different conditions.

Next, TFs can regulate genes for other TFs. This can set up a variety of network motifs. It can create cascades in which a series of regulatory events ultimately determines the expression of a non-TF target gene. For example, a dimeric TF called FlhDC activates the expression of a σ-factor called FliA. FliA then regulates the expression of a host of genes that allow the bacterial cell to move in particular ways. But then, the cascade is not purely linear along a single path. FlhDC, in addition to regulating the gene encoding the σ-factor FliA, also directly activates genes that allow the cell to move. This creates what is called a feed-forward loop: TF *A* regulates the expression of TF *B*, and *A* and *B* together regulate the expression of a non-transcription-factor target gene *C*. Depending on whether the effect of *A* on *C* is the same as the composite effect of *A* and *B* on *C*, the feed-forward loop is either coherent or incoherent.

Most feed-forward loops in the *E. coli* regulatory network appear to be coherent.[30] FliA, being a σ-factor, activates all its targets and FlhDC activates its target genes. Therefore, both routes to the motility-determining genes are activating, so this represents a coherent feed-forward loop. The two activating regulatory arms leading to *C* may represent an AND gate, in which both arms are required for the full expression of *C*, or they may form a SUM gate in which the effect on C is the sum of the effects of the two individual arms. The regulatory system for motility forms a SUM gate.[31] The arabinose system also includes an AND feed-forward loop in which CRP sits as a top-level TF that regulates the expression of the arabinose-responsive TF as well as that of the enzymes that metabolise arabinose. OR gates between the two arms of a feed-forward loop are also possible. For example, a coherent feed-forward loop forming an OR gate regulates the expression of a negative regulator of an adhesive structure called holdfast in a bacterium *Caulobacter crescentus*.[32] Whereas the activating SUM input of a coherent feed-forward loop keeps the expression of motility genes on for a long time, a coherent OR input structure helps to decrease the effect of environmental fluctuations on the expression of the ultimate target gene. Other types of local network structures include one in which the same TF regulates multiple genes, but with varying binding affinities such that some targets are prioritised for regulation over others.[33]

29 A. Ishihama, 'Prokaryotic genome regulation: a revolutionary paradigm', *Proceedings of the Japan Academy B* 88 (2012), 485–508. https://doi.org/10.2183/pjab.88.485

30 S.S. Shen-Orr, R. Milo, S. Mangan, and U. Alon, 'Network motifs in the transcriptional regulation network of Escherichia coli', *Nature Genetics* 31 (2002), 64–68. https://doi.org/10.1038/ng881

31 S. Kalir, S. Mangan, and U. Alon, 'A coherent feed-forward loop with a SUM input function prolongs flagella expression in *Escherichia coli*', *Molecular Systems Biology* 1 (2005), 2005.0006. https://doi.org/10.1038/msb4100010

32 M. McLaughlin, D.M. Hershey, L.M.R. Ruiz, A. Fiebig, and S. Crosson, 'A cryptic TF regulates Caulobacter adhesin development', *PLoS Genetics* 18 (2022), e1010481. https://doi.org/10.1371/journal.pgen.1010481

33 A. Zaslaver, A.E. Mayo, R. Rosenberg, P. Bashkin, H. Sberro, M. Tsalyuk, M.G. Surette, and U. Alon,

Finally, feedback loops also occur. If one were to include the signal molecule, such as lactose (or allolactose), in our representation of the regulatory network, the regulation of the lactose operon would be an example of a positive feedback loop. In this positive feedback loop, activation of the lactose operon in a cell by a small quantity of the signal would result in more of the signal molecule being taken up by this cell. When lactose is limiting such that not all cells see lactose first, a bistable state, in which two sub-populations of cells with very distinct levels of expression of the *lac* operon will be established. The lucky few cells that came into contact with lactose early would be primed to take up most of the lactose available in the local environment, whereas others would be oblivious to the availability of this sugar. Here we immediately notice a situation where genetically identical cells display two distinct traits because of how their regulatory network has interpreted the environment. This highlights the important argument that the genome sequence is not a dictatorial directive but, as the science writer Philip Ball puts it, it helps to establish some "flexible rules" from which life can emerge.[34] Negative feedback loops also exist, and there are also instances where one TF regulates another and the latter returns the favour. In other words, even if we do not include the signal molecule in our representation of the regulatory network, feedback loops involving TFs and no other types of molecules exist.[35] Finally, TFs can auto-regulate their own expression, often negatively, but also positively. Whereas the former helps to reduce response time and decrease fluctuations, the latter does the opposite. Thus, each type of network motif has its own kinetic properties.[36] Taken together, the take home message from this short discussion is that the transcriptional regulatory network is a complex set of highly interconnected nodes that form a variety of converging, diverging, and even circular loops.

Back to global regulators: do some regulators with a large number of targets possess additional properties that distinguish them from local TFs that regulate a small set of genes on demand? Agustino Martinez-Antonio and Julio Collado-Vides found that some TFs with a large number of targets share certain properties which together qualify them as global regulators. Firstly, they regulate genes belonging to distinct functions.[37] For example, CRP regulates genes involved in carbohydrate metabolism as well as the

---

'Just-in-time transcription program in metabolic pathways', *Nature Genetics* 36 (2004), 486–91. https://doi.org/10.1038/ng1348

34 P. Ball, 'How life really Works', *Nautilus*, 6 November 2023. https://nautil.us/how-life-really-works-435813/.

35 J.A. Freyre-Gonzalez, J.A. Alonso-Pavon, L.G. Trevino-Quintanilla, and J. Collado-Vides, 'Functional architecture of *Escherichia coli*: new insights provided by a natural decomposition approach', *Genome Biology* 9 (2008), 2008. https://doi.org/10.1186/gb-2008-9-10-r154

36 These have been reviewed elsewhere. See, for example, U. Alon, 'Network motifs: theory and experimental approaches', *Nature Reviews Genetics* 8 (2007), 450–461. https://doi.org/10.1038/nrg2102
I have also summarised some of this in chapter 5 of my previous book. A.S.N. Seshasayee, *Bacterial Genomics: Genome organisation and gene expression tools* (Cambridge: Cambridge University Press, 2016). https://doi.org/10.1017/cbo9781139942225.005

37 A. Martinez-Antonio and J. Collado-Vides, 'Identifying global regulators in transcriptional regulatory networks in bacteria', *Current Opinion in Microbiology* 6 (2003), 482–489. https://doi.org/10.1016/j.mib.2003.09.002

genes encoding FlhDC, the master regulator of motility, among others. In contrast, TFs with only a few targets often regulate genes which all or mostly belonging to a single pathway or function. For instance, the TF TyrR regulates a small number of operons, all encoding genes involved in the metabolism of aromatic amino acids. In a more recent work, Julio Freyre-Gonzalez and colleagues suggested that global regulators integrate multiple *modules* within the transcriptional regulatory network.[38] Modules are groups of highly interconnected nodes, such that nodes within a module are more connected to each other than nodes from across modules. Modules are often determined by one or more TFs with local scope whereas global regulators sit astride multiple modules, presumably priming the expression of a large and diverse set of genes. Each module would, in some ways, represent a functionally coherent set of genes, and therefore Freyre-Gonzalez and colleagues' work might be taken as an independent validation of Martinez-Antonio's suggestion that global TFs regulate genes from multiple functions.

The *E. coli* genome encodes seven σ-factors. Each σ-factor helps to transcribe its own set of genes. Though there is some overlap between the sets of genes regulated by different σ-factors, one can say that the transcriptional space of *E. coli*, or for that matter that of many bacteria with large genomes, is *partitioned* among σ-factors.[39] Global TFs often regulate genes from different σ-factor partitions. According to the data analysed by Martinez-Antonio and co-workers, CRP regulates genes from as many as four σ-factor partitions! Global TFs do not usually regulate a gene as its sole regulator; they often act in concert with other TFs. Again, the regulation by CRP of its targets in sugar metabolism in concert with local TFs is a good example. Global TFs often regulate other TFs, something that TFs with a local scope rarely do. Finally, global TFs are also active in multiple conditions, whereas local regulators are usually activated by highly specific signals. Based on these parameters, Martinez-Antonio and Collado-Vides concluded that the *E. coli* genome encodes seven global TFs, and that a majority of target genes have at least one of these seven TFs as their regulators.

Years ago, I was involved in a piece of work that attempted to study how genes involved in metabolism are regulated in the transcriptional regulatory network of *E. coli*, and how different segments of the metabolic network might be regulated differently by global and local TFs.[40] The metabolic network can be visualised as an hourglass. A great diversity of nutrient breakdown pathways converge down to what is called central metabolism, which eventually produces energy. And a variety of biosynthetic pathways diverge away from these central metabolic pathways. Usually, nutrient breakdown pathways are regulated by TFs that respond to the nutrient itself, i.e., these are regulated by supply levels. On the other hand, biosynthetic pathways are regulated

38 Freyre-Gonzalez et al., 2008.

39 T.M. Gruber and C.A. Gross, 'Multiple sigma subunits and the partitioning of bacterial transcription space', *Annual Review of Microbiology* 57 (2003), 441–466. https://doi.org/10.1146/annurev.micro.57.030502.090913

40 A.S.N. Seshasayee, G.M. Fraser, M.M. Babu, and N.M. Luscombe, 'Principles of transcriptional regulation and evolution of the metabolic system in *E. coli*', *Genome Research* 19 (2009), 79–91. https://doi.org/10.1101/gr.079715.108

by TFs that bind to final product molecules, i.e., these are controlled by demand. These TFs are often local. Central metabolic pathways, in contrast, are regulated by a multitude of global TFs. This makes sense in light of Martinez-Antonio and Collado-Vides' work showing that global TFs respond to a wide range of environmental and cellular conditions, as well as the notion that central metabolic pathways are the culmination of a whole range of metabolic cues.[41] Whereas biosynthetic pathways are usually regulated by a single TF, the regulation of breakdown pathways often involves combinations of global and local TFs, as exemplified by CRP acting as a major node regulating a whole variety of breakdown pathways in concert with local TFs responding to particular nutrient molecules.

Finally, global TFs are usually expressed at high levels in the cell, and this presumably is to ensure their availability in sufficient concentrations to bind to all their targets. In contrast, local TFs are expressed at low levels; in fact, there is a positive correlation between the expression level of a TF and the number of targets it regulates. As shown by Grigory Kolesov and colleagues, for a TF present at low concentrations, finding one or a few target binding sites in the cell can be inefficient.[42] To counter this, local TFs are often encoded adjacently to their target sites. This logic works for bacteria in which transcription and translation occur more or less simultaneously, but not for eukaryotic cells in which transcription occurs within the nucleus whilst proteins are synthesised outside and so the TF will have to be transported back into the nucleus before it can bind to and regulate its target. There is also an evolutionary explanation for why local TFs are encoded close to their cognate targets. Horizontal transfer of a stretch of DNA carrying genes for a metabolic pathway is more likely to be successful if it also carries its own TF!

In summary, the transcriptional regulatory network—as exemplified in *E. coli*—is a complex structure with highly interconnected regulators and their targets. This structure ultimately determines which genes are expressed in the cell and when. Many regulators act in an intuitive manner, responding to a signal and regulating genes that should respond precisely to the inducing signal. In contrast, there are global TFs that integrate multiple segments of the metabolic network, and whose significance is a lot harder to understand and rationalise. In the next few sections, we will describe and try to understand the functioning of two regulatory networks determined by global TFs.

5.3. Driving the stress response: σ^S and its competition with σ^D

In *E. coli*, the major σ-factor σ^D regulates the expression of most genes involved in growth. It is the most abundant σ-factor protein in the cell and, out of the seven σ-factors encoded by the *E. coli* genome, it has the highest affinity for binding to the core RNAP. Its function contrasts with that of σ^S, the stress responsive σ-factor, whose production

41 Martinez-Antonio and Collado-Vides, 2003.

42 G. Kolesov, Z. Wunderlich, O.N. Laikova, M.S. Gelfand, and L.A. Mirny, 'How gene order is influenced by the biophysics of transcription regulation', *Proceedings of the National Academy of Sciences USA* 104 (2007), 13948–53. https://doi.org/10.1073/pnas.0700672104

increases as nutrients deplete and growth starts to cease. σ^S is also expressed during growth, but under conditions that are not ideal for growth of the bacterium. These include conditions of suboptimal osmolarity, temperature, and pH.[43] Like the global regulator CRP, σ^S does not respond nor activate cognate responses to specific stresses but plays a "preventive"[44] role, priming the cell to tolerate a wide range of stresses. Given its ability to contribute positively to cellular responses to such stresses, it is not surprising that the gene for σ^S was discovered independently multiple times by several researchers in the 1980s–early 1990s[45] before it was recognised that all these researchers had been referring to the same protein doing its job in different contexts![46]

The expression of σ^S is regulated by a plethora of mechanisms at every conceivable step in gene expression, from transcription through translation to protein stability.[47] Transcription of *rpoS*, the gene encoding σ^S, increases during down-shifts in growth, most notably as an *E. coli* culture transitions from exponential growth to a stationary phase.[48] There is evidence that the global TF CRP is involved in the up-regulation of σ^S during the stationary phase. The RegulonDB database for transcriptional regulatory interactions in *E. coli* also includes an acid stress regulator called GadX as a regulator of the σ^S gene. The small molecule guanosine tetraphosphate, which is produced during transition to states of starvation, also up-regulates the transcription of the σ^S gene. The concentration of the mRNA for σ^S also decreases in the absence of the cytosine methyltransferase Dcm (see Chapter 4). Various other regulators of σ^S expression have been described in other bacteria as well. However, the amount of σ^S mRNA is not a good predictor of the amount of σ^S protein: σ^S protein has been reported to be hardly detectable under some conditions in which the σ^S mRNA is abundant.[49] This suggests regulation of this gene beyond transcription. Some RNA binding proteins such as Hfq play roles in enabling translation of the σ^S mRNA in concert with other proteins and RNA that act as specific stress signals.[50] Even the protein HU, best known as a non-

43 R. Hengge-Aronis, 'Signal Transduction and Regulatory Mechanisms Involved in Control of the σS (RpoS) Subunit of RNAP', *Microbiology and Molecular Biology Reviews* 66 (2002), 373–395. https://doi.org/10.1128/mmbr.66.3.373-395.2002

44 Ibid., p. 374.

45 For example, P.C. Loewen and B.L. Triggs, 'Genetic mapping of katF, a locus that with katE affects the synthesis of a second catalase species in *Escherichia coli*', *Journal of Bacteriology* 160 (1984), 668–675. https://doi.org/10.1128/jb.160.2.668-675.1984; E. Touati, E. Dassa, and P.L. Bouquet, 'Pleiotropic mutations in appR reduce pH 2.5 acid phosphatase expression and restore succinate untilization in CRP-deficient strains of *Escherichia coli*', *Molecular and General Genetics* 202 (1986), 257–64. https://doi.org/10.1007/bf00331647

46 R. Lange and R. Hengge-Aronis, 'Identification of a central regulator of stationary-phase gene expression in Escherichia coli', *Molecular Microbiology* 5 (1991), 49–59. https://doi.org/10.1111/j.1365-2958.1991.tb01825.x

47 Hennge-Aronis, 2002.

48 Recall from Chapter 4 that a small inoculum of *E. coli* cells in fresh media, after a brief period of adaptation would start growing 'exponentially', or double in number at regular intervals until nutrient exhaustion and accumulation of toxic byproducts of growth metabolism cause cessation of cell multiplication in what is referred to as 'stationary phase'. During the stationary phase, any low rate of population growth is offset by cell death.

49 Hengge-Aronis, 2002.

50 D.D. Sledjeski and C. Whitman, 'Hfq is necessary for regulation by the untranslated RNA DsrA',

specific DNA binding protein, may bind to the σ^S mRNA and affect its translation.[51] The stability of σ^S protein is also regulated by complex signal cascades involving multiple proteins. Protein stability is a pivot point in σ^S expression that is targeted by adaptive strategies that use genetic evolution of σ^S, which we will return to later in this chapter. Thus, regulation at multiple stages of gene expression and protein stability seems to play a role in determining σ^S levels under different conditions. Whereas regulation at the level of transcription seems to predominate during steady transition to slow growth rates, stresses such as high osmolarity and low temperatures seem to particularly stimulate σ^S translation.

Harald Weber and colleagues identified genes regulated by σ^S in *E. coli* growing in three different conditions (stationary phase, osmotic, and acid stress) by measuring changes in the levels of the mRNA of all genes encoded by the genome in the presence and absence of an intact *rpoS* gene.[52] Of the ~500 genes that changed in expression in a σ^S-dependent manner in at least one of the three conditions, a third were defined as the 'core' σ^S *regulon*, being responsive to σ^S in all three conditions. The members of the core σ^S regulon typically contained an extended −10 element in their promoters, whereas the rest did not and might be expressed from sub-optimal promoter elements by σ^S-containing RNAP holoenzyme, in concert with other stress-responsive regulatory proteins. Byung-Kwan Cho and colleagues, while assembling a network of regulatory interactions between all σ-factors and their targets in *E. coli* (Fig. 5.5), showed that σ^S was bound to over 1,000 promoters in *E. coli*, but a majority of these interactions did not result in a change in gene expression when σ^S was removed.[53] The role of these binding interactions, if any, is yet to be understood. Whereas most of the genes whose expression changes in response to σ^S are activated by the σ-factor, the rest are in fact less expressed when σ^S is present. How could this be possible? Cho et al. showed that in many of these genes, the presence of σ^S reduced the binding of σ^D, the major housekeeping σ-factor. This suggests that competition between σ-factors, through which the presence of one σ-factor affects the influence of another, plays a role in determining the mRNA levels of several genes.

Journal of Bacteriology 183 (2001), 997–2005. https://doi.org/10.1128/jb.183.6.1997-2005.2001

51 A. Balandina, L. Claret, R. Hengge-Aronis, and J. Rouviere-Yaniv, 'The *Escherichia coli* histone-like protein HU regulates rpoS translation', *Molecular Microbiology* 39 (2001), 1069–1079. https://doi.org/10.1046/j.1365-2958.2001.02305.x

52 H. Weber, T. Polen, J. Heuveling, V.F. Wendisch, and R. Hengge, 'Genome-wide analysis of the general stress response network in Escherichia coli: σS-dependent genes, promoters, and sigma factor selectivity', *Journal of Bacteriology* 187 (2005), 1591–1603. https://doi.org/10.1128/jb.187.5.1591-1603.2005

53 B.-K. Cho, D. Kim, E.M. Knight, K. Zengler, and B.O. Palsson, 'Genome-scale reconstruction of the sigma factor network in *Escherichia coli*: topology and functional states', *BMC Biology* 12 (2014), 4. https://doi.org/10.1186/1741-7007-12-4

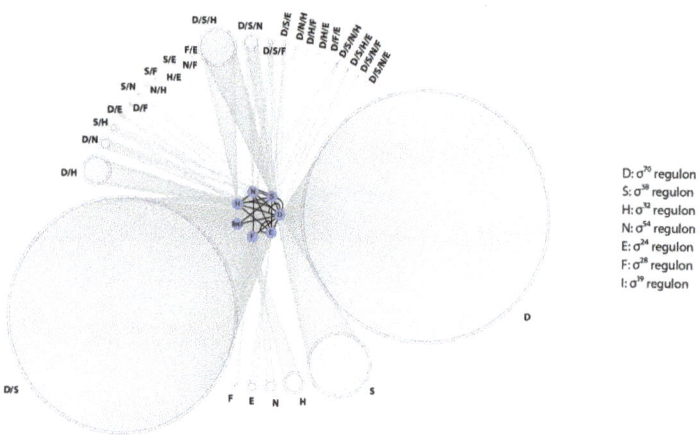

Fig. 5.5. Targets of various σ-factors in *E. coli*. This figure shows the sizes of various regulons (sets of targets of a regulatory protein) for *E. coli* σ-factors. σ^{38} is an alternative name for σ^S, and σ^{70} for σ^D. Originally published as Figure 2D in B.-K. Cho, D. Kim, E.M. Knight, K. Zengler, and B.O. Palsson, 'Genome-scale reconstruction of the sigma factor network in *Escherichia coli*: topology and functional states', *BMC Biology* 12 (2014), 4, CC BY 2.0.

For a σ-factor to activate the transcription of its targets, it first has to outcompete with other σ-factors for binding to the core RNAP; only then do holoenzymes bind to their respective promoters. Competition between σ-factors for a limited core RNAP—particularly the RNAP between σ^D and σ^S, both of which regulate large numbers of genes—can present some interesting problems. Though the concentration of σ^S increases as *E. coli* cells transition from exponential growth to a stationary phase, the absolute concentration of σ^S stays well below that of σ^D. In addition, the affinity of σ^S to the core RNAP is much less than that of σ^D. Therefore, even in stationary phase, σ^S, on its own, would be hard-pressed to compete effectively with σ^D to form the RNAP holoenzyme. According to calculations using the known total concentrations of σ^S, σ^D, and the RNAP in free and complexed forms, as well as the affinities of σ^S and σ^D to the RNAP, the concentration of the σ^S-holoenzyme (henceforth called $E\sigma^S$, with 'E' standing for the RNAP core enzyme) would be much smaller than that of $E\sigma^D$ even in stationary phase.[54] This immediately suggests the need for additional players that modulate σ-factor competition to favour the formation of $E\sigma^S$ in the stationary phase. Indeed, several such factors have been identified and described. One such protein, Crl, binds to σ^S and increases its affinity for the core RNAP, thus favouring the formation of $E\sigma^S$. This promotes the transcription of several σ^S-dependent genes.[55] There is evidence again that the small molecule guanosine tetraphosphate,

54 Comparing numbers from M. Mauri and S. Klumpp, 'A model for sigma factor competition in bacterial cells', *PLoS Computational Biology* (2014) e1003845, and those from Lal et al. 2018 quoted below. https://doi.org/10.1371/journal.pcbi.1003845

55 A. Typas, C. Barembruch, A. Possling, and R. Hengge, 'Stationary phase reorganisation of the *Escherichia coli* transcription machinery by Crl protein, a fine-tuner of sigmas activity and levels',

which signals starvation, also favours alternative σ-factors including σ^S in its competition with σ^D for holoenzyme formation.[56]

Two other molecules influence σ-factor competition by negatively targeting σ^D and $E\sigma^D$ activity. One is the protein Rsd, whose level increases during slow growth,[57] including in the stationary phase.[58] This protein binds to σ^D and sequesters it away from the core RNAP, thus removing a proportion of σ^D from the σ-factor competition. This should favour $E\sigma^S$ formation during the stationary phase. The other molecule influencing σ-factor competition is a non-coding RNA called 6S RNA.[59] 6S RNA is produced by the transcription of a gene called *ssrS*, but the RNA is not translated into protein. The 6S RNA adopts a looped structure that forms a base-paired motif resembling a standard *E. coli* promoter. This attracts $E\sigma^D$, thus reducing its availability for transcription while also removing part of σ^D during the stationary phase when the 6S RNA level is at its highest. Thus, while Rsd removes σ^D from the equation, 6S RNA reduces the availability of $E\sigma^D$ for transcription. The effect of 6S RNA therefore would also reduce the amounts of core RNAP available for σ-factors to bind to. Avantika Lal in my lab asked what effect each of these two methods of modulating the partitioning of transcription space across different $E\sigma$ holoenzymes would have on global gene expression in *E. coli*.[60] She used a combination of genome-scale gene expression measurements and mathematical modelling of σ-factor competition to approach this.

Experiments measuring mRNA expression levels of genes in *E. coli* cells lacking Rsd, in comparison with those that have the protein, showed that very few genes changed substantially in their expression between these two conditions. However, many σ^S target genes showed small but consistent decreases in expression levels in the absence of Rsd. Though Rsd operates by sequestering σ^D, σ^D targets did not show any change in their mRNA levels when Rsd was removed from the system. On the other hand, deletion of 6S RNA resulted in large changes in the expression levels of several genes, but the sets of genes responding to 6S RNA availability depended on the growth phase that the cells were in. But like the Rsd deletion, removal of 6S RNA also caused decreases in the expression of many σ^S target genes. In contrast to the Rsd deletion, removal of 6S RNA caused an increase in the expression of several σ^D target genes. Many of these σ^D targets showing elevated expression in the absence of 6S RNA were expressed at below average levels in the presence of 6S RNA. This suggests that one role for 6S RNA is in suppressing

EMBO Journal 26 (2007), 1569–1578. https://doi.org/10.1038/sj.emboj.7601629

56 M. Jishage, K. Kvint, V. Shingler, and T. Nystrom, 'Regulation of sigma factor competition by the alarmone ppGpp', *Genes and Development* 16 (2002), 1260–1270. https://doi.org/10.1101/gad.227902

57 R. Balakrishnan, M. Mori, I. Segota, Z. Zhang, R. Aebersold, C. Ludwig, and T. Hwa, 'Principles of gene regulation quantitatively connect DNA to RNA and proteins in bacteria', *Science* 378 (2022), eabk2066. https://doi.org/10.1126/science.abk2066

58 M. Jishage and A. Ishihama, 'Transcriptional organization andin vivo role of the Escherichia coli rsd gene, encoding the regulator of RNAP sigma D', *Journal of Bacteriology* 181 (1999), 3768–3776. https://doi.org/10.1128/jb.181.12.3768-3776.1999

59 K.M. Wassarman and G. Storz, '6S RNA regulates E. coli RNAP activity', *Cell* 101 (2000), 613–623. https://doi.org/10.1016/s0092-8674(00)80873-9

60 A. Lal, S. Krishna, and A.S.N. Seshasayee, 'Regulation of Global Transcription in *Escherichia coli* by Rsd and 6S RNA', *Genes Genomes Genetics* 8 (2018), 2079–2089. https://doi.org/10.1534/g3.118.200265

the expression of genes with relatively weak promoters. The removal of both 6S RNA and Rsd produced much greater effects on gene expression than would be expected from the product of these individual deletions.

These findings raised some important questions. First, Rsd acts by sequestering σ^D and any effect it has on σ^S function should be indirect. Yet, the aforementioned gene expression experiments showed that the removal of Rsd has an effect on the expression of σ^S targets but little, if any, on genes regulated by σ^D. How does this work? To answer this, Lal and colleagues used a mathematical model of holoenzyme formation and transcription that incorporated core RNAP, σ^D, σ^S, Rsd, 6S RNA, and DNA (Fig. 5.6). This model showed that in the presence of 6S RNA, increasing Rsd concentration results in the freeing up of $E\sigma^D$ from its complex with 6S RNA, leading to the release of some core polymerase which can then bind to σ^S and form $E\sigma^S$. The end result is an increase in transcription of $E\sigma^S$-dependent promoters. The model also showed that an increase in Rsd concentration also causes a decrease in transcription of σ^D target genes, something that was not apparent in the gene expression data. The gene expression experiments however showed that Rsd regulates 6S RNA: when Rsd is removed, there is a ~2.5-fold increase in the expression of 6S RNA. This could potentially decrease the availability of $E\sigma^D$ for transcription. Incorporation of this regulation of 6S RNA by Rsd into the mathematical model showed that it preserves the effect of Rsd on σ^S-dependent transcription but abolishes its effect on σ^D-dependent gene expression.

The second question that arises pertains to the effect of 6S RNA on transcription. 6S RNA not only sequesters σ^D but, by binding to $E\sigma^D$, also reduces the amount of the core RNAP available for σ^S to interact with. This should result in an overall decrease in transcription, though this effect would be more pronounced for σ^D-dependent genes. The theoretical model also supports this view. How then does the removal of 6S RNA selectively decrease σ^S-dependent transcription while increasing the expression of many σ^D-dependent genes? The gene expression data showed that the absence of 6S RNA reduced the expression of Rsd and increased that of σ^S itself. Further, it also decreased the expression of the catalytic subunit of the RNAP, which reduces the overall availability of this enzyme. These effects together help reduce σ^S-dependent transcription; in fact, a ~20% reduction in Rsd appears to be sufficient to reduce σ^S-dependent transcription in the absence of 6S RNA. 6S RNA also appears to affect the expression of other regulators of σ-factor competition such as Crl, all of which should contribute to σ^S competing effectively with σ^D for binding to core RNAP.

Thus, Rsd and 6S RNA regulate each other and also other players involved in transcription and σ-factor competition. These combined effects appear to be necessary for these molecules to modulate σ-factor competition in a manner that favours σ^S-dependent gene expression. Why such a complex web of interactions to modulate the interplay between σ^D and σ^S? We will try to answer this question when we explore how transcription regulatory networks evolve a little later in this chapter.

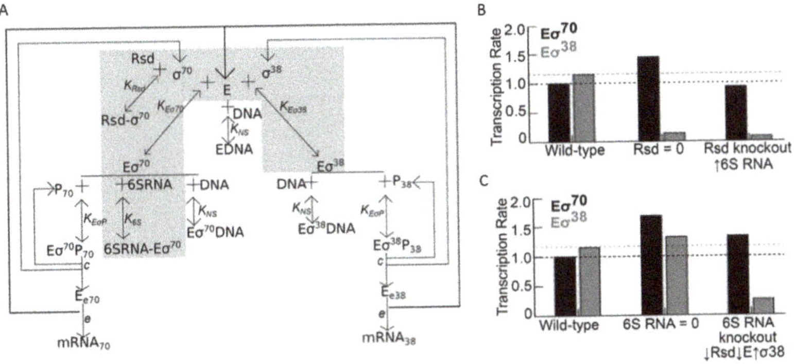

Fig. 5.6. Complex regulation of σS-σD competition. (A) A schematic showing the set of interactions involved in the regulation of σ^S-σ^D competition by the protein Rsd and the non-coding RNA 6S RNA. (B) Figure shows calculated transcription rate when Rsd is removed and when 6S RNA levels are increased in the absence of Rsd. (C) Figure shows calculated transcription rate when 6S RNA is removed, and the effect of the observed decrease in Rsd and core RNAP and increase in σ^S in the absence of 6S RNA. (B) and (C) show how functional connections between 6S RNA and Rsd appear to be important in determining the outcome of the competition between σ^S and σ^D. Originally published as part of Figure 6 in A. Lal, S. Krishna, and A.S.N. Seshasayee, 'Regulation of Global Transcription in *Escherichia coli* by Rsd and 6S RNA', *Genes Genomes Genetics* 8 (2018), 2079–2089, CC BY 4.0.

5.4. Managing the costs of horizontally acquired DNA: the 'genome sentinel'[61]

Horizontal transfer is a major mode of gene acquisition in bacteria (see Chapter 4), leading to genome expansion. As with any other segment of DNA, whether a piece of horizontally acquired DNA is maintained in a genome is a function of the selective pressure in its favour. This is especially so in bacteria with such large population sizes that the cost of merely maintaining and expressing a non-functional or neutral piece of DNA is enough for this DNA to be lost from the population (see Chapter 3). As described in Chapter 4, the selective pressure in favour of maintaining a piece of DNA in a genome could be conventional in that the DNA includes genes that enhance the growth and survival of the organism in its niche—for example, by allowing it to resist antibiotics in an antibiotic-rich environment. Or selection could arise from addiction, in which the loss of a piece of DNA once acquired proves toxic to the cell—a concept demonstrated by what are called toxin-antitoxin systems, which are often horizontally acquired. The cost that a piece of DNA presents to its host arises from the metabolic expense of maintaining it, as well as the possibility that its expression could prove

61 C.J. Dorman, 'H-NS, the genome sentinel', *Nature Reviews Microbiology* 5 (2007), 157–161. https://doi.org/10.1038/nrmicro1598

toxic to the cell or interfere with the functioning of molecules already well-established in the host. For example, Rotem Sorek and colleagues analysed gene fragments, from nearly 80 prokaryotic genomes, that could not be successfully transferred into *E. coli*.[62] They presented evidence that the expression of such genes, even at low levels, could not be tolerated by the host cell, arguing that these genes are toxic to *E. coli*. This toxicity prevents their successful establishment in the *E. coli* genome. For other genes, their failure to transfer into *E. coli* appeared to be best explained by increased dosage—or, in other words, their high expression levels. That high gene expression is a barrier to horizontal transfer of some genes was reaffirmed very recently by Rama Bhatia and colleagues, who studied the transfer of genes from *E. coli* into the closely-related *Salmonella*.[63] Thus, both toxicity and inappropriately high expression levels can act as barriers to horizontal gene transfer.

Many organisms encode dedicated mechanisms to control or even 'silence' horizontally-acquired genes at the level of their expression. A prominent example among eukaryotes is the silencing of selfish transposable elements by small regulatory RNA molecules in many plants. An example in bacteria involves a TF called H-NS, which in *E. coli* and related bacteria represses the expression of a variety of horizontally-acquired genes. H-NS, best known for its DNA-binding activities,[64] recognises A+T-rich sequences and binds extensively to such stretches of DNA. While doing so, it can form highly rigid or tightly looped structures that can either block the binding of RNAP to promoters or the relative movement of DNA and RNAP when the latter is already bound to a promoter. Now, it turns out that many horizontally-acquired genes in *E. coli* and related free-living bacteria tend to be A+T-rich. The genomes of *E. coli* and many related bacteria are usually nearly 50% A+T on average. However, the distribution of the A+T content of genes in these genomes is skewed towards the right. This means that very few genes are G+C-rich, but a larger proportion are A+T-rich, or more A+T-rich than the mean. Many such genes are believed, with good reasons, to have been acquired horizontally. These are often poorly conserved even across closely-related strains and species, and also include genes of bacteriophage origin. The high A+T content of many horizontally-acquired genes makes them attractive to H-NS for binding. Where H-NS binds, it represses transcription. Therefore, H-NS, which preferentially binds A+T-rich genes, emerges as a 'silencer' or repressor of the transcription of horizontally-acquired genes.

H-NS, as a protein that modifies the structure of DNA and also regulates gene

62 R. Sorek, Y. Zhu, C.J. Creevey, F.M. Pilar, P. Bork, and E. Rubin, 'Genome-wide experimental determination of barriers to horizontal gene transfer', *Science* 318 (2007), 1449–1452. https://doi.org/10.1101/2022.06.29.498157

63 R.P. Bhatia, H.A. Kirit, C.M. Lewis Jr, K. Sankaranarayanan, J.P. Bollback, 'Evolutionary barriers to horizontal gene transfer in macrophage-associated Salmonella', *Evolution Letters* 7 (2023), 227–239. https://doi.org/10.1093/evlett/qrad020

64 There is some evidence that H-NS can also bind to RNA. See C.C. Brescia, M.K. Kaw, and D.D. Sledjeski, 'The DNA binding protein H-NS binds to and alters the stability of RNA in vitro and in vivo', *Journal of Molecular Biology* 339 (2004), 505–514. https://doi.org/10.1016/s0022-2836(04)00382-1

expression, has been known since the 1980s.[65] However, its major role as a silencer of horizontally-acquired genes was not recognised until 2006, when two papers described the binding of H-NS to the chromosome of *Salmonella*—a close relative of *E. coli*—and the impact this binding has on gene expression on a genomic scale. The two pieces of work, one by William Navarre et al.[66] and the other by Sacha Lucchini et al.,[67] showed—by isolating and identifying chromosomal DNA regions bound by H-NS in *Salmonella* cells—that H-NS binds to hundreds of genes, including many coding for proteins that help the bacteria cause disease (Fig. 5.7). When H-NS was removed from cells by genetic means, genes bound by the protein greatly increased in expression, showing that H-NS represses the expression of genes it binds to. Many genes bound and regulated by H-NS are poorly conserved, are often specific to *Salmonella*, and show higher A+T-content than is typical of the average gene in this bacterium. Many of these genes have a role to play in the virulence of *Salmonella* and are normally expressed only during specific stages of infection and not during normal growth. The uncontrolled expression of these virulence-associated genes in the absence of H-NS is detrimental to the host bacterium. In fact, in the absence of additional, compensating mutations in the stress responsive σ-factor σ^S, the removal of H-NS is lethal to *Salmonella*. These findings clearly emphasised the importance of gene silencing to the fitness and evolutionary success of these bacteria.

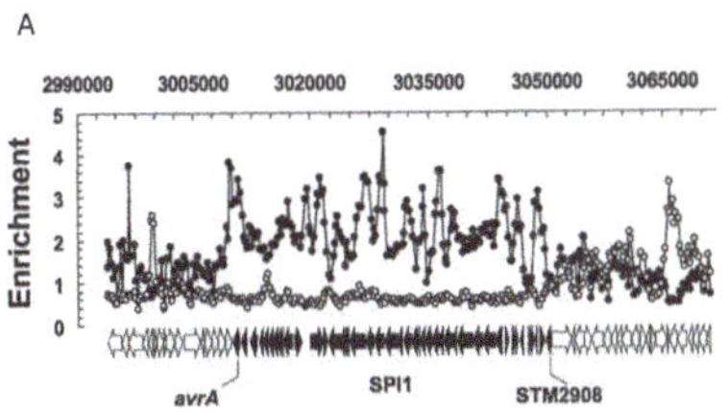

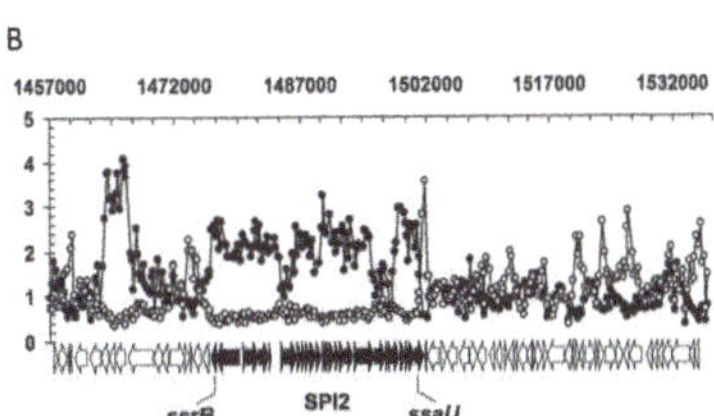

Fig. 5.7. Binding of H-NS to virulence genes. This figure shows the binding of H-NS (enrichment, on the y-axis) to pathogenicity-determining genes belonging to (A) SPI-1 and (B) SPI-2 pathogenicity islands in *Salmonella enterica* Typhimurium. Originally published as Figures 3D and 3E in S. Lucchini, G. Rowley, M.D. Goldberg, D. Hurd, M. Harrison, and J.C. Hinton, 'H-NS Mediates the Silencing of Laterally Acquired Genes in Bacteria', *PLoS Pathogens* 2 (2006), e81, Creative Commons Attribution License.

65 For an early review of H-NS, see C.F. Higgins, J.C. Hinton, C.S. Hulton, T. Owen-Hughes, G.D. Pavitt, and A. Seirafi, 'Protein H1: a role for chromatin structure in the regulation of bacterial gene expression and virulence?', *Molecular Microbiology* 4 (1990), 2007–2012. https://doi.org/10.1111/j.1365-2958.1990.tb00559.x

66 W.W. Navarre, S. Porwollik, Y. Wang, M. McClelland, H. Rosen, S.J. Libby, and F.C. Fang, 'Selective Silencing of Foreign DNA with Low GC Content by the H-NS Protein in Salmonella', *Science* 313 (2006), 236–238. https://doi.org/10.1126/science.1128794

67 S. Lucchini, G. Rowley, M.D. Goldberg, D. Hurd, M. Harrison, and J.C. Hinton, 'H-NS Mediates the Silencing of Laterally Acquired Genes in Bacteria', *PLoS Pathogens* 2 (2006), e81. https://doi.org/10.1371/journal.ppat.0020081

A few years later, Christina Kahramanoglou and colleagues, in a piece of work I was involved in, showed—using genome-scale techniques again—that H-NS binds to A+T-rich horizontally-acquired genes in *E. coli* and represses their expression.[68] However, unlike typical TFs, H-NS binding regions on the DNA extend over long stretches. The longer the stretch of DNA bound by H-NS, the more likely that a gene located at or proximal to the bound DNA would be transcriptionally silenced. Thus, the silencing of horizontally-acquired genes by H-NS requires many molecules of the protein to bind adjacently, essentially covering long stretches of DNA like beads on a string.

Studies of H-NS in *E. coli*, unlike those in *Salmonella*, rarely reported lethality when the protein was removed. Some studies observed a slight reduction in growth rates, and others none. Despite performing similar functions in *Salmonella* and *E. coli*, H-NS is essential for bacterial survival in one and not the other. These observations suggest that there is something else in the genome of *E. coli*, presumably differing in some way from the contents of the *Salmonella* genome, that minimises the impact of the loss of H-NS on survival. What might this be? The *E. coli* genome encodes a protein called StpA, which is similar in sequence to H-NS. StpA also binds to A+T-rich DNA sequences, but is produced by *E. coli* cells in much smaller quantities than H-NS. Removing StpA from *E. coli* cells has little, if any, impact on growth, at least in laboratory conditions. Ebru Uyar and colleagues identified sites on the chromosome within *E. coli* cells that are bound by H-NS and StpA in the presence and absence of the other protein.[69] They first found that the binding of H-NS to the *E. coli* chromosome is unaffected by the presence or absence of StpA. StpA binds to the same locations as H-NS when the latter is also present. However, when H-NS is removed from cells, StpA loses its ability to bind to as many as two-thirds of its sites. Uyar and co-workers further suggest that this reduction in binding of StpA to the chromosome in the absence of H-NS probably reflects the intrinsic binding properties of StpA. They also propose that H-NS can induce changes in the structure of the DNA that it binds to, which might enable StpA to bind to these regions of the chromosome. The loss of such DNA structural features in the absence of H-NS might reduce the ability of StpA to bind to it.

Does the binding of StpA to a small subset of its targets in the absence of H-NS help maintain the repression of these genes? If so, is there some basis to *which* subset of H-NS regulated genes are 'chosen' for repression by the StpA-dependent backup regulatory system? Rajalakshmi Srinivasan in my lab attempted to address this question. She first asked what effect the removal of one or both proteins will have on the expression levels of horizontally-acquired genes in *E. coli*.[70] Consistent with the

---

68 C. Kahramanoglou, A.S. Seshasayee, A.I. Prieto, D. Ibberson, S. Schmidt, J. Zimmermann, V. Benes, G.M. Fraser, and N.M. Luscombe, 'Direct and indirect effects of H-NS and Fis on global gene expression control in Escherichia coli', *Nucleic Acids Research* 39 (2011), 2073–2091. https://doi.org/10.1093/nar/gkq934

69 E. Uyar, K. Kurokawa, M. Yoshimura, S. Ishikawa, N. Ogasawara, and T. Oshima, 'Differential binding profiles of StpA in wild-type and hns mutant cells: a comparative analysis of cooperative partners by chromatin immunoprecipitation-microarray analysis', *Journal of Bacteriology* 191 (2009), 2388–2391. https://doi.org/10.1128/jb.01594-08

70 R. Srinivasan, D. Chandraprakash, R. Krishnamurthi, P. Singh, V. Scolari, S. Krishna, and A.S.N.

findings of Uyar and colleagues, she found that loss of StpA affects the expression of very few genes. However, when H-NS is removed, the expression of many of its targets—including horizontally-acquired genes—greatly increases. Most importantly, when both H-NS and StpA are removed, many other genes—including those bound by H-NS but unaffected by the loss of H-NS alone—increase in expression. Srinivasan and colleagues compared their gene expression data with Uyar and colleagues' chromosome binding data for H-NS and StpA. In doing so, they found that genes whose expression increases when StpA is removed from *E. coli* already lacking H-NS are often those to which StpA remains bound in the absence of H-NS.

In an earlier study, Blair Gordon and colleagues had measured the affinity of H-NS to tens of thousands of 8-mer DNA sequences.[71] Using these data, Srinivasan and colleagues noted that regions to which StpA binds in the absence of H-NS tend to have a high density of 8-mers that display high-affinity binding to H-NS—and, by inference, to StpA. This finding offers a biophysical rationale for how StpA is able to retain its ability to bind to some but not all its wildtype sites. Srinivasan and co-workers also found that while the loss of H-NS alone has very little effect on growth of *E. coli* under the conditions they had tested in the lab, the loss of both H-NS and StpA resulted in a large growth impairment. Thus, StpA binds to a subset of H-NS-repressed, horizontally-acquired genes in the absence of H-NS and dampens their over-expression when H-NS is lost from the system. This backup function of StpA also helps soften the adverse effect of the loss of H-NS on bacterial growth. Thus, these results lead to the suggestion that keeping horizontally-acquired genes—to which StpA binds in the absence of H-NS—transcriptionally silent is important to ensure that the loss of H-NS on its own does not severely impair the growth of *E. coli*.

Srinivasan and co-workers also observed that genes that are repressed by StpA in the absence of H-NS: (a) are expressed at very low levels, lower than genes repressed by H-NS but not by StpA in the absence of H-NS; (b) transition to very high expression levels when H-NS and StpA are removed from the system. These two findings suggest that these horizontally-acquired genes that are silenced by both H-NS and StpA have a high intrinsic ability for transcription. Transcription at these genes may not even produce full length mRNA.[72] Instead, the high A+T content of these genes, in the absence of H-NS/StpA, exposes many promoter-like elements within gene sequences. This attracts RNAP, causing it to waste resources by performing useless transcription. Going by the bioenergetic cost calculations by Lynch and Marinov that we discussed in Chapter 3, these genes—if left unregulated—would carry a very large negative

---

Seshasayee, 'Genomic analysis reveals epistatic silencing of "expensive" genes in Escherichia coli K-12', *Molecular Biosystems* 9 (2013), 2021–2033. https://doi.org/10.1039/c3mb70035f

71 B.R. Gordon, Y. Li, A. Cote, M.T. Weirauch, P. Ding, T.R. Hughes, W.W. Navarre, B. Xia, and J. Liu, 'Structural basis for recognition of AT-rich DNA by unrelated xenogeneic silencing proteins', *Proceedings of the National Academy of Sciences USA* 108 (2011), 10690–10695. https://doi.org/10.1073/pnas.1102544108

72 S.S. Singh, N. Singh, R.P. Bonocora, D.M. Fitzgerald, J.T. Wade, and D.C. Grainger, 'Widespread suppression of intragenic transcription initiation by H-NS', *Genes and Development* 28 (2014), 214–219. https://doi.org/10.1101/gad.234336.113

selection coefficient ($s << 0$) and be detrimental to the cell if not eliminated. And the role of the silencing system orchestrated by H-NS and StpA is to ensure that they do not get expressed inappropriately when not required. In fact, as demonstrated by Marie Doyle and co-workers, some horizontally-acquired plasmids—accessory genetic elements found in many copies in cells—encode their own version of H-NS to ensure that the expression of plasmid-borne genes is kept under control without syphoning off the host cell's endogenous H-NS reserves.[73] This function, by ensuring that the use of H-NS to silence plasmid-encoded genes does not compromise its function on chromosomal DNA, enables the maintenance of the plasmid in its host.

Though many horizontally-acquired genes may be detrimental to the cell if allowed to be transcribed under favourable growth conditions, there is evidence that some of these genes might in fact be beneficial ($s >> 0$) under stress. For example, several genes normally repressed by H-NS in *Salmonella* are required for virulence, but can negatively impact growth when expressed under benevolent laboratory conditions. Positive s under certain conditions could ensure that these genes are maintained in the bacterial population. Additional regulatory mechanisms, such as anti-H-NS proteins that displace H-NS from the DNA, can relieve repression by H-NS and StpA precisely when necessary.[74]

Salmonella, in which the deletion of H-NS is lethal, also encodes an StpA. However, it does not seem to be capable of supporting bacterial survival and growth in the absence of H-NS. Why would this be so? Sabrina Ali and colleagues allowed *Salmonella* lacking H-NS[75] to grow in the laboratory in such a way that these slow-growing populations could accumulate additional mutations.[76] Some of these mutations would be adaptive, allowing the bearer to grow faster than its parent. Such adaptive mutations would soon dominate in the population, as predicted by natural selection. This would then allow investigators to discover mechanisms by which *Salmonella* can compensate for growth defects caused by the loss of H-NS. Ali et al. found that the loss of pathogenicity islands—which are usually kept silent by H-NS but are expressed at high levels in an inappropriate manner in the absence of the repressor—allows *Salmonella* lacking H-NS to adapt to the loss of the repressor. This finding reinforces the idea that improper expression of horizontally-acquired virulence genes, under conditions in which virulence has no role to play in the organism's lifestyle, can be costly. It also shows that evolution would quickly result in the loss of such expensive pieces of DNA if they happened to reside and be expressed inside bacterial cells when not required. In addition, these researchers found that mutations in StpA also allowed

73 M. Doyle, M. Fookes, A. Ivens, M.W. Mangan, J. Wain, and C.J. Dorman, 'An H-NS-like stealth protein aids horizontal DNA transmission in bacteria', *Science* 315 (2007), 251–252. https://doi.org/10.1126/science.1137550

74 D.M. Stoebel, A. Free, and C.J. Dorman, 'Anti-silencing: overcoming H-NS-mediated repression of transcription in Gram-negative enteric bacteria', *Microbiology* 154 (2008), 2533–2545. https://doi.org/10.1099/mic.0.2008/020693-0

75 As well as σS. *Salmonella* lacking H-NS can survive in the absence of σS activity.

76 S.S. Ali, J. Soo, C. Rao, A.S. Leung, D.H. Ngai, A.W. Ensminger, and W.W. Navarre, 'Silencing by H-NS potentiated the evolution of Salmonella', *PLoS Pathogens* 10 (2014), e1004500. https://doi.org/10.1371/journal.ppat.1004500

the H-NS-negative *Salmonella* to adapt to the absence of H-NS (Fig. 5.8A). StpA in *Salmonella* is not identical in sequence to that in *E. coli*. A quick look at the two StpA sequences showed that some of the mutations which allowed *Salmonella* to adapt to the loss of H-NS targeted amino acid positions at which the *Salmonella* StpA differed from the *E. coli* StpA.[77] An open question from this analysis is whether the Salmonella *StpA*, in its normal form, is unable to act as an effective backup for H-NS, and whether the mutations discovered by Ali and co-workers allow it to do so!

Rajalakshmi Srinivasan in my lab performed an experiment similar to that by Sabrina Ali and colleagues, but for *E. coli* lacking both H-NS and StpA.[78] She expected to see losses of segments of horizontally-acquired DNA in populations displaying adaptation to the absence of H-NS and StpA. However, this did not happen. Instead, she first observed that mutations that inactivate σS emerged; this was not surprising in light of prior evidence linking H-NS and σS. Yet, this did highlight an important point which we will examine shortly: that mutations which perturb portions of the transcriptional regulatory network can be adaptive.

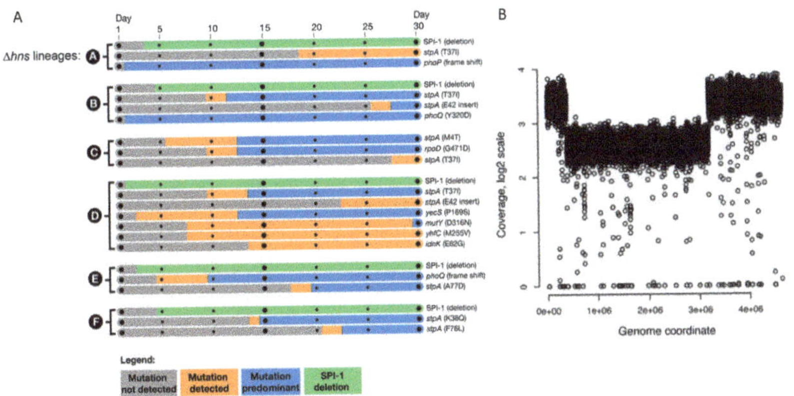

Fig. 5.8. Compensation for the loss of the H-NS gene silencing system. (A) This figure shows that mutations that change StpA and those that delete clusters of pathogenicity-related genes compensate for the loss of H-NS in *Salmonella*. Originally published as Figure 3 in S.S. Ali, J. Soo, C. Rao, A.S. Leung, D.H. Ngai, A.W. Ensminger, and W.W. Navarre, 'Silencing by H-NS potentiated the evolution of Salmonella', *PLoS Pathogens* 10 (2014), e1004500, Creative Commons Attribution License. (B) This figure shows that a duplication of nearly 40% of the chromosome centred around *ori* (which is located at ~3.9e06 on the x-axis) partially compensated for the lack of H-NS and StpA in *E. coli*. Originally published as Figure 3E in R. Srinivasan, V.F. Scolari, M.C. Lagomarsino, and A.S.N. Seshasayee, 'The genome-scale interplay amongst xenogene silencing, stress response and chromosome architecture in *Escherichia coli*', *Nucleic Acids Research* 43 (2005), 295–308, CC BY 4.0.

77 An analysis I had quickly performed when Ali et al., 2014 was published.

78 R. Srinivasan, V.F. Scolari, M.C. Lagomarsino, and A.S.N. Seshasayee, 'The genome-scale interplay amongst xenogene silencing, stress response and chromosome architecture in *Escherichia coli*', *Nucleic Acids Res.* 43 (2015), 295–308. https://doi.org/10.1093/nar/gku1229

A second mutation that emerged, partly compensating for the loss of H-NS and StpA, was a duplication of nearly 40% of the chromosome (Fig. 5.8B). This mutation, while increasing the expression of many genes in the duplicated segment of the chromosome, also caused a strong reduction in the expression of horizontally-acquired genes that had been derepressed by the loss of H-NS and StpA. This underlined the idea that rearrangements of large parts of the chromosome can also be adaptive, and can compensate in part for the loss of a global regulatory network. Keeping in mind the fact that the H-NS-StpA system primarily represses horizontally-acquired genes, we can now ask the following questions: are genes of different functions and of different evolutionary origins positioned differently on the chromosome, and how does this arrangement interplay with gene expression? We will address these questions in the final section of this book. But before that, we make a detour and ask how transcriptional regulatory networks evolve.

5.5. Evolving regulation

Evolution, via changes in the sequence of the genome, is central to adaptation and to the continued existence of life on our planet. Over shorter timescales, gene regulation is a physiological response that allows an organism to react to fast-changing circumstances. The repertoire and function of the machinery responsible for gene expression and its regulation are also subject to change through genetic evolution, even as organisms explore new niches and lifestyles. Even the RNAP, despite being a highly conserved and essential multi-subunit protein, shows variation in the sequences of its component subunits across bacteria, and some of these variations are adaptive. Even closely-related bacteria, and members of the same species, show such variations!

As discussed in Chapter 4, mutations in the main enzymatic subunit of the RNAP confer resistance to the antibiotic rifampicin, most notably in the pathogen *Mycobacterium tuberculosis*. A catalogue of known antibiotic resistance mutations in this pathogen, made publicly available by the World Health Organisation on their website,[79] includes nearly 25 entries (plus several hundred more whose significance for resistance is unclear) for the core subunits of the RNAP. These mutations, in all likelihood, inhibit the binding of the antibiotic to its target protein, the catalytic component of the RNAP, or make the RNAP impervious to interactions with the antibiotic. In addition, the presence of such mutations, at least in one example, is associated with the elevated expression of transporter proteins that throw the antibiotic out of the cell.[80] It is not by any means clear that this effect on gene expression is a direct and specific consequence of the mutation in RNAP and not merely a feedback mechanism arising from extended exposure of these resistant bacteria to the antibiotic. Nevertheless, this does support a role for gene expression changes, which

79 https://www.who.int/publications/i/item/9789240028173

80 G.J. de Knegt, O. Bruning, M.T. ten Kate, M. de Jong, A. van Belkum, H.P. Endtz, T.M. Breit, I.A. Bakker-Woudenberg, and J.E. de Steenwinkel, 'Rifampicin-induced transcriptome response in rifampicin-resistant Mycobacterium tuberculosis', *Tuberculosis* 93 (2013), 96–101. https://doi.org/10.1016/j.tube.2012.10.013

may or may not be directly linked to mutations in the RNAP, in resistance to rifampicin.

In *E. coli*, mutations in the core or the σ subunits of the RNAP are not uncommon. Yasmin Cohen and Ruth Hershberg found that *E. coli* populations adapting to new, uncomfortable environments—most commonly exposure to antibiotics or high temperatures—found mutations in core RNAP subunits (Fig. 5.9).[81] These mutations often affected amino acid residue positions that are otherwise conserved across RNAPs from thousands of different *E. coli* types. These residues also happen to be present close to the active centre of the RNAP enzyme. With conserved amino acid residues usually being important for some crucial aspect of protein function, one can assume with reasonable confidence that most of these RNAP core mutations change RNAP activity in some way, and presumably in a manner that allows the bacterium to improve in the environment concerned. A particular example of a circumstance in which mutations in both the core and the σ subunits of the RNAP confer a selective advantage on the bacterium is late in the stationary phase, during which the environment is inimical to bacterial population growth.

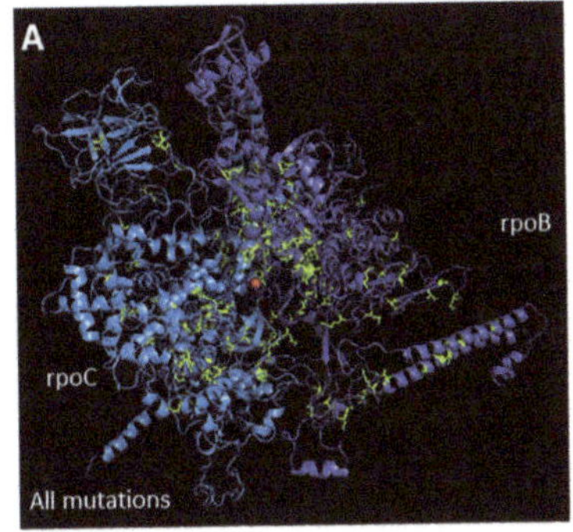

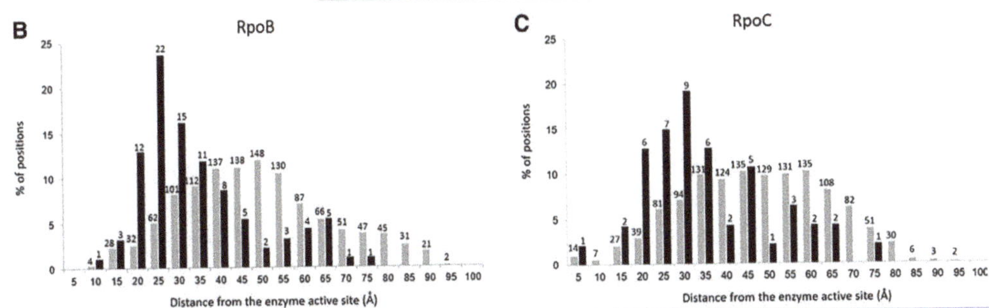

Fig. 5.9. Adaptive mutations in the RNAP. (A) This figure shows the structure of the RNAP, highlighting sites showing adaptive mutations in light green. (B) and (C) These figures show that adaptive mutations in the RNAP occur close to the active site of the enzyme, which is involved in performing the transcription reaction. Originally published as Figure 2 in Y. Cohen and R. Hershberg, 'Rapid Adaptation Often Occurs through Mutations to the Most Highly Conserved Positions of the RNAP Core Enzyme', *Genome Biology and Evolution* 14 (2022), evac105, CC BY 4.0.

81 Y. Cohen and R. Hershberg, 'Rapid Adaptation Often Occurs through Mutations to the Most Highly Conserved Positions of the RNAP Core Enzyme', *Genome Biology and Evolution* 14 (2022), evac105. https://doi.org/10.1093/gbe/evac105

Many bacteria survive long in the stationary phase, and many do so by developing into 'spores' or forms that are dormant but hardy and ready to explode into normal, rapid growth when the environment changes. *E. coli*, despite lacking the ability to form spores, can survive for years in a stationary phase! A few days after entering stationary phase, non-growing *E. coli* cells start to die. However, this phase of death does not fully wipe out the population. Some 1% of the population manages to survive. Maria Zambrano and co-workers[82] showed that after 10 days of stationary phase, the survivors include mutants of the stress-responsive σ-factor σ^S, a mutation we will refer to here as σ^S -GASP1.[83] The mutation is a duplication of a portion of the gene for σ^S, which produced a variant protein with much lower expression and, therefore, host cells with reduced σ^S activity. The activity of σ^S-GASP1 in these cells, though reduced, was not fully abolished. Bacteria carrying this mutation multiplied in the nutrient-depleted stationary phase environment and outcompeted their parents lacking the mutation. Thus, the variant σ^S-GASP1 protein allowed the desperate *E. coli* population to adapt to its trying circumstances. The phenomenon in which a newly emerging variant/mutant outcompetes its parent while growing its population deep in the stationary phase has been termed *GASP*, an acronym for Growth Advantage in Stationary Phase.

Deactivating mutations in σ^S also emerge in *E. coli* lacking H-NS and StpA introduced earlier in this chapter, and partially offset the growth defect caused by the loss of these proteins. In fact, the lethality of the loss of H-NS in *Salmonella* can be reduced by de-activating mutations in σ^S. Thus, σ^S mutations appear to emerge and provide adaptive benefits to bacteria in multiple contexts, often those that cause prolonged slow, very suboptimal, growth.[84] As argued by Thomas Ferenci, such circumstances call for a delicate balance between growth-promoting and stress-responsive functions,[85] which are enabled by opposing σ-factor activities—σ^D and σ^S respectively, as described earlier. Full σ^S activity may not permit even the slow growth that a non-benevolent environment might still allow. Complete abolition of σ^S activity, on the other hand, might make *E. coli* extremely vulnerable to the hazards presented by the same environment. Thus, it might be careful fine-tuning of σ^S activity and balance in the competition between σ^D and σ^S that is called for under such circumstances. This might very well occur in *E. coli* populations entering a deep stationary phase.[86]

Considerable work, published in the years following the publication of Zambrano and colleagues' work, showed that as the stationary phase progresses, fresh variants

---

82 M.M. Zambrano, D.A. Siegele, M. Almiron, A. Tormo, and R. Kolter, 'Microbial competition: *Escherichia coli* mutants that take over stationary phase cultures', *Science* 259 (1993), 1757–1760. https://doi.org/10.1126/science.7681219

83 This mutation is often referred to in the published literature as *rpoS819*.

84 T. Ferenci, 'What is driving the acquisition of mutS and rpoS polymorphisms in Escherichia coli?', *Trends in Microbiology* 11 (2003), 457–461. https://doi.org/10.1016/j.tim.2003.08.003

85 T. Ference, 'Maintaining a healthy SPANC balance through regulatory and mutational adaptation', *Molecular Microbiology* 57 (2005), 1–8. https://doi.org/10.1111/j.1365-2958.2005.04649.x

86 P. Nandy, 'The role of sigma factor competition in bacterial adaptation under prolonged starvation', *Microbiology* 168 (2022), 001195. https://doi.org/10.1099/mic.0.001195

keep emerging, each trying to thrive while outcompeting their parents and driving them to extinction.[87] Thus, the stationary phase for *E. coli* is not as stationary as its name would indicate. Instead, it is characterised by the aggregate lives of ever-emerging and disappearing variants of *E. coli* engaging in various interactions among themselves while adapting, surviving, and growing in the dynamic, yet increasingly forbidding environment. Till recently, however, the genetic composition or diversity of the *E. coli* population inhabiting this changing environment remained uncharacterised. Savita Chib and Farhan Ali in my lab sought to fill this gap.[88]

Chib and Ali sequenced the genomes of five *E. coli* populations maintained in stationary phase for four weeks at several time points over the lifetime of these populations.[89] They identified several mutations appearing in these populations and found that most waxed and waned over these four weeks. All in all, the genetic diversity of these populations—defined by the number of different mutations seen in a population—increased consistently over time. This increase in genetic diversity was not random and occurred under selection. A more recent study by Sophia Katz and colleagues, interrogating the genomes of multiple *E. coli* populations kept in stationary phase for a staggering three years, also showed that some populations acquired the ability to mutate at a faster rate over time.[90] Both studies found that the same mutations appeared across multiple populations. This suggested that the same genetic strategies enable adaptation to deep stationary phases in independent populations, pointing to the repeatability of evolutionary strategies in these environments. Importantly in the context of the present discussion, both studies identified several mutations in the RNAP subunits emerging in these populations. In fact, Katz et al. showed that over 90% of all genomes sequenced in their study, across multiple independent populations over several years, carried a mutation in a core RNAP subunit.

Pabitra Nandy in my lab, following up on the work by Chib and Ali, noticed that *E. coli* with a mutation in core RNAP had appeared after around three weeks of maintained stationary phase in the population.[91] This mutation was found alongside another in σ^S, which we will call σ^S-GASP2 for it is a variant of σS-GASP1 described above. This mutant was hardy and slow-growing, and yet able to outcompete its faster-growing (in rich media) ancestors in highly limiting stationary phase media—but not in media from, say, *E. coli* cultures grown for only a few days in stationary

87 Reviewed in S.E. Finkel, 'Long-term survival during stationary phase: evolution and the GASP phenotype', *Nature Reviews Microbiology* 4 (2006), 113–120. https://doi.org/10.1038/nrmicro1340

88 S. Chib, F. Ali, and A.S.N. Seshasayee, 'Genomewide mutational diversity in Escherichia coli population evolving in prolonged stationary phase', *mSphere* 2 (2017), e00059-17. https://doi.org/10.1128/msphere.00059-17

89 Ibid.

90 S. Katz, S. Avrani, M. Yavneh, S. Hilau, J. Gross, and R. Hershberg, 'Dynamics of adaptation during three years of evolution under long-term stationary phase', *Molecular Biology and Evolution* 38 (2021), 2778–2790. https://doi.org/10.1093/molbev/msab067

91 P. Nandy, S. Chib, and A. Seshasayee, 'A Mutant RNA Polymerase Activates the General Stress Response, Enabling *Escherichia coli* Adaptation to Late Prolonged Stationary Phase', *mSphere* 5 (2020), e00092-20. https://doi.org/10.1128/msphere.00092-20

phase. The σS-GASP2 mutant showed higher σ^S activity than σ^S-GASP1, as measured by the extent to which genes known to be expressed under the action of RNAP- σ^S holoenzyme changed in expression between the two σ^S mutants. Now, the RNAP core mutation somehow enhanced the degree to which known σ^S targets are expressed, irrespective of whether σ^S-GASP1 or σ^S-GASP2 was present. This ability of the RNAP core mutation, however, required some residual σ^S activity and was lost in the complete absence of σ^S. This suggested that the RNAP core mutation differentially affected gene expression via σ^S, presumably by tilting the σ^S–σ^D competition in favour of σ^S. The RNAP mutation did not clearly enhance the stationary phase growth of *E. coli* carrying σ^S-GASP2; however, the σ^S-GASP1 + RNAP core double mutant considerably outperformed the σ^S-GASP1 single mutant, suggesting that the RNAP core mutation might have emerged in a σ^S-GASP1 background and that σ^S-GASP2 developed later. Together, these findings present the argument that the balance between growth and stress response shifts in favour of the latter as stationary phase progresses, and that this balance can be tuned by mutations not only in the relevant σ-factor genes but also those in the core RNAP.

The bacterial regulatory network, though constrained by the capabilities of the RNAP and the σ-factors, is driven by a plethora of regulatory proteins such as TFs and their interactions with the DNA. Mutations in any of these components can influence gene expression and, in some instances, do so in an adaptive manner. A few mutations in a TF can easily change the binding properties of a TF, as demonstrated by Ryan Schultzaberger and colleagues using artificially-generated sequence variants of a bacterial TF.[92] Let us, to begin with, examine the phenomenon of GASP once more. Chib and Ali, while analysing their data on mutations emerging over four weeks in stationary phase *E. coli* populations, found several mutations in proteins involved directly or indirectly in the regulation of gene expression. This was in addition to mutations in core RNAP and σ-factors. They found that regulatory proteins were in fact more likely to be altered by adaptive mutations than non-regulatory proteins. These included an enzyme responsible for the degradation of cyclic AMP, a small molecule whose levels depend on the availability of glucose. Mutations in this gene were observed in several independent stationary phase *E. coli* populations, pointing to an important role for this modification in stationary phase growth. Cyclic AMP binds to and activates the global TF CRP, which in turn regulates the expression of hundreds of genes. CRP, being required for the metabolism of many unusual carbon sources, can be expected to play a role in gene expression affecting stationary phase survival. It is therefore not unreasonable to expect that this mutation in the enzyme that degrades cyclic AMP will result in an increase in the levels of this small molecule and will thereby change the expression of some genes under the control of CRP. In fact, a very recent study by Shira Zion and

92 R.K. Shultzaberger, S.J. Maerkl, J.F. Kirsch, and M.B. Eisen, 'Probing the informational and regulatory plasticity of a transcription factor DNA-binding domain', *PLoS Genetics* 8 (2012), e1002614. https://doi.org/10.1371/journal.pgen.1002614

colleagues reported that multiple *E. coli* populations gain mutations in CRP, which further potentiate the emergence of many other secondary mutations, deep into stationary phase.[93] Consistent with the findings of Chib and Ali, Nicole Ratib and co-workers also found many mutations in regulatory proteins in *E. coli* maintained in stationary phase for nearly three years.[94]

In an early work, Erik Zinser and Roberto Kolter showed that a mutation in the global TF Lrp provides a growth advantage during stationary phase.[95] Lrp is primarily a regulator of genes involved in amino acid metabolism.[96] In particular, it represses the expression of genes that help in the breakdown of certain amino acids, and those involved in the uptake of peptides (short chains of amino acid residues). The GASP mutation in Lrp, which abolishes the DNA binding activity of the protein, may enable increased growth in nutrient-deprived stationary phase conditions by increasing amino acid breakdown, leading to their increased utilisation as sources of nutrition and energy instead of mere building blocks of proteins.

In a more recent piece of work, Savita Chib and Subramony Mahadevan showed that a mutation in H-NS, the silencer of horizontally-acquired genes introduced earlier, conferred the GASP phenotype.[97] The mutation they discovered reduced the activity of H-NS. Among genes whose expression was derepressed by decreased H-NS activity are those involved in the utilisation of unusual sugars as carbon sources. These sugars may be unusual in standard laboratory media, but may become available as cells metabolise and excrete their way into a deep stationary phase. Taken together, mutations in Lrp and in H-NS allow the simultaneous expression of genes that help cells find unusual sources of nitrogen and carbon respectively, something that they would not do during normal, rapid growth, and are selected for during deep stationary phases. Since these genes are typically under the control of σ^D, their activation is further enabled in σ-factor mutations that tune the balance between growth-promoting nutrient utilisation programmes and stress responses.

Mutations in TFs enable adaptation in other circumstances as well, such as in antibiotic resistance. A commonly cited example of a TF with a role in antibiotic resistance is MarR.[98] MarR indirectly, through the action of another TF whose

93 S. Zion, S. Katz, and R. Hershberg, '*Escherichia coli* adaptation under prolonged resource exhaustion is characterized by extreme parallelism and frequent historical contingency', *PLoS Genetics* 20 (2024), e1011333. https://doi.org/10.1101/2024.03.21.586114

94 N.R. Ratib, F. Seidl, I.M. Ehrenreich, and S.E. Finkel, 'Evolution in Long-Term Stationary-Phase Batch Culture: Emergence of Divergent Escherichia coli Lineages over 1,200 Days', *mBio* 12 (2021), e03337–20. https://doi.org/10.1128/mbio.03337-20

95 E.R. Zinser and R. Kolter, 'Prolonged stationary phase incubation selects for lrp mutations in *Escherichia coli* K12', *Journal of Bacteriology* 182 (2000), 4361–4365. https://doi.org/10.1128/jb.182.15.4361-4365.2000

96 J.M. Calvo and R.M. Matthews, 'The leucine responsive regulatory protein, a global regulator of metabolism in *E. coli*', *Microbiology Reviews* 1994 (1994), 466–490. https://doi.org/10.1128/mmbr.58.3.466-490.1994

97 S. Chib and S. Mahadevan, 'Involvement of the global regulator H-NS in the survival of *Escherichia coli* in stationary phase', *Journal of Bacteriology* 194 (2012), 5285–5293. https://doi.org/10.1128/jb.00840-12

98 G.A. Beggs, R.G. Brennan, and M. Arshad, 'MarR family proteins are important regulators of clinically relevant antibiotic resistance', *Protein Science* 29 (2020), 647–653. https://doi.org/10.1002/pro.3769

expression it directly controls, represses the expression of efflux pumps, which eject antibiotics and other toxic molecules out of the cell with broad specificity. Tens of mutations that reduce MarR activity to different extents have been associated with resistance to unrelated antibiotics such as ciprofloxacin, trimethoprim, tetracycline, and several more, in the laboratory as well as in clinically-relevant contexts.

Mutations in TFs mediate antibiotic resistance, albeit indirectly by affecting the expression of some other protein which may then eliminate the antibiotic in some manner. In contrast, high levels of resistance to antibiotics such as ciprofloxacin can be readily attained via precise mutations in the protein that the antibiotic targets, resulting in reduced binding of the antibiotic to its target. We have already noted how mutations in RNAP can confer resistance to rifampicin, which acts by binding to RNAP. The set of such mutations is small however, for these affect highly conserved, essential proteins. Mutations that inadvertently reduce the activity of such an essential target protein while protecting it from an antibiotic could have severe consequences for the cell. This again does not mean that the alternative mutations—which affect efflux pump expression by inactivating a TF—are without a cost, though the number of mutations that have the same outcome of inactivating a protein can be quite large and therefore these mutations can be discovered more easily. As discussed earlier in this text, inappropriate expression of any given gene can potentially be expensive to bacterial lineages that are part of large populations. Thus, it is often *combinations* of mutations walking the tightrope between antibiotic resistance and general fitness that establish themselves in a population. For instance, Lisa Alzrigat and colleagues showed that clinical samples of *E. coli* resistant to ciprofloxacin carry a combination of mildly-inactivating mutations in MarR in addition to mutations in ciprofloxacin's direct target, DNA gyrase.[99] The conclusion arising from this study is that constitutive, high expression of efflux pumps may impose a cost that effectively selects against strongly inactivating mutations in the TF MarR.

One can surmise that environments which present a range of diverse toxic substances would favour robust inactivation of TFs such as MarR, for high, persistent expression of a broad-specificity efflux pump may be a more efficient solution to the problem presented by such environments than mutations in all possible target proteins. However, an alternative possibility is that large genetic elements carrying multiple antibiotic resistance genes are acquired horizontally. The very evolution, let alone the spread, of such a complex series of genes would require persistent selection by consistent exposure to multiple antimicrobial agents. Unfortunately, in response to antibiotic abuse that has created environments replete with antibiotics and other toxins, these genetic modules are becoming increasingly common in bacterial populations.

Richard Lenski's pioneering long-term experimental evolution (LTEE, see Chapter 4), during which several independent populations of *E. coli* grew and evolved

99 L.P. Alzrigat, D.L. Huseby, G. Brandis, and D. Hughes, 'Fitness cost constrains the spectrum of marR mutations in ciprofloxacin-resistant *Escherichia coli*', *Journal of Antimicrobial Agents and Chemotherapy* 72 (2017), 3016–3024. https://doi.org/10.1093/jac/dkx270

for tens of thousands of generations, has also provided us with instances of TF evolution. An example involves the global TF and chromosome shaping protein FIS. FIS is expressed at high levels as cells transition from a period of physiological adaptation to a new medium to one of exponential population multiplication, providing a boost to the transcription of growth-enabling genes such as those involved in protein synthesis. Estelle Crozat and colleagues noticed that FIS, alongside the topoisomerase protein topoisomerase 1, had mutated in several LTEE populations.[100] These mutations altered chromosome supercoiling states. This finding suggested that mutations affecting chromosome structure can be adaptive.

In a later study, Crozat again—with other co-workers—discovered that a FIS mutation that had emerged during LTEE had additional, more subtle effects relating to its function as a TF rather than as a chromosome-shaping protein.[101] They found that FIS activates the transcription of a protein—a *porin*, that forms pores on the cell surface, allowing the exchange of material between a cell and its environment—called OmpF. This regulatory interaction had not been described previously. In fact, this interaction is known to be absent in some other varieties of *E. coli*. These *E. coli* varieties have a second porin OmpC in addition to OmpF. The overlapping nature of the functions of OmpC and OmpF means that the high level of OmpF achieved by the activation of its transcription by FIS is not necessary in these *E. coli*. It can be hypothesised that an ancestor of the *E. coli* variety that was used in LTEE lost OmpC and instead co-opted FIS as a transcriptional activator of OmpF. However, this FIS-dependent activation of OmpF was being compromised by the FIS mutation that had evolved during LTEE. Despite this FIS mutation, the levels of OmpF did not quite decrease in the evolved lines, suggesting that additional compensatory mutations elsewhere had kicked in, reducing the dependence of OmpF on FIS. In addition, the expression level of FIS itself had decreased over time during LTEE, suggesting that evolution had reduced the requirement of FIS for the rapid growth of *E. coli*. Why might this be the case? FIS is an extraordinarily abundant TF. At its peak, there are some 60,000 molecules of FIS in the *E. coli* cell, which is several times more than the expression levels of other global TFs such as CRP and H-NS, and comparable to the levels of non-specific DNA binding proteins such as HU that coat the chromosome. Are evolving populations of *E. coli* attempting to minimise the cost of expressing FIS to such high levels by discovering combinations of mutations, in FIS and in other parts of the genome, that compensate for reduced FIS availability in other ways?

It has been assumed in the past that if an ortholog of a TF and that of its target gene in one organism is present in another organism, the regulatory interaction between

---

100 E. Crozat, N. Philippe, R.E. Lenski, J. Geiselmann, and D. Schneider, 'Long-Term Experimental Evolution in *Escherichia coli*. XII. DNA topology as a key target of selection', *Genetics* 169 (2005), 523–532. https://doi.org/10.1534/genetics.104.035717

101 E. Crozat, T. Hindre, L. Kuhn, J. Garin, R.E. Lenski, and D. Schneider, 'Altered regulation of the OmpF porin by Fis in *Escherichia coli* during an evolution experiment and between B and K-12 strains', *Journal of Bacteriology* 193 (2011), 429–440. https://doi.org/10.1128/jb.01341-10

the TF and the target gene is also conserved.[102] However, the FIS example shows that orthologous TFs can regulate different genes in different organisms. Is this a rule or an exception? Morgan Price and co-workers attempted to answer this question using the regulatory network of *E. coli* as a reference to predict gene expression patterns in *E. coli* as well as in other bacteria.[103] Their work was based on the premise that genes regulated by the same set of TFs will be expressed in the same manner across conditions. As mentioned earlier, this is a good assumption that holds in many cases, despite the incompleteness of the known regulatory network, as demonstrated in the first instance by Price and co-workers, and can lead to useful inferences.[104]

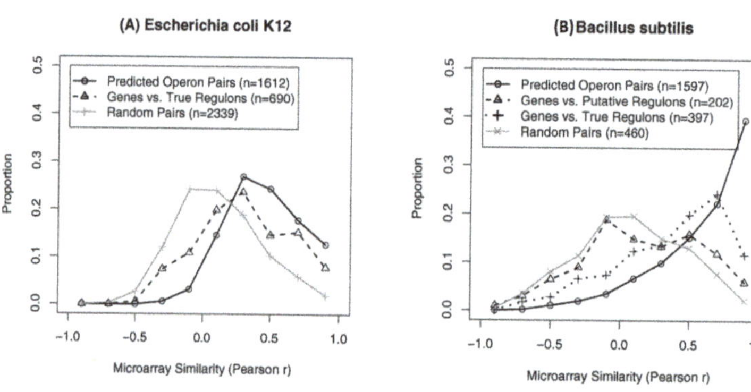

Fig. 5.10 Transcriptional regulatory interactions are not conserved. (A) This figure shows correlation in gene expression between pairs of genes belonging to the same operon and therefore expected to be co-regulated, those regulated by the same TFs, and random pairs; (B) This figure measures the same correlations as A, but for *B. subtilis*. Compare the distribution of co-expression measures between genes that are known to be regulated by the same TFs in *B. subtilis* ('true regulons') and those that are predicted to be co-regulated based on the *E. coli* regulatory network ('putative regulons'). Originally published as Figures 5A and 5E in M. Price, P.S. Dehal, and A.P. Arkin, 'Orthologous transcription factors in bacteria have different functions and regulate different genes', *PLoS Computational Biology* 3 (2007), e175, Creative Commons Public Domain declaration.

Price and colleagues identified orthologs of *E. coli* TFs and their target genes in the evolutionarily distant bacterium *Bacillus subtilis* and asked if pairs of genes[105] predicted to be co-regulated in the latter based on data from the former showing

102 M.M. Babu, S.A. Teichmann, and L. Aravind, 'Evolutionary dynamics of prokaryotic transcriptional regulatory networks', *Journal of Molecular Biology* 358 (2006), 614–633. https://doi.org/10.1016/j.jmb.2006.02.019

103 M. Price, P.S. Dehal, and A.P. Arkin, 'Orthologous Transcription Factors in Bacteria Have Different Functions and Regulate Different Genes', *PLoS Computational Biology* 3 (2007), e175. https://doi.org/10.1371/journal.pcbi.0030175

104 And by other work such as Balazsi et al., 2005.

105 This is a simplified description of Price et al., 2007's work. What they did is the following: for each gene belonging to a *regulon*, i.e., the set of genes regulated by the same TF, they measured the correlation in its expression with the average expression across genes in the same regulon.

correlated expression patterns. They found, across ~200 examples, that predicted groups of co-regulated genes in *B. subtilis* were not necessarily co-expressed, implying that regulatory interactions between TFs and target genes in *E. coli* are not conserved in evolutionarily distant bacteria (Fig. 5.10). This held true even when regulatory interactions known in *E. coli* were used to predict expression patterns in more closely related bacteria such as *Vibrio cholerae*. Thus, orthologs of both a TF and its target gene may be present in another organism, but the regulatory interaction between the two need not be. This is exemplified by the FIS-OmpF pair. Both FIS and OmpF are conserved in two different strains of *E. coli*, whilst FIS regulates the transcription of OmpF in one strain but not the other. Price and colleagues' work suggests that this may not be an exception. Note here though that such differences in regulatory networks can arise not only through mutations in TFs, which can alter the DNA recognition specificity and/or activity of TFs, but also mutations in inter-genic DNA sequences to which these TFs bind.

We have seen several examples of adaptive mutations in TFs. The fact that some TFs regulate several genes implies that mutations in such TFs would affect the expression of all or most of these targets. If the adaptation driven by a TF mutation is determined by only a small subset of the TF's targets, then the change in expression of all the other genes that the TF regulates should be considered as a side-effect and potentially one with negative consequences. Thus, consequential mutations in TFs can be a double-edged sword. Though genetic alteration of TF function might offer early adaptation, would it be maintained over longer timescales? One can surmise that the side-effects of an otherwise beneficial TF mutation may cause such mutations to be selected against as evolution progresses, allowing populations to discover and select for other beneficial combinations of mutations with fewer detrimental side-effects.

To test this, Farhan Ali in my lab investigated sequence variation in *E. coli* TFs at two distinct timescales:[106] a very short timescale represented by ~30 years of LTEE and a longer timescale covering ~100 million years that have elapsed since the divergence of *E. coli* and its relative *Salmonella*. Novel mutations in TFs, especially those that regulate a large number of genes, emerged early during the LTEE. As time progressed and as *E. coli* populations adapted to the environment imposed on them by the experiment, the frequency of new mutations in TFs declined. At the ~100 million year timescale represented by the diversity of *E. coli*, TFs show significantly smaller sequence diversity than the genes that they regulate. This was especially apparent in TFs regulating many genes. Thus, the pleiotropic nature of TF mutations might, over the time of divergence and diversification of a single bacterial species, reduce the extent of variation in TF sequences (Fig. 5.11).

106 F. Ali and A.S.N. Seshasayee, 'Dynamics of genetic variation in transcription factors and its implications for the evolution of regulatory networks in Bacteria', *Nucleic Acids Research* 48 (2020), 4100–4114. https://doi.org/10.1093/nar/gkaa162

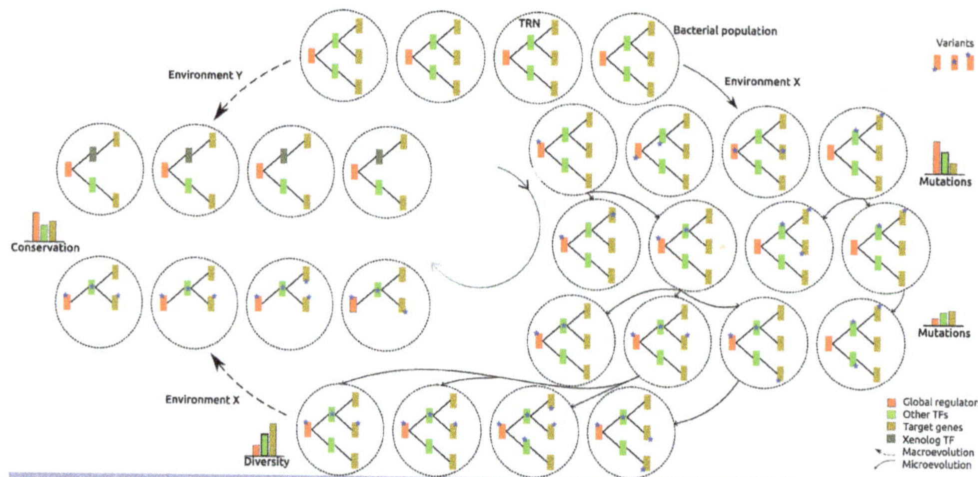

Fig. 5.11 TF evolution in bacteria. A population of bacteria beginning to adapt to a new environment accumulates mutations in TFs, especially global TFs. Over longer timescales, of the order of divergence of a whole species like *E. coli*, TFs show low sequence diversity. As the niche changes to *Y*, adaptation may proceed through TF repertoire changes created by gene loss and horizonal gene transfer. Originally published as Figure 8 in F. Ali and A.S.N. Seshasayee, 'Dynamics of genetic variation in transcription factors and its implications for the evolution of regulatory networks in Bacteria', *Nucleic Acids Research* 48 (2020), 4100–4114, CC BY 4.0.

In the examples we have seen so far, regulatory changes were effected by mutations in TFs. Even if a mutation in a TF is strong enough to inactivate it completely, the gene for the TF is still present. In such cases, there is a chance for TF activity to recover quickly even in the absence of horizontal re-acquisition of a lost TF gene. For example, much of σ^S activity lost in σ^S-GASP1 was recovered in σ^S-GASP2. In many cases, such as TF σ^S,[107] mutations in GASP, or the FIS mutations in LTEE, mutations do not fully abolish the protein's activity and will instead subtly alter it. However, over longer evolutionary distances, even within the period required for the diversification of a species, whole TF genes can be gained or lost. Ali found that TFs showing high sequence divergence within *E. coli* are often lost in related species.[108] These may represent examples of TFs that are lost in some lineages as a result of the inexorable process of sequence decay by mutation. Or these might merely be TFs with high tolerance for mutation that were acquired specifically in *E. coli* and related lineages.

Bacterial genomes encode fairly large numbers of TFs. The numbers of TFs coded for by a bacterial genome increases with increase in genome size, or more precisely, the total number of genes. However, the relationship is not linear, but is closer to being quadratic. Very small bacterial genomes such as those of obligate parasites and endosymbionts code for hardly any TFs and only a single σ-factor. Bacteria like *E. coli* with, say, 4,000 genes in total, contain ~300 TFs—similar if not more than the number of TFs encoded by the genome of the eukaryote budding yeast. Large bacterial genomes

107 We assume that σ-factors are TFs here, although I usually do not assume this.

108 Ali and Seshasayee, 2020.

with ~10,000 genes code for as many as ~1,000 TFs, a number that is comparable to that for higher eukaryotes. The lack of TFs in parasitic or endosymbiotic genomes likely arises from a preferential loss of TFs. As genomes grow, where do new TFs come from? The DNA binding portion of bacterial TFs, called the DNA-binding *domain* (DBD), belong to a fairly small set of sequence *families* that in many cases are variants within a common theme known as the helix-turn-helix (HTH) motif. These DBDs are often found alongside other domains that may help to sense a signal. Proteins belonging to the same family show fairly high sequence similarities with each other. When a family of TFs shows an expansion—i.e., an increase in the number of protein sequences belonging to the family—in a clade of bacteria, gene duplications initiated from a progenitor family member become available as an option. Or, new family members may be gained independently of pre-existing relatives through horizontal gene transfer. Which of these two is more predominant? This is merely a special case of the broader duplication vs horizontal gene transfer debate we had examined in Chapter 4.

While horizontal gene acquisition is believed to be the more dominant force responsible for bacterial genome/gene repertoire expansion overall, arguments in favour of both forces have been advanced in the special case of TFs. Early work by Sarah Teichmann and Madan Babu suggested that duplication dominates TF evolution.[109] Many proteins are modular, comprising of multiple 'domains'. Each domain is usually responsible for one function. For example, a bacterial TF may have one domain that binds to the DNA and another that binds to a signal molecule that directs it to bind to the DNA. Teichmann and Madan Babu assumed that any pair of proteins with the same 'domain architecture'—namely the same set of domains arranged in a particular order from one end of the sequence to the other—are likely to have arisen by duplication. Thus, they concluded that a majority of TFs in *E. coli* have arisen by duplication. Unlike other studies that have compared rates of duplication and horizontal gene transfer in other contexts (see Chapter 4), this study did not compare *E. coli* TF sequences with those from other organisms.

A few years later, Morgan Price and co-workers disagreed with Teichmann and Madan Babu. They built phylogenetic trees of *E. coli* TFs, comparing sequences from *E. coli* with those of similar proteins from other bacterial species.[110] They compared these trees with species trees. Based on the incongruity between species trees and gene trees, they concluded that nearly two-thirds of all *E. coli* TFs had been acquired by horizontal gene transfer (Fig. 5.12). Farhan Ali had also noticed that several TFs with low sequence divergence in *E. coli* are poorly conserved in related species; these are probably useful TFs horizontally acquired in the *E. coli* lineage.[111] Horizontal gene transfer appears

109 S.A. Teichmann and M.M. Babu, 'Gene regulatory network growth by duplication', *Nature Genetics* 36 (2004), 492–496. https://doi.org/10.1038/ng1340

110 M.N. Price, P.S. Dehal, and A.P. Arkin, 'Horizontal gene transfer and the evolution of transcriptional regulation in *Escherichia coli*', *Genome Biology* 9 (2008), R4. https://doi.org/10.1186/gb-2008-9-1-r4

111 Ali and Seshasayee, 2020.

to be rare for global TFs with a large number of targets, but common for TFs that are encoded adjacently to their target genes on the genome. The latter group of TFs might have been transferred into *E. coli* as modules comprising of the TF as well as its target genes. Thus, TF repertoires do change over certain phylogenetic distances. These may reflect niche divergence between the species concerned, and the primary means by which TF repertoires expand is likely to be horizontal gene acquisition.

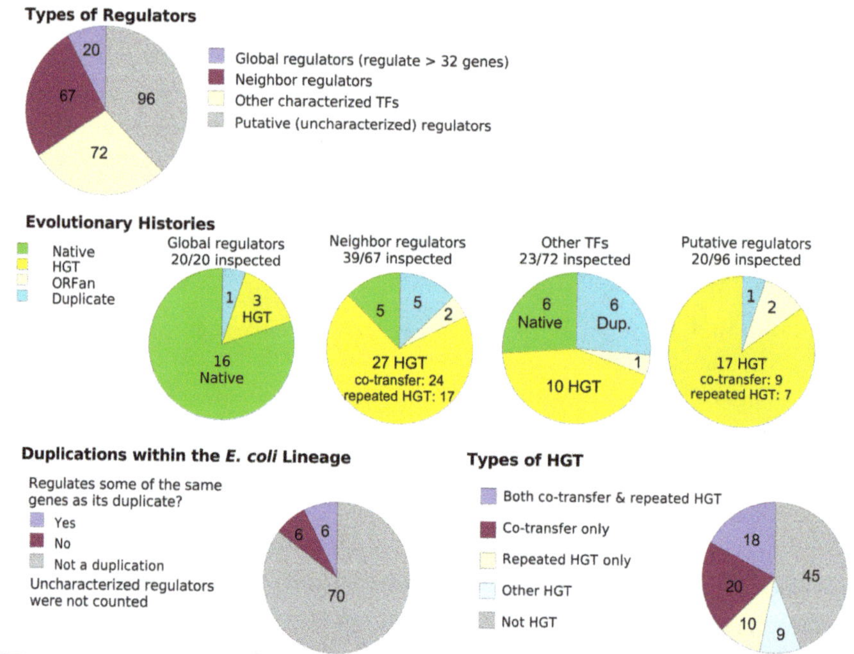

Fig. 5.12 Evolution of TFs by horizontal gene transfer. This figure shows that a majority of TFs in *E. coli* were likely acquired by horizontal gene transfer. This does not hold true for global regulators, most of which are 'native' or not recently acquired. This was published as Figure 6 in Price et al. (2008). *Genome Biology*, CC BY 2.0.

Acquiring even a single regulatory system can have major consequences for an organism's lifestyle. This is exemplified by some beautiful work done some time ago by Mark Mandel and colleagues.[112] They investigated the biology of two varieties of the bacterial species *Vibrio fischeri*. One variety is a symbiont of fish whereas the other colonises squid. Comparing the genomes of the two symbionts revealed that the fish symbiont, which was unable to colonise squid, lacked a protein called RscS that was encoded in the genome of the squid symbiont. RscS is a protein that activates a TF by modifying it,[113] which in turn activates the expression of genes that help the

112 M.J. Mandel, M.S. Wollenberg, E.V. Stabb, K.L. Visick, and E.G. Ruby, 'A single regulatory gene is sufficient to alter bacterial host range', *Nature* 458 (2009), 215–218. https://doi.org/10.1038/nature07660

113 RscS is what is called a sensor kinase that phosphorylates a response regulator protein. The response regulator is a TF.

bacterium colonise surfaces inside squid. The fish symbiont, when engineered by the addition of the gene for RscS, gained the ability to colonise squid. Further analysis of a collection of *V. fischeri* isolates from fish and squid showed that all squid colonisers encoded RscS, whereas only a subset of fish symbionts did, and the RscS found in fish symbionts was very divergent in sequence from that encoded by squid colonisers. The phylogeny of *V. fischeri* suggested that the ancestor of this species likely lacked the ability to colonise squid, and that the acquisition of RscS allowed its descendants to do so. If this acquisition was achieved through horizontal transfer, its source remains unknown. The fact that the addition of a single regulator allowed a bacterium to access a new niche indicated that genes responsible for this colonisation were already part of the ancestral bacterium. The addition of a single master regulator, an organiser or pied-piper, was sufficient to activate these genes in a coordinated manner, enabling the colonisation of a new environment.

We have so far examined how TFs evolve by mutation and how the repertoire of TFs encoded in a genome can change. We finally ask how TFs evolved in the first place. What is the ultimate origin of transcription regulation by TFs? This question is motivated by the argument, articulated by Sandhya Visweswaraiah and Stephen Busby, that "transcription regulation is a 'luxury' for a bacterium."[114] No TF, unless one considers the major σ-factor σD a TF, is part of the hypothetical minimal bacterial genome (Chapter 3). Bacteria with small genomes code for few TFs, and endosymbiont genomes encode hardly any that we know of. As we had briefly mentioned earlier, properties inherent to the chromosome such as DNA supercoiling can regulate the expression of genes,[115] and can probably play a key role in gene regulation in bacteria with highly reduced genomes such as the *Mycoplasma*. The so-called contingency loci in bacteria such as *Helicobacter pylori* and *Campylobacter jejuni* can mutate, reversibly, at rates as high as ~1/20 and help to achieve rapid phenotypic switches at a population level.[116] Thus, TF-based regulation might become important only as bacteria evolve to adopt 'complex' lifestyles—complexity defined as, say, the diversity of environmental and cellular situations the organism has to deal with.

In contrast to transcription regulation, DNA shaping and compaction are likely to be fundamental for survival to even a primitive cell. Even a genome coding for ~100 genes will need to be compacted by two orders of magnitude to be packed inside a tiny cell. Unlike TFs, which recognise particular sequence motifs on the DNA to make sequence-specific interactions with operator sites, most chromosome shaping proteins bind non-specifically. Often, both sequence-specific and nonspecific DNA binding

114 S. Visweswaraiah and S.J.W. Busby, 'Evolution of bacterial transcription factors: how proteins take on new tasks, but do not always stop doing the old ones', *Trends in Microbiology* 23 (2015), 463–467, p. 465. https://doi.org/10.1016/j.tim.2015.04.009

115 For a specific example, see W. Zhang and J.B. Baseman, 'Transcriptional regulation of MG_149, as osmoinducible lipoprotein gene from *Mycoplasma genitalium*', *Molecular Microbiology* 81 (2011), 327–339. https://doi.org/10.1111/j.1365-2958.2011.07717.x

116 J. Parkhill, B.W. Wren, K. Mungall, J.M. Ketley, C. Churcher, et al., 'The genome sequence of the food-borne pathogen *Campylobacter jejuni* reveals hypervariable sequences', *Nature* 403 (2000), 665–668. https://doi.org/10.1038/35001088

proteins adopt similar structural folds, belonging to the same protein 'superfamily', and likely share a common ancestor.[117] At the level of protein function, the boundary between a chromatin-shaping protein (called nucleoid-associated proteins in bacteria) and a TF is not all that clear.[118] For example, the protein FIS, which we had encountered earlier, is not only a TF regulating the expression of many genes by making precise contacts with operator sites near promoters, it also affects topological properties of the DNA such as supercoiling. Several TFs, including FIS and CRP—the latter being an important subject of extended discussion, in Visweswaraiah and Busby's opinion—bind to thousands of sites on the genome, a vast majority of which appear to have nothing to do with the regulation of transcription of a proximal gene. Do these other interactions have anything to do with functions such as chromosome shaping? Likely yes for FIS and maybe for CRP. Maybe chromosome shaping interactions represent the 'original' function of these proteins, with features required for transcription regulation evolving subsequently? These excess interactions, however, are not non-specific contacts per se but are likely sequence-specific contacts with lower affinity targets. Yet this begs the following extrapolation: did an ancestral non-specific or weakly-specific DNA binding protein gain the ability of sequence-specific DNA recognition to evolve into a TF? We can only guess at the moment, and explore whether sequence data publicly available for extant TFs and nucleoid-associated proteins allows us to answer this question systematically.

The evolution of a TF involves much more than just evolving DNA recognition specificity. Take, for example, the protein CRP. This protein not only has the ability to recognise a target motif and bind to it, it also has the capacity to bind to the signal molecule cyclic AMP and then present interfaces that attract the RNAP.[119] Though the DNA-binding and the cyclic AMP-binding domains are distinct and lie in different segments of the protein sequence, one residue in the latter contributes to DNA binding. The binding of cyclic AMP somewhere on the protein should translate to modulation of its DNA binding properties which are encoded elsewhere on the same protein. These point to some correlated evolution between the two domains. Further, CRP has the ability to activate transcription by more than one mechanism depending on which part of the RNAP it interacts with. That means that this protein has evolved multiple RNAP binding interfaces. Assuming that TFs evolved from a nucleoid-associated protein ancestor, all these elements had to have evolved on a non-specific DNA binding protein backbone from scratch. This would have involved a combination of domain sequence evolution and the fusion of distinct domains. The evolutionary

117 N.M. Luscombe, S.E. Austin, H.M. Berman, and J.M. Thornton, 'An overview of the structures of protein-DNA complexes', *Genome Biology* 1 (2000), REVIEWS001. https://doi.org/10.1186/gb-2000-1-1-reviews001

118 C.J. Dorman, M.A. Schumacher, M.J. Bush, R.G. Brennan, and M.J. Buttner, 'When is a transcription factor a NAP?', *Current Opinion in Microbiology* 55 (2020), 26–33. https://doi.org/10.1016/j.mib.2020.01.019

119 Visweswaraiah and Busby, 2015; A. Kolb, S. Busby, H. Buc, A. Garges, and S. Adhya, 'Transcriptional regulation by cAMP and its receptor protein', *Annual Review of Biochemistry* 62 (1993), 749–795. https://doi.org/10.1146/annurev.bi.62.070193.003533

trajectories leading to the evolution of such a complex set of coordinated activities remain to be explored.

In summary, there exists an intriguing hypothesis that proteins involved in fundamental processes of chromosome shaping and compaction might have evolved the ability to bind to specific sites on the DNA and regulate the transcription of nearby genes. The repertoire of TFs is in flux and presumably responds to selection imposed by the environment and is probably driven, in large part, by genome reduction and horizontal gene acquisition. Finally, evolution often allows mutations that alter TF activity early in adaptation; the pleiotropic nature of TF mutations, which can potentially cause much collateral damage, might eventually select against sequence variation in conserved TFs.

5.6. Building to read and reading to build

Evolving a bacterial genome is complicated business. Its construction reflects the fine balance of a host of selection pressures within a relatively short stretch of DNA. This is unlike the genomes of higher eukaryotes, in which relaxed selection allows the accommodation of whole stretches of non-functional or 'junk' DNA (see Chapter 3). In bacteria, selection appears to decide not only the gene repertoire but also where each gene is positioned on the chromosome. We will call this 'gene order', 'gene organisation', or 'chromosome organisation' interchangeably. Here we will discuss the interplay between transcription and gene organisation; how gene organisation helps enable efficient transcription, and how transcription drives gene organisation. Note here that transcription is by no means the only driver behind chromosome organisation, but in the context of this book and this chapter it is merely the most relevant factor.

But first, a brief reiteration and description of some features of a bacterial genome. Most known bacteria contain a single circular chromosome. There are several exceptions to both 'single' and 'circular'. Bacteria such as *Vibrio cholerae* carry more than one chromosome and members of *Streptomyces* have linear chromosomes. Our discussion will apply to the single, circular bacterial genome and some conclusions might apply to any chromosome identifiable as primary in bacteria with multiple chromosomes.

The bacterial genome encodes genes for a variety of functions, starting from those that are required minimally for any cell to function to those needed for defence against the most unusual threats. Many of these functions emerge or become better represented in the genetic repertoire in larger bacterial genomes than in smaller ones, whereas others show no such relationship.[120] For example, the number of genes coding for proteins that are part of the ribosome would be more or less constant irrespective of genome size, for the bacterial ribosome is a highly conserved structure that is grossly

120 E. van Nimwegen, 'Scaling laws in the functional content of genomes', *Trends in Genetics*. 19 (2009), 479–484. https://doi.org/10.1007/0-387-33916-7_14

the same for most, if not all, bacteria. On the other hand, the number of genes involved in small molecule metabolism—the breakdown of nutrients, biosynthesis of monomeric building blocks among other molecules, and energy generation—increases linearly with genome size. This suggests that an increase in genome size in bacteria reflects an expansion of metabolic capabilities, which in turn is in response to an increase in the complexity and diversity of its habitats. Curiously, as we had briefly discussed in the previous section, the number of TF-encoding genes (and other regulatory protein genes) increases more or less quadratically with genome size. This brings to the table the idea that as genome size and gene function repertoire increase, the regulatory 'overhead' for ensuring their optimal function increases more than linearly—to the extent that Juan Ranea and colleagues argued that regulatory cost can impose a ceiling on how large a bacterial genome can grow.[121]

Genes representing this vast diversity of functions are arranged fairly tightly along the bacterial chromosome, with very little intergenic DNA separating adjacent genes. Adjacent genes can be transcribed in the same direction, being encoded on the same strand of DNA. They can be convergent, with the end of one gene being next to that of the other. Or they can be divergent, with the starts of the two genes adjacent to each other. Both convergent and divergent pairs of genes are encoded on opposite strands of DNA. As mentioned earlier, groups of co-directional, adjacent genes are often organised as operons. Genes forming part of the same operon are transcribed together from a single promoter as a single mRNA, with each protein-coding gene within the operon translated from its own ribosome loading site.

When genes are close to each other, the transcription of one gene can affect that of neighbouring genes. This arises from the interplay between DNA supercoiling and transcription. When RNAP is transcribing a gene, the mechanics of the process is such that the DNA in front is overwound, or positively supercoiled, while the DNA behind the RNAP is underwound, or hyper-negatively supercoiled (Fig. 5.13).[122] The progress of the RNAP, and therefore transcription, would require topoisomerases to act and stabilise supercoiling states. This imposes additional constraints that can impede idealised, smooth progress of transcription. A recent study by Ihab Boulas et al. showed that, in artificial DNA constructs introduced into a bacterial cell, the expression of a downstream gene decreases when that of another gene upstream increases, unless an 'insulator'—an element that blocks the diffusion of supercoiling states along the length of a DNA molecule—is introduced between the two genes.[123]

Patrick Sobetzko, in an earlier study, had performed an analysis of the *E. coli* genome and asked how genes whose expression is sensitive to supercoiling states are organised

121 J.A. Ranea, A. Grant, J.M. Thornton, and C.A. Orengo, 'Microeconomic principles explain an optimal genome size in bacteria', *Trends in Genetics* 21 (2005), 21–25. https://doi.org/10.1016/j.tig.2004.11.014

122 C.J. Dorman, 'DNA supercoiling and transcription in bacteria: a two-way street', *BMC Molecular Cell Biology* 20 (2019), 26. https://doi.org/10.1186/s12860-019-0211-6

123 I. Boulas, L. Bruno, S. Rimsky, O. Espeli, I. Junier, and O. Rivoire, 'Assessing in vivo the impact of gene context on transcription through DNA supercoiling', *Nucleic Acids Research* 51 (2023), 9509–9521. https://doi.org/10.1093/nar/gkad688

on the chromosome in terms of their adjacency properties.[124] He found that genes that respond to hyper-negative supercoiling are enriched among divergently oriented gene pairs. When two genes are divergently encoded, the transcription of one would leave the other highly negatively supercoiled. In contrast, genes whose expression is favoured by less negatively supercoiled DNA are encoded in a convergent manner. RNAP activity at one gene would cause the other gene to be more tightly wound and would thus favour its expression. This work suggested that local DNA organisation is such that it enables transcription in the face of constraints imposed by transcription-induced DNA supercoiling.

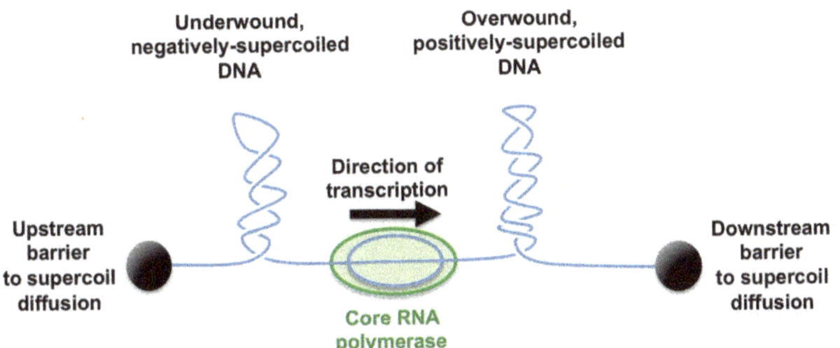

Fig. 5.13 Model of transcription and supercoiling. This figure shows the process of transcription, highlighting positive supercoils forming in front of and negative supercoils behind of the elongating RNAP. Originally published as Figure 1 in C.J. Dorman, 'DNA supercoiling and transcription in bacteria: a two-way street', *BMC Molecular Cell Biology* 20 (2019), 26, CC BY 4.0.

Transcription and gene organisation appear intertwined even if we are to zoom out and take a bird's-eye view of large chunks of the bacterial chromosome. To investigate this, we must first understand the effect of DNA replication on gene dosage, i.e., the number of copies of each gene present in the cell as a result of chromosome replication.[125] The typical bacterial chromosome has a single origin of replication (*ori*). This is the site at which DNA polymerase (DNAP), the enzyme that replicates the chromosome, binds. Replication proceeds bidirectionally outwards from *ori* and ends at a series of terminus sites (*ter*) located more or less at a diametrically opposite location on the circular chromosome. Consider the chromosome as a perfect circle and draw its diameter from *ori* to *ter*. We will call this line the *ori-ter axis* and the two semicircles thus formed as *replichores*. The two replichores, on either side of the *ori-ter* axis, would be nearly equal in length. The DNA polymerase in *E. coli* replicates the DNA at the rate of ~1,000 bp per second. Assuming that the average *E. coli* chromosome is ~5 Mbp long, and that the two replichores are being replicated simultaneously, it will take over 40 minutes for the chromosome to be replicated completely. If, in a minimal growth medium, *E.*

124 P. Sobetzko, 'Transcription-coupled DNA supercoiling dictates the chromosomal arrangement of bacterial genes', *Nucleic Acids Research* 44 (2016), 1514–1524. https://doi.org/10.1093/nar/gkw007

125 E.P.C. Rocha, 'The replication-related organization of bacterial genomes', *Microbiology* 150 (2004), 1609–1627. https://doi.org/10.1099/mic.0.26974-0

coli populations double, say, every hour, then we can expect *ori*-proximal genes to be present in two copies for at least two-thirds of a bacterium's life cycle. On the other hand, *ter*-proximal genes would be replicated, producing a second copy, only a short while before the cell divides. Imagine the situation of *E. coli* growing in rich media that supports the population's doubling every 20–30 minutes on average, much less than the time required to make two copies of the chromosome to be partitioned between the two daughter cells. In such situations, replication initiates at *ori* more than once per life cycle such that the DNA copy number or dosage between *ori* and *ter*-proximal genes can be much higher than two—say, four, or possibly even eight.

The more copies of a gene, the higher the availability of promoters for its transcription. This is especially true when a gene promoter in one copy of the chromosome is saturated with RNAP. In such a situation, creating a second copy would pretty much be the only way to increase transcription even if the cell's physiology absolutely requires it. Thus, this replication-dependent difference in gene dosage between *ori*- and *ter*-proximal genes can make additional promoters of *ori*-proximal genes available for RNAP to access and bind to. To what extent does this aspect of the interplay between replication and transcription affect gene organisation?

Patrick Sobetzko and colleagues, by analysing gene organisation in the *E. coli* genome, showed that genes under σD control are relatively more frequently encoded proximally to *ori* whereas those regulated by σS are located closer to *ter*[126]. This applies equally well to both replichores. σD regulated genes are expressed primarily during exponential growth during which chromosome replication prominently occurs. One can hypothesise that some part of the increased expression of σD-regulated genes during exponential growth is facilitated by their higher dosage, which in turn is a consequence of their presence in *ori*-proximal regions of the chromosome and ongoing DNA replication. Rajalakshmi Srinivasan and co-workers in my lab found further evidence that gene expression coherence extends well beyond the confines of the operon.[127] A gene encoded in one *half* of the chromosome, centred around *ori* or *ter* and thus comprising one half of each replichore, is more likely to be expressed under similar conditions as another present in the same half than with one found in the opposite half. Further, if a gene in one half of the chromosome is activated in one condition, then a gene from the other half tends to be repressed in the same condition. Thus, genes that are expressed together tend to be found in the same half of the chromosome whereas mutually exclusive or antagonistic pairs of genes are encoded in opposite halves. This could well be explained at least in part by the differential localisation of σD- and σS-regulated genes in *ori*-proximal and *ter*-proximal parts of the chromosome respectively.

126 P. Sobetzko, A. Travers, and G. Muskhelishvili, 'Gene order and chromosome dynamics coordinate spatiotemporal gene expression during the bacterial growth cycle', *Proceedings of the National Academy of Sciences USA* 109 (2011), E42–E50. https://doi.org/10.1073/pnas.1108229109

127 Srinivasan et al., 2015.

In the results published by Sobetzko et al., the relative preference of σD genes to be encoded in *ori-proximal-* over *ter*-proximal regions of the chromosome was relatively small. This suggests that only a small, though significant, subset of σD genes are preferentially encoded close to *ori*. Is this a random subset or does it represent particular sets of gene functions? The answer to this question has already been provided by Etienne Couturier and Eduardo Rocha, who asked where genes involved in different functions and expressed at different levels are encoded on a couple of hundred different bacterial genomes, and how these patterns might change across bacteria capable of growing at different rates.[128] We know that the growth rate of a bacterial population is determined to a large extent by the nutrition available to it. Nutrient availability is dynamic, but gene organisation is not nearly as dynamic. So, what is the rationale behind trying to correlate the two? It may well be the case that richer media conditions support faster growth, but this relationship cannot hold indefinitely. The genomic content of a bacterium would impose a ceiling on how fast its population can grow. After a point, one can keep adding better and better nutrients, but growth rates would saturate. This ceiling is a product of evolution and probably a reflection of the bacterium's ecology. Couturier and Rocha used an experimentally determined dataset of the highest known growth rates for ~200 bacteria and made the reasonable assumption that these correspond to the maximum growth rate possible for these bacteria. Given that this is a product of evolutionary optimisation, they asked whether gene organisation is in any way linked to maximum growth rate as contained in data they had assembled. As we had discussed earlier, a higher growth rate can result in higher copy number differences between *ori* and *ter* and presumably stronger selective pressures arising from such a difference. In fact, instead of growth rate, Couturier and Rocha sought to find the relationship between gene organisation and a measure of the gene dosage difference between *ori* and *ter*, which can be estimated from growth rates and the rate of progress of replication.

Couturier and Rocha first predicted highly expressed protein-coding genes from their sequence.[129] A prediction, as opposed to an experimentally-determined set of highly expressed genes, was necessitated by the fact that appropriate experimental data were available for only a small set of bacteria and not across the broad phylogenetic spread these researchers studied. The prediction was based on the degeneracy of the genetic code. Most amino acids are encoded by multiple codons, and in most organisms one or a smaller subset of codons for each amino acid is preferentially used. This may reflect the relative availability of the different types of tRNAs, each of which recognises a particular codon and brings its respective amino acid to the translating ribosome. The presence of a rare codon results in translation slowing down, because the appropriate tRNA is not immediately available. Thus, highly expressed protein-

128 E. Couturier and E.P.C. Rocha, 'Replication-associated gene dosage effects shape the genomes of fast-growing bacteria but only for transcription and translation genes', *Molecular Microbiology* 59 (2006), 1506–1518. https://doi.org/10.1111/j.1365-2958.2006.05046.x

129 Ibid.

coding genes can be expected to often use a preferred codon for most amino acids. This tendency of any gene to utilise preferred codons can be measured by calculating what is known as the '*codon adaptation index*' (CAI) from its sequence. We expect genes with high CAI to be expressed at high levels.

Couturier and Rocha[130] found that genes with high CAI are preferentially encoded closer to *ori*, but only in fast-growing bacteria. In particular, highly expressed genes responsible for fast growth, including those encoding RNAP, rRNA, and ribosomal proteins, are encoded in *ori*-proximal regions in fast-growing bacteria. rRNA genes are of particular relevance here. In fast-growing bacteria, some 80–90% of all transcription is diverted to the synthesis of rRNA. In fact, growth rate and rRNA expression levels are tightly correlated. For the same bacteria, increasing growth rate—for example by the provision of better nutrition—is associated with an increase in rRNA expression. Across bacterial species, the number of rDNA copies per chromosome increases with increasing maximum growth rates. Overall, high translation supplies fast growth. This requires both rRNA and ribosomal proteins. The supply of the latter is provided by a product of transcription and translation. The former, however, lack the luxury of amplification provided by translation and should be entirely supplied by transcription. The high levels of rRNA transcription required for fast growth cannot be provided by a single gene copy. Thus, fast growing bacteria code for multiple copies of rRNA genes per chromosome. Their being encoded near *ori* will further increase their gene dosage, making more of their promoters available for transcription, during rapid growth. The limiting factor then is the supply of RNAP. The encoding of RNAP genes close to the *ori*, closer than rDNA, should help the cell beat this constraint. Thus, genes required at high levels for fast growth are encoded in *ori*-proximal regions in bacteria capable of rapid population growth.

In a much more recent work, Supriya Khedkar in my lab reiterated the findings of Couturier and Rocha showing that essential genes encoding proteins involved in translation are present in *ori*-proximal regions in fast-growing bacteria (Fig. 5.14A).[131] She also showed that horizontally-acquired genes are depleted from regions close to *ori* in both fast- and slow-growing bacteria, but more so in the former (Fig. 5.14B). There are two possible explanations for this. The first is that regions around the *ori* are rich anyway in essential genes involved in crucial information processes such as transcription and translation. In gene-rich bacterial genomes, a random insertion of a horizontally-acquired gene will more often than not split and disrupt a gene already present in the chromosome. The successful maintenance or loss of such a disruptive insertion will depend on the relative contributions of the inserted and the disrupted gene to growth and survival. When this occurs in *ori*-proximal regions, the chance that it will disrupt an essential gene and cause lethality is relatively high.

130 Ibid.

131 S. Khedkar and A.S.N. Seshasayee, 'Comparative genomics of interreplichore translocations in bacteria: a measure of chromosome topology?', *Genes, Genomes, Genetics* 6 (2016), 1597–1606. https://doi.org/10.1534/g3.116.028274

Such insertions will be purged out by selection, leading to the under-representation of horizontally-acquired genes in regions near *ori* in many extant bacterial genomes. A second explanation, not mutually exclusive with the first, is that some aspect of chromosome structure, such as the protection of bound DNA by nucleoid-associated proteins, disallows insertions in regions near *ori* in the first place. The evidence for this is complex.

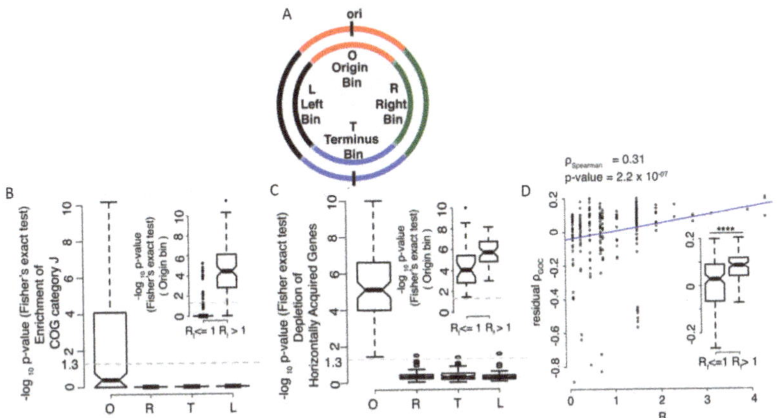

Fig. 5.14 Gene organisation in fast- and slow-growing bacteria. (A) This figure shows the division of the bacterial chromosome into four bins referred to in B and C. (B) This figure shows the enrichment of translation related genes (referred to as 'COG category J') in the *ori*-proximal region across bacterial genomes; inset shows the difference between fast-growing ($R > 1$) and slow-growing ($R <= 1$) bacteria. (C) As in B, but this figure shows the depletion of horizontally-acquired genes in *ori*-proximal regions in both fast- and slow-growing organisms but more so in the former. (D) This figure shows that gene order conservation (ρ_{GOC}) between pairs of closely-related bacteria after correcting for phylogenetic relatedness is correlated with the growth rate of bacteria. Originally published as part of S. Khedkar and A.S.N. Seshasayee, 'Comparative genomics of interreplichore translocations in bacteria: a measure of chromosome topology?', *Genes, Genomes, Genetics* 6 (2016), 1597–1606, CC BY 4.0.

In a recent study, Malikmohammed Yousuf and colleagues found that insertions of transposons may occur more frequently near *ori* than *ter*.[132] Thus, it is reasonable to posit that the lack of horizontally-acquired genes near *ori* is more likely a result of selection than any insertion bias. Yousuf et al. also noticed that insertion sites are enriched in locations bound by H-NS. Though H-NS binding sites are more often in the *ter*-half of the chromosome, this relationship between H-NS binding sites and insertion was not strong enough to overcome the overall balance favouring insertions in the *ori*-half. Because Yousuf et al. did not look for insertions in *E. coli* lacking H-NS, it is unclear whether the protein influences where insertions happen. Transposon insertions are not entirely random, and many prefer inserting in A+T-rich loci. This can also explain

132 M. Yousuf, I. Iuliani, R.T. Veetil, A.S.N. Seshasayee, B. Sclavi, and M.C. Lagomarsino, 'Early fate of exogenous promoters in *E. coli*', *Nucleic Acids Research* 48 (2020), 2348–2356. https://doi.org/10.1093/nar/gkz1196

why insertions are more common in H-NS binding sites, which themselves are A+T-rich. It is probably more likely that H-NS inhibits integration. N. Sharadamma and colleagues had shown—using purified protein from *Mycobacterium tuberculosis* and chemically synthesised DNA—that H-NS inhibits biochemical processes underlying integration.[133] Further, a genome-wide study of transposon insertion in *Vibrio cholerae* by Satoshi Kimura and co-workers showed that H-NS binding sites were depleted for such insertions, and this was no longer observed in bacteria lacking H-NS.[134] Thus, the balance of evidence is in favour of H-NS, which helps to keep horizontally-acquired DNA transcriptionally silent, operating one step earlier by discouraging the insertion of foreign DNA. As a result, mechanistic processes can create non-uniformities in the insertion of foreign DNA, but the evidence presented here suggests that these would not block insertions near *ori*, thus further strengthening the case of selection acting to minimise insertions in *ori*-proximal DNA.

Khedkar also showed that gene order is usually more stable in fast-growing bacteria,[135] consistent with similar findings made earlier by Couturier and Rocha using a different analytical approach (Fig. 5.14C).[136] In other words, the replication-dependent dosage of a gene is better conserved in fast-growing than in slow-growing bacteria. Khedkar was, in particular, interested in measuring long-range translocations in bacterial genomes—i.e., is a gene located at a position p in one bacterial genome positioned elsewhere at q, distant from p, in another, related genome? Genes encoded in *ori*-proximal regions rarely translocate to distant *ter*-proximal parts of the genome. However, genes encoded on one replichore in one bacterium have in several instances moved to the opposite replichore in a related bacterium. These translocations from one replichore to the other do not usually disrupt gene dosage; in other words, the distance of a translocated gene from *ori* remains more or less the same, within reasonable limits, in whichever replichore it is found in (Fig. 5.15A). Or, inter-replichore translocations are often symmetric about the *ori-ter* axis, thus conserving distance from the *ori*. This helped to generalise findings made years earlier based on very few genomes.[137] Assuming translocations can occur randomly (symmetrically or asymmetrically at more or less equal frequencies), then selection imposed by the gene dosage gradient can eliminate a good proportion of asymmetric inter-replichore translocations.

An alternative explanation for the predominance of symmetric translocations is that asymmetric translocations that disrupt gene dosage do not happen at all, or happen at

133 N. Sharadamma, Y. Harshavardhana, P. Singh, and K. Muniyappa, 'Mycobacterium tuberculosis nucleoid-associated DNA-binding protein H-NS binds with high-affinity to the Holliday junction and inhibits strand exchange promoted by RecA protein', *Nucleic Acids Research* 38 (2010), 3555–3569. https://doi.org/10.1093/nar/gkq064

134 S. Kimura, T.B. Hubbard, B.M. Davis, and M.K. Waldor, 'The Nucleoid Binding Protein H-NS Biases Genome-Wide Transposon Insertion Landscapes', *mBio* 7 (2016), e01351–16. https://doi.org/10.1128/mbio.01351-16

135 Khedkar and Seshasayee, 2016.

136 Couturier and Rocha, 2006.

137 M. Suyama and P. Bork, 'Evolution of prokaryotic gene order: genome rearrangements in closely related species', *Trends in Genetics* 17 (2001), 10–13. https://doi.org/10.1016/s0168-9525(00)02159-4

very low frequencies. Multiple factors may contribute to the symmetry of translocations. One is that these events often require single-stranded DNA. This happens in replication forks—places where the DNAP is replicating the chromosome. The two replication forks, one on each replichore, move at more or less the same speed. If such forks promote translocations, then the manner in which replication occurs can ensure that translocations are usually symmetric around the *ori-ter* axis. Further, in some bacteria, the chromosome is structured in 3D space such that the two replichores are intertwined about each other, more or less symmetrically about the *ori-ter* axis.[138] Translocations would require the two regions of the chromosome to lie in close proximity. Khedkar showed that in one such bacterium, *C. crescentus*, translocation events appear to have occurred more frequently between pairs of regions that are often in contact with one another. But any such effect is likely to be amplified by selection that ensures that disadvantageous translocations are lost. Therefore, long-range translocations of bacterial genes are not uncommon, but these, along with other mechanisms of gene rearrangements, minimally disrupt gene dosage. This, in part, reflects selection acting to eliminate rearrangements that detrimentally disrupt gene dosage.

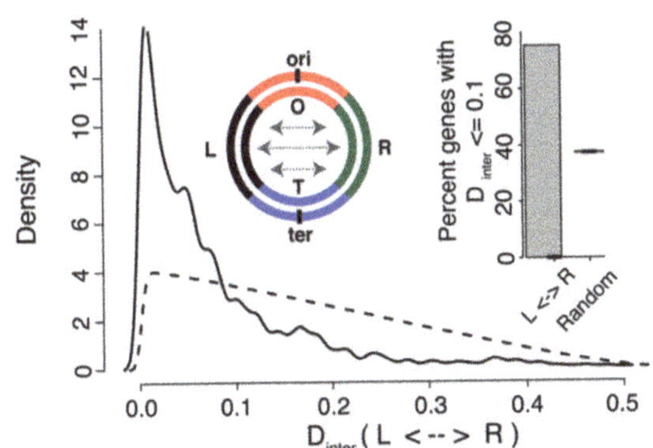

Fig. 5.15 Chromosome rearrangements maintaining gene dosage along the *ori-ter* axis. (A) This figure shows that inter-replichore translocations tend to be symmetric about the *ori-ter* axis. D_{inter} refers to the distance of the translocated pair of genes on either replichore from *ori*. The smaller the distance, the more symmetric the translocation. The dotted line shows what would be expected if translocations occurred between random sites across replichores. Originally published as Figure 5A in S. Khedkar and A.S.N. Seshasayee, 'Comparative genomics of interreplichore translocations in bacteria: a measure of chromosome topology?', *Genes, Genomes, Genetics* 6 (2016), 1597–1606, CC BY 4.0.

In another work published shortly after Khedkar's study, Jelena Repar and Tobias Warnecke showed that the tendency for inter-replichore translocations to be symmetric was not uniform across bacterial clades.[139] In different clades, different

138 T.B. Le, M.V. Imakaev, L.A. Mirny, and M.T. Laub, 'High- resolution mapping of the spatial organization of a bacterial chromosome', *Science* 342 (2013), 731–734. https://doi.org/10.1126/science.1242059

139 J. Repar and T. Warnecke, 'Non-random inversion landscapes in prokaryotic genomes are shaped by heterogeneous selection pressures', *Molecular Biology and Evolution* 8 (2017), 1902–1911. https://doi.org/10.1093/molbev/msx127

selection pressures seemed to be associated with symmetric translocations. In some, it was the presence of translation-associated genes close to *ori*, which can be interpreted as a measure of growth rate. In other clades, this relationship did not hold. Instead, correlations with what is known as '*gene strand bias*' (*GSB*) were observed. What is GSB and how is this related to replication?

Replication and transcription are two essential processes that engage the bacterial chromosome simultaneously. While replication is ongoing, RNAP is also going about doing its job transcribing genes. It therefore becomes inevitable that the two polymerases with collide, either codirectionally or in a head-on fashion. The addition of a new nucleotide to a growing DNA chain during replication is directional; the phosphate group of an incoming nucleotide is attached to a hydroxyl group at the end of the growing, nascent DNA chain. One strand, called the *leading strand*, is that which is synthesised continuously, in the direction in which new nucleotides are added to the growing DNA chain. The opposite strand, or the lagging strand, is synthesised discontinuously as fragments that are later glued together. Transcription, because it produces single stranded RNA, is free from such concerns despite being just as directional as replication. A gene that is encoded on the leading strand is transcribed in the same direction as replication. Therefore, DNAP and RNAP move codirectionally over a leading strand gene. Note that when we say that a gene is encoded on the leading strand, we mean that this strand acts as the coding strand whose sequence is identical[140] to that of the RNA chain being synthesised; for such a gene, the *lagging strand* serves as the template for transcription. Any meeting between the DNAP and RNAP transcribing a lagging strand gene will be head-on.

We now understand that codirectional collisions between RNAP and DNAP are largely inconsequential, whereas head-on collisions lead to several problems, from the mere slowing down of transcription[141] to genotoxicity arising from the stalling of replication.[142] In particular, head-on conflicts between the two polymerases increase local mutation rates around the collision site.[143]

One can expect that highly expressed genes would usually be encoded on the leading strand to minimise the chance of detrimental head-on collisions between RNAP and DNAP. A gene that is highly transcribed is more likely to see an RNAP meet a DNAP than one that is less transcribed, with replication being equal for all genes. As pointed out earlier, over 80% of all transcription in growing cells is reserved for rRNA synthesis. Thus, even the most highly expressed mRNA would not be as highly transcribed as rRNA genes, and so rRNA genes—across bacteria—are almost always

140 But for uracil replacing thymine.

141 B. Liu and B.M. Alberts, 'Head-on collision between a DNA replication apparatus and RNA polymerase transcription complex', *Science* 267 (1995), 1131–1137. https://doi.org/10.1126/science.7855590

142 E.V. Mirkin and S.M. Mirkin, 'Mechanisms of Transcription-Replication Collisions in Bacteria', *Molecular and Cellular Biology* 25 (2005), 888–895. https://doi.org/10.1128/mcb.25.3.888-895.2005

143 S. Paul, S. Million-Weaver, S. Chattopadhyay, E. Sokurenko, and H. Merrikh, 'Accelerated gene evolution through replication–transcription conflicts', *Nature* 495 (2013), 512–515. https://doi.org/10.1038/nature11989

encoded on the leading strand. More broadly, in most bacteria, a majority of genes are encoded on the leading strand. Now, are highly expressed genes preferentially encoded on the leading strand? For example, if 55% of all genes in a bacterial genome are present on the leading strand, are, say, 75% or 80% of highly expressed genes so encoded? Eduardo Rocha and Antoine Danchin showed that it is the essentiality of a gene, and not its expression level as measured by its CAI, that determines which strand a gene is encoded on.[144] Whereas a large proportion of essential genes are encoded on the leading strand irrespective of whether they are highly or less expressed in *Bacillus subtilis* and *E. coli*, the same proportion for non-essential genes drops ~20% again independently of expression level.

In a more recent study, Christopher Merrikh and Houra Merrikh showed that, in contrast to essential genes, genes involved in processes like antibiotic resistance and virulence are encoded on the lagging strand and show elevated mutation rates.[145] It is well established that there are more guanines than cytosines on the leading strand, and by definition the reverse is true for the lagging strand. This bias can be measured by what is called as '*GC skew*' ((G-C)/(G+C)). This value is positive for the leading strand. The skew arises from differences in mutational pressures between the two strands. It is a reflection of the long-term evolutionary history of which strand the piece of DNA sequence—for which the skew is measured—has been encoded on. However, not every stretch of leading strand DNA sequence exhibits a positive skew, and not every segment of lagging strand DNA exhibits a negative skew. Local variations in GC skew along a strand can tell us whether a segment of DNA in an extant genome under study has stayed on the same strand over long periods or has switched strands in more recent times. Using this reasoning, Merrikh and Merrikh showed that several highly mutable genes have inverted or switched from the leading to the lagging strand recently. This led them to propose that these genes—often involved in antibiotic resistance and virulence—might undergo accelerated evolution by deploying head-on DNAP-RNAP collisions to cause mutations.

Though certain genes may undergo high rates of evolution through head-on polymerase collisions, it is known that a majority of genes in most bacteria are encoded on the leading strand. For example, most of the inter-replichore translocations present in Khedkar's analysis are inversions. In an inversion, a gene that is on one strand of the DNA switches to the other. Now, because of the bidirectional nature of replication, the leading strand on one replichore becomes lagging in the other. Thus, inversions *within* a replichore would flip a leading strand gene to the lagging strand and vice-versa. On the other hand, inversions across replichores preserve gene strandedness. Thus, inter-replichore translocations analysed by Khedkar not only kept gene dosage disruptions to a minimum, but also maintained GSB. In a very recent study, Malhar

144 E.P.C. Rocha and A. Danchin, 'Essentiality, not expressiveness, drives gene-strand bias in bacteria', *Nature Genetics* 34 (2003), 377–378. https://doi.org/10.1038/ng1209

145 C.N. Merrikh and H. Merrikh, 'Gene inversion potentiates bacterial evolvability and virulence', *Nature Communications* 9 (2018), 4662. https://doi.org/10.1038/s41467-018-07110-3

Atre and colleagues performed a detailed analysis of inversions and GSB in over 2,000 bacterial genomes.[146] Consistent with the idea that intra-replichore inversions disrupt GSB whereas inter-replichore inversions do not, the latter are more common in most bacterial genomes.

Now, GSB is not uniform across bacterial clades. Some genomes code for only a slight excess of leading strand genes whereas in others, as many as 80–90% of all genes are on the leading strand. This indicates that different bacterial clades have different mechanisms to resolve head-on DNAP-RNAP conflicts and thus manage them differently. For example, if the mechanism of replication itself causes high rates of detrimental head-on DNAP-RNAP conflicts, it stands to reason that selection would keep a large proportion of genes in such bacteria to be encoded on the leading strand. There is a correlation between GSB and the nature of DNAP utilised by a bacterium for replication.[147] Again, an evolutionary argument for the variation in GSB is that different bacteria have selective pressures, independent of any differences in mechanisms of replication, against the detrimental effects of such collisions. An example would be growth rate: Anjana Srivatsan and colleagues demonstrated in *B. subtilis*, a genome with high GSB, that a large inversion near the *ori*—which causes many rRNA genes to shift from the leading to the lagging strand—has a stronger negative effect on growth, specifically during fast growth.[148] Atre and co-workers discovered that genomes with high GSB display very low frequencies of inversions overall. Further, in high GSB genomes, whatever inversions there are tend to be of the non-disruptive inter-replichore type. The authors argue that differences in inversion frequencies and type may be a factor underlying variation in GSB. Alternatively, if differences in the DNAP cause genomes with high GSB to be less tolerant of head-on DNAP-RNAP conflicts, then disruptive inversions would be much more strongly selected in such genomes. Thus, there is a strong mechanistic basis for the variation in GSB, and any evolutionary factor may arise from these mechanistic differences. Growth rate may also be a player, operating at a level distinct from the mechanistic factor, but its strength across bacteria needs to be clarified.

Chromosome rearrangements such as duplications, deletions, and inversions are often facilitated by repetitive sequences or just repeats. Bacterial genomes, unlike eukaryotic genomes, are relatively poor in repeats but are not entirely devoid of them. Pairs of repeats are called *direct repeats* when they are encoded in the same orientation. Rearrangements mediated by interactions between direct repeats are duplications and deletions. On the other hand, repeat pairs that are coded for on opposite strands are

---

146 M. Atre, B. Joshi, J. Babu, S. Sawant, S. Sharma, and T.S. Sankar, 'Origin, evolution and maintenance of gene-strand bias in bacteria', *Nucleic Acids Research* 52 (2024), 3493–3509. https://doi.org/10.1093/nar/gkae155

147 Ibid.; E.P.C. Rocha, 'Is there a role for replication fork asymmetry in the distribution of genes in bacterial genomes?', *Trends in Microbiology* 10 (2002), 393–395. https://doi.org/10.1016/s0966-842x(02)02420-4

148 A. Srivatsan, A. Tehranchi, D.M. MacAlpine, and J.D. Wang, 'Co-orientation of replication and transcription preserves genome integrity', *PLoS Genetics* 6 (2010), e1000810. https://doi.org/10.1371/journal.pgen.1000810

called *inverted repeats*, and these promote inversions. Repetitive sequences are usually horizontally acquired, and in many instances are transposable elements that can jump or copy themselves around the chromosome. One can expect these sequences to be randomly distributed, subject to constraints arising from the mechanisms by which they are generated. Nitish Malhotra in my lab showed that repeats are non-randomly distributed across bacterial genomes, especially in fast-growing bacteria.[149] He found that inverted repeats are more commonly present inter-replichore than intra-replichore, in fast-growing bacteria in particular (Fig. 5.16A). Thus, inversions mediated by a majority of inverted repeat pairs would be inter-replichore and would not affect GSB. Further, inter-replichore inverted repeat pairs tend to be positioned more or less symmetrically about the *ori-ter* axis, implying that inversions that they promote would disrupt gene dosage of inverted genes to a reduced extent. This was again prominent in fast-growing bacteria (Fig. 5.16B). Direct repeat pairs are often intra-replichore and are positioned closer to each other than would be predicted by random chance. This ensures that deletions and duplications caused by such repeats affect only relatively short stretches of DNA. Direct repeats are usually generated by duplications that create tandem copies of the duplicated sequence.[150] Thus, one can expect the mere processes that generate repeats to keep direct repeat pairs closer to each other than expected by chance. However, Malhotra also noted that the distance between direct repeat pairs was significantly shorter in fast-growing than in slow-growing bacteria, thus invoking an argument in favour of selection against large deletions and duplications that distant direct repeat pairs might cause. Therefore, the organisation of repetitive elements—among the drivers of chromosome rearrangements—on the chromosome is non-random and probably set up such that the rearrangements which can promoted do not drastically alter favourable gene organisation.

Finally, given that the location of *ori* on the genome is a central player in gene organisation, what will be the consequences for bacterial growth and gene organisation of the *ori* shifting elsewhere on the same chromosome? Xindan Wang and colleagues engineered *E. coli* to carry an origin of replication[151] at a non-native site, and removed the native *ori*. They called the new origin of replication *oriZ*. This newly engineered *oriZ* was placed about 1 Mbp away from the native *ori*. They found that *oriZ* was fully functional, causing initiation of chromosome replication normally, in this engineered *E. coli*. They also noticed that the initiation of replication from *oriZ*, instead of from the native *ori*, had a minimal effect on time to cell doubling. This is a curious observation. The sequence between *oriZ* and where the native *ori* originally

149 N. Malhotra and A.S.N. Seshasayee, 'Replication-Dependent Organization Constrains Positioning of Long DNA Repeats in Bacterial Genomes', *Genome Biology and Evolution* 14 (2022), evac102. https://doi.org/10.1093/gbe/evac102

150 G. Achaz, E.P.C. Rocha, P. Netter, and E. Coissac, 'Origin and fate of repeats in bacteria', *Nucleic Acids Research* 30 (2002), 2987–2994. https://doi.org/10.1093/nar/gkf391

151 X. Wang, C. Lesterlin, R. Reyes-Lamothe, G. Ball, and D.G. Sherratt, 'Replication and segregation of an *Escherichia coli* chromosome with two replication origins', *Proceedings of the National Academy of Sciences USA* 108 (2011), E243–E250. https://doi.org/10.1073/pnas.1100874108

was includes several highly expressed rRNA genes. These rRNA genes, which would have been on the leading strand in relation to replication initiated from the native *ori*, are now on the lagging strand. Thus, DNAP initiating replication from the newly-introduced *oriZ* would engage in head-on conflicts with RNAP transcribing rRNA genes. So how come bacteria facing such detrimental conflicts are replicating and growing just fine?

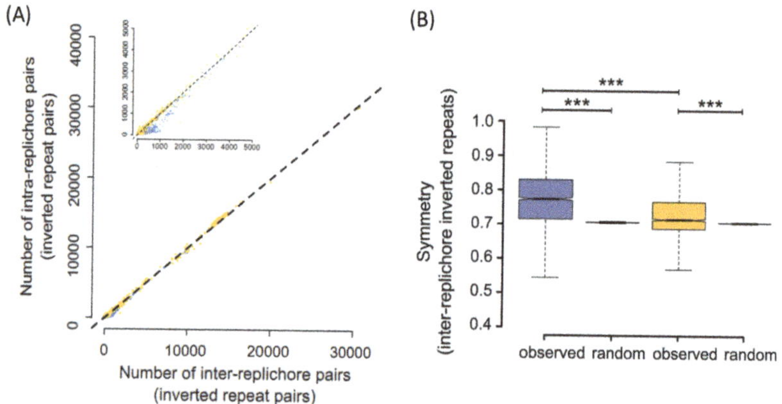

Fig. 5.16 Non-random organisation of repeats in bacterial genomes. (A) This figure shows that fast growing genomes contain more inter-replichore inverted repeats. This is especially clear in the zoomed-in version shown in the inset. Blue dots are for fast-growing bacteria and yellow for slower growing ones. Originally published as Figure 6E in N. Malhotra and A.S.N. Seshasayee, 'Replication-Dependent Organization Constrains Positioning of Long DNA Repeats in Bacterial Genomes', *Genome Biology and Evolution* 14 (2022), evac102, CC BY-NC 4.0; copyright held by the author of this book. (B) Inter-replichore inverted repeat pairs are more symmetric about the *ori-ter* axis in fast-growing bacteria. Originally published as Figure 7H in Malhotra and Seshasayee (2022).

Ivanova and co-workers answered this question.[152] They created an *E. coli* strain similar to that engineered by Wang and colleagues, but—unlike Wang and co-workers—found that this experienced a severe growth defect when grown in rich media. They also observed that problems in replication arose near rRNA genes, suggesting that head-on DNAP-RNAP conflicts contribute to the growth defect these bacteria suffer from. They provided further evidence for this argument by showing that a mutation in RNAP that allows it to bypass head-on conflicts alleviated the growth defect. They also noticed that the *E. coli* strain generated by Wang et al. had acquired a chromosome rearrangement that masked the growth defect that replication initiating at *oriZ* would have otherwise caused. This solution is simple. A fairly long stretch of DNA containing several rRNA genes had simply inverted, thus returning them onto the leading strand given replication initiation from the new *ori*.

152 D. Ivanova, T. Taylor, S.L. Smith, J.U. Dimude, A.L. Upton, et al., 'Shaping the landscape of the *Escherichia coli* chromosome: replication-transcription encounters in cells with an ectopic replication origin', *Nucleic Acids Research* 43 (2015), 7865–7877. https://doi.org/10.1093/nar/gkv704

Reshma Veetil and others in my lab found a similar adaptation evolving in an *E. coli* mutant which was defective in initiating replication at the native *ori* but managed to do so from elsewhere on the chromosome.[153] They found that these *E. coli* initiated replication from a site called *oriX*, 0.4–0.7 Mbp away from the native *ori*, creating a situation similar to the strains used by Wang et al. and Ivanova and colleagues. Again, an inversion of a stretch of DNA including several rRNA genes was one adaptive strategy discovered by these *E. coli* under selection to multiply faster. Such inversions are easy to achieve. rRNA genes are themselves repetitive elements, and a pair of such repeats on either side of the native *ori* would form an inter-replichore inverted repeat pair, which would promote inversions of the intervening DNA. If such inversions were to happen in natural isolates of *E. coli* replicating from the native *ori*, they would maintain strandedness, being catalysed by a pair of inter-replichore inverted repeats. However, this inversion in the context of replication from *oriX* caused several essential genes to switch from the leading to the lagging strand. This is probably an acceptable compromise. As emphasised earlier, rRNA synthesis accounts for the bulk of transcription. Ensuring that the highly transcribed rRNA genes stay on the leading strand even at the cost of other essential mRNA genes switching to the lagging strand is a fair bargain. Given enough time, *E. coli* would probably discover additional mutations that help to manage head-on DNAP-RNAP collisions even at essential genes.

Taken together, the way in which a bacterial chromosome replicates establishes a difference in gene dosage between *ori*- and *ter*-proximal regions of the chromosome in a growth rate-dependent manner. The selection arising from this contributes to the evolution of gene organisation, especially in fast-growing organisms. Conflicts between DNAP and RNAP add a further layer of constraint on the genome, determining how many and which types of genes are encoded on which strand of DNA, and how this can vary across clades of bacteria.

Thus, transcription is a crucial first step in the reading of the genome. The regulation of transcription is intricate, involving a vast network of regulators and the regulated. This helps bacteria to adapt to external environments as well as to changing genetic circumstances, such as the introduction and integration of a potentially expensive horizontally-acquired gene. Physiological adaptation through the regulation of gene expression meets genetic evolution when regulators evolve, and this appears to be a common phenomenon—especially early during adaptation to a new environment. Adaptation by mutations of regulators will have to balance the advantage such mutations provide against the collateral damage that they can cause by altering the expression of genes unrelated to the present adaptive challenge. Transcription not only provides fast physiological adaptations, but—along with chromosome replication—is also a factor determining the manner in which genes are organised around the bacterial chromosome.

---

153 R.T. Veetil, N. Malhotra, A. Dubey, and A.S.N. Seshasayee, 'Laboratory Evolution Experiments Help Identify a Predominant Region of Constitutive Stable DNA Replication Initiation', *mSphere* 5 (2020), e00939–19. https://doi.org/10.1128/msphere.00939-19

Afterword

Two years ago, even as the whole world and India—which had reeled from the effects of the so-called second wave the year before—was recovering from the impact of Covid, I got an invitation from the magazine *Current Science* to review the 2021 issue of the *Annual Review of Microbiology*. I had no idea how one could review a book that is not always a coherent whole but rather is a collection of authoritative reviews on disparate aspects of microbiology. But it turned out this was a special issue, celebrating 75 years of this important publication, which helped me to find a unifying framework for reviewing the book. And that issue of the Annual Review was headlined by a historical perspective on microbiology by none other than Roberto Kolter,[1] a microbiologist I have never met but have learnt a lot from! His conceptualisation of microbiology as an integrated discipline that marries biology, chemistry, and whatever it takes to understand the place of this fascinating group of organisms on Earth took hold in my mind. His dissatisfaction over how 'schisms' in microbiology had split this integrated field into isolated islands that rarely talked to each other resonated with my own philosophical musings, which at that time were suffused with Covid-induced pessimism.

Kolter's article also did light a spark, and I realised that studying bacterial adaptation does reunite these islands. It involves the molecular component of how genes mutate, evolve, and how they are transcribed, and the ecological aspects (some relevant to disease processes) of how these molecular processes operate under selection and enable adaptation. I felt chuffed that I had been working on, or thinking about, this very area for years, and a bit chagrined that I had not looked at it from the obvious (only in retrospect) perspective so beautifully articulated by Kolter.

My lab was also undergoing a transformation. From being a lab that did both experimental and computational work on bacteria and their genomes, we had—just before the arrival of SARS-CoV2 and nearly two years of stay-at-home requirements—opted to return to my roots in computational genomics. Personnel were changing, with some key members finishing their PhD theses and leaving. Our research interests were also shifting from regulatory networks in *E. coli* to the evolution of transcription regulation more broadly in bacteria and to some extent in eukaryotes as well. Somehow, these circumstances had led me to want to write my second book and, while I was thinking about an organising theme for this work, Kolter's piece suddenly showed up

1 Kolter, 2021.

and gave me a platform to launch from—a fulcrum to anchor the book's contents. Two years later, I hope I have done justice to the question of bacterial adaptation and how the interplay of genetic evolution and physiological adaptation enables this.

A good chunk of the book, on history and population genetics, deals with topics that are quite outside my comfort zone. It was both challenging and entertaining to try and learn as much of these fields as I could, to think about it and put it into writing. These are things I would not have done had I not opted to write this book.

Many may question why I have not emphasised the burgeoning field of microbiome and metagenome biology in this book—a field that Kolter, rightly, has argued reunites the multiple strands of microbiology. My answer is that while these studies have led us to appreciate the hitherto underappreciated diversity of bacteria and have also contributed to making fundamental changes to how we see the tree of life, they haven't—in my view—fundamentally altered our understanding of how bacterial genomes are organised, how they evolve, and how they are read. These concepts are better illustrated by studies based on single, complete bacterial genomes. But at the same time, I believe that the benefits of an ecological approach towards the study of microbiology run as an undercurrent throughout the book.

Selected bibliography

Chapter 1

Fleming, A., *Nobel Lecture* (1945). https://www.nobelprize.org/uploads/2018/06/fleming-lecture.pdf

Kolter, R., 'The History of Microbiology—A Personal Interpretation', *Annual Review of Microbiology* 75 (2021), 1–17. https://doi.org/10.1146/annurev-micro-033020-020648

Lane, N., 'The unseen world: reflections on Leeuwenhoek (1677) "Concerning little animals"', *Phil Trans Royal Soc. B.* 370 (2015), 20140344. https://doi.org/10.1098/rstb.2014.0344

Moberg, C.L., 'René Dubos, a harbinger of microbial resistance to antibiotics', *Perspectives in Biology and Medicine* 42 (1999), 559–580. https://doi.org/10.1089/mdr.1996.2.287

Salyers, A.A., and D.D. Whitt, *Revenge of the Microbes: How Bacterial Resistance is Undermining the Antibiotic Miracle* (Washington, DC: ASM Press, 2005). https://doi.org/10.1128/9781555817602

Whitman, W.B., D.C. Coleman, and W.J. Wiebe, 'Prokaryotes: the unseen majority', *Proc. Natl. Acad. Sci. USA* 95 (1998), 6578–6583. https://doi.org/10.1073/pnas.95.12.6578

Chapter 2

Cobb, M., 'Who discovered messenger RNA?', *Current Biology* 25 (2015), R525–R532. https://doi.org/10.1016/j.cub.2015.05.032

Dubos, R., 'Oswald Theodore Avery, 1877–1955', *Biographical Memoirs of the Fellows of the Royal Society* 2 (1956), 35–48. https://doi.org/10.1098/rsbm.1956.0003

Duckworth, D., 'Who discovered bacteriophage?', *Bacteriological Reviews* 40 (1976), 793–802. https://doi.org/10.1128/mmbr.40.4.793-802.1976

Hayes, W., 'Genetic transformation: a retrospective appreciation', *Journal of General Microbiology* 45 (1966), 385–397. https://doi.org/10.1099/00221287-45-3-385

Hershey, A.D., 'The injection of DNA into cells by phage', *Phage and the Origins of Molecular Biology* (Plainview, NY: Cold Spring Harbor Laboratory Press, 1966).

Hesse, S., and S. Adhya, 'Phage Therapy in the Twenty-First Century: Facing the Decline of the Antibiotic Era; Is It Finally Time for the Age of the Phage?', *Annual Review of Microbiology* 73 (2019), 155–174. https://doi.org/10.1146/annurev-micro-090817-062535

Stanier, R., and C.B. van Niel, 'The concept of a bacterium', *Archiv für Mikrobiologie* 42 (1962), 17–35. https://doi.org/10.1007/bf00425185

Woese, C.R., 'Bacterial evolution', *Microbiology and Molecular Biology Reviews* 51 (1987), 221. https://doi.org/10.1128/mmbr.51.2.221-271.1987

Chapter 3

Doolittle, W.F., 'Is junk DNA bunk? A critique of ENCODE', *Proceedings of the National Academy of Sciences USA* 110 (2013), 5294–5300. https://doi.org/10.1073/pnas.1221376110

Doolittle, W.F., and C. Sapienza, 'Selfish genes, the phenotype paradigm and genome evolution', *Nature* 284 (1980), 601–603. https://doi.org/10.1038/284601a0

Gil, R., F.J. Silva, J. Pereto, and A. Moya, 'Determination of the core of a minimal bacterial gene set', *Microbiology and Molecular Biology Reviews* 68 (2004), 518–537. https://doi.org/10.1128/mmbr.68.3.518-537.2004

Koonin, E.V., and Y.I. Wolf, 'Genomics of bacteria and archaea: the emerging dynamic view of the prokaryotic world', *Nucleic Acids Research* 36 (2008), 6688–6719. https://doi.org/10.1093/nar/gkn668

Lynch, M., and G. Marinov, 'The bioenergetic costs of a gene', *Proceedings of the National Academy of Sciences USA* 112 (2015), 15690–15695. https://doi.org/10.1073/pnas.1514974112

Romero, M.L.R., A. Rabin, and D. Tawfik, 'Functional Proteins from Short Peptides: Dayhoff's Hypothesis Turns 50', *Angewandte Chemie International Edition* 55 (2016), 15966–15971. https://doi.org/10.1002/anie.201609977

Thomas, C.A. Jr, 'The genetic organization of chromosomes', *Annual Review of Genetics 5* (1971), 237–256. https://doi.org/10.1146/annurev.ge.05.120171.001321

Wu, D., S.C. Daugherty, S.E. Van Aken, G.H. Pai, K.L. Watkins, H. Khouri, L.J. Tallon, J.M. Zaborsky, H.E. Dunbar, P.L. Tran, N.A. Moran, and J.A. Eisen, 'Metabolic complementarity and genomics of the dual bacterial symbiosis of sharpshooters', *PLoS Biology* 4 (2006), e188. https://doi.org/10.1371/journal.pbio.0040188

Chapter 4

Doolittle, W.F., 'Phylogenetic classification and the universal tree', *Science* 284 (1999), 2124–2129. https://doi.org/10.1126/science.284.5423.2124

Gómez-Valero, L., E.P.C. Rocha, A. Latorre, and F.J. Silva, 'Reconstructing the ancestor of Mycobacterium leprae: The dynamics of gene loss and genome reduction', *Genome Research* 17 (2007), 1178–1185. https://doi.org/10.1101/gr.6360207

Hood, D., and R. Moxon, 'Gene variation and gene regulation in bacterial pathogenesis', in D.A. Hodgson and C.M. Thomas (ed.) *Signals, Switches, Regulons and Cascades* (Cambridge: Cambridge University Press, 2002).

Mira, A., H. Ochman, and N.A. Moran, 'Deletional bias and the evolution of bacterial genomes', *Trends in Genetics* 17 (2001), 589–596. https://doi.org/10.1016/s0168-9525(01)02447-7

Nilsson, A.I., S. Koskiniemi, S. Eriksson, E. Kugelberg, J.C.D. Hinton, and D.I. Andersson, 'Bacterial genome size reduction by experimental evolution', *Proceedings of the National Academy of Sciences USA* 102 (2005), 12112–12116. https://doi.org/10.1073/pnas.0503654102

Puigbó, P., A.E. Lobkovsky, D.M. Kristensen, Y.I. Wolf, and E.V. Koonin, 'Genomes in turmoil: quantification of genome dynamics in prokaryote supergenomes', *BMC Biology* 12 (2014), 66. https://doi.org/10.1186/s12915-014-0066-4

Puigbo, P., Y.I. Wolf, and E.V. Koonin, 'The Tree and Net Components of Prokaryote Evolution', *Genome Biology and Evolution* 2 (2010), 745–756. https://doi.org/10.1093/gbe/evq062

Tenaillon, O., J.E. Barrick, N. Ribeck, D.E. Deatherage, J.L. Blanchard, et al., 'Tempo and mode of genome evolution in a 50,000-generation experiment', *Nature* 536 (2016), 165–170. https://doi.org/10.1038/nature18959

Treangen, T.J., and E.P.C. Rocha, 'Horizontal transfer, not duplication, drives the expansion of protein families in prokaryotes', *PLoS Genetics* 7 (2011), e1001284. https://doi.org/10.1371/journal.pgen.1001284

Waldor, M.K. and J.J. Mekalanos, 'Lysogenic conversion by a filamentous phage encoding cholera toxin', *Science* 272 (1996), 1910–1914. https://doi.org/10.1126/science.272.5270.1910

Chapter 5

Browning, D.F., and S.J.W. Busby, 'The regulation of bacterial transcription initiation', *Nature Reviews Microbiology* 2 (2004), 57–65. https://doi.org/10.1038/nrmicro787

Cohen, Y., and R. Hershberg, 'Rapid Adaptation Often Occurs through Mutations to the Most Highly Conserved Positions of the RNAP Core Enzyme', *Genome Biology and Evolution* 14 (2022), evac105. https://doi.org/10.1093/gbe/evac105

Dorman, C.J., 'H-NS, the genome sentinel', *Nature Reviews Microbiology* 5 (2007), 157–161. https://doi.org/10.1038/nrmicro1598

Dorman, C.J., 'DNA supercoiling and transcription in bacteria: a two-way street', *BMC Molecular and Cell Biology* 20 (2019), 26. https://doi.org/10.1186/s12860-019-0211-6

Ferenci, T., 'What is driving the acquisition of mutS and rpoS polymorphisms in Escherichia coli?', *Trends in Microbiology* 11 (2003), 457–461. https://doi.org/10.1016/j.tim.2003.08.003

Gruber, T.M., and C.A. Gross, 'Multiple sigma subunits and the partitioning of bacterial transcription space', *Annual Review of Microbiology* 57 (2003), 441–466. https://doi.org/10.1146/annurev.micro.57.030502.090913

Lucchini, S., G. Rowley, M.D. Goldberg, D. Hurd, M. Harrison, and J.C. Hinton, 'H-NS Mediates the Silencing of Laterally Acquired Genes in Bacteria', *PLoS Pathogens* 2 (2006), e81. https://doi.org/10.1371/journal.ppat.0020081

Price, M.N., P.S. Dehal, and A.P. Arkin, 'Horizontal gene transfer and the evolution of transcriptional regulation in *Escherichia coli*', *Genome Biology* 9 (2008), R4. https://doi.org/10.1186/gb-2008-9-1-r4

Ptashne, M., *A Genetic Switch: Phage Lambda Revisited* (Plainview, NY: Cold Spring Harbor Laboratory Press, 2004).

Rocha, E.P.C., 'The replication-related organization of bacterial genomes', *Microbiology* 150 (2004), 1609–1627. https://doi.org/10.1099/mic.0.26974-0

Srinivasan, R., V.F. Scolari, M.C. Lagomarsino, and A.S.N. Seshasayee, 'The genome-scale interplay amongst xenogene silencing, stress response and chromosome architecture in *Escherichia coli*', *Nucleic Acids Res.* 43 (2015), 295–308. https://doi.org/10.1093/nar/gku1229

Srivatsan, A., A. Tehranchi, D.M. MacAlpine, and J.D. Wang, 'Co-orientation of replication and transcription preserves genome integrity', *PLoS Genetics* 6 (2010), e1000810. https://doi.org/10.1371/journal.pgen.1000810

van Nimwegen, E., 'Scaling laws in the functional content of genomes', *Trends in Genetics* 19 (2009), 479–484. https://doi.org/10.1007/0-387-33916-7_14

Visweswaraiah, S., and S.J.W. Busby, 'Evolution of bacterial transcription factors: how proteins take on new tasks, but do not always stop doing the old ones', *Trends in Microbiology* 23 (2015), 463–467. https://doi.org/10.1016/j.tim.2015.04.009

Index

About the Team

Alessandra Tosi was the managing editor for this book.

Annie Hine and Lucy Barnes proof-read this manuscript.

The cover was designed by Jeevanjot Kaur Nagpal, and produced in InDesign using the Fontin and Calibri fonts.

Jeremy Bowman typeset the book in InDesign and produced the paperback, hardback and EPUB editions. The text font is TeX Gyre Pagella; the heading font is Californian FB.

Cameron Craig produced the PDF and HTML editions. The conversion was performed with open-source software and other tools freely available on our GitHub page at https://github.com/OpenBookPublishers.

Raegan Allen was in charge of marketing.

This book was peer-reviewed. Experts in their field, our readers give their time freely to help ensure the academic rigour of our books. We are grateful for their generous and invaluable contributions.

This book need not end here...

Share

All our books — including the one you have just read — are free to access online so that students, researchers and members of the public who can't afford a printed edition will have access to the same ideas. This title will be accessed online by hundreds of readers each month across the globe:
why not share the link so that someone you know is one of them?

This book and additional content is available at
https://doi.org/10.11647/OBP.0446

Donate

Open Book Publishers is an award-winning, scholar-led, not-for-profit press making knowledge freely available one book at a time. We don't charge authors to publish with us: instead, our work is supported by our library members and by donations from people who believe that research
shouldn't be locked behind paywalls.

Join the effort to free knowledge by supporting us at
https://www.openbookpublishers.com/support-us

You may also be interested in:

Introduction to Systems Biology

Workbook for Flipped-classroom Teaching

Thomas Sauter and Marco Albrecht

https://doi.org/10.11647/OBP.0291

Writing and Publishing Scientific Papers

A Primer for the Non-English Speaker

Gábor L. Lövei

https://doi.org/10.11647/OBP.0235

Advanced Problems in Mathematics

Preparing for University

2nd Edition

Stephen Siklos

https://doi.org/10.11647/OBP.0181

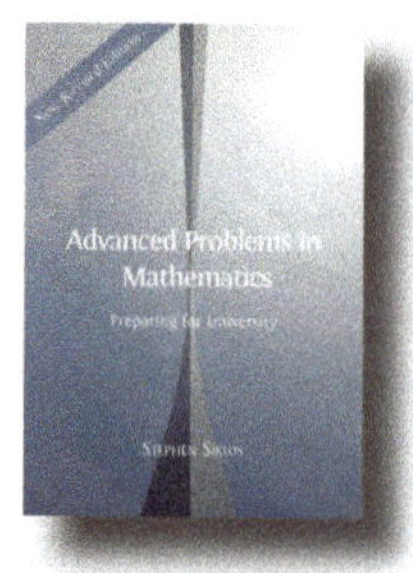

Animals and Medicine

The Contribution of Animal Experiments to the Control of Disease

Jack Botting (author), Regina Botting (editor) and Adrian R. Morrison (foreword by)

https://doi.org/10.11647/OBP.0055

www.ingramcontent.com/pod-product-compliance
Lightning Source LLC
Chambersburg PA
CBHW050236220326

41598CB00044B/7414